Visual Astronomy of the Deep Sky

Visual Astronomy of the Deep Sky

Roger N. Clark, Ph.D.

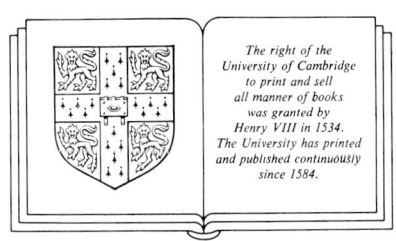

CAMBRIDGE UNIVERSITY PRESS

Cambridge

New York   Port Chester   Melbourne   Sydney

&

SKY PUBLISHING CORPORATION

Cambridge, Massachusetts

Published by Sky Publishing Corporation
49 Bay State Road, Cambridge, Massachusetts, 02238 – 1290
and by the Press Syndicate of the University of Cambridge CB2 1RP
The Pitt Building, Trumpington Street, Cambridge CB2 1RP
40 West 20th Street, New York, NY 10011, USA
10 Stamford Road, Oakleigh, Melbourne 3166, Australia

© Roger N. Clark 1990

First published 1990

Printed in the United States of America

British Library cataloguing in publication data
Clark, Roger N.
Visual Astronomy of the Deep Sky.
1. Astronomical bodies. Observation -
Amateurs' manual
I. Title
522

Library of Congress cataloguing in publication data
Clark, Roger N. (Roger Nelson)
Visual Astronomy of the Deep Sky / Roger N. Clark
p. cm.
Bibliography: p.
Includes Index.
ISBN 0 933 346 54 9 (Sky Publishing Corporation)
ISBN 0 521 36155 9 (Cambridge University Press)
1.Astronomy – Amateurs' manuals. 4. Nebulae – Atlases –
Amateurs' manuals. I. Title
QB64.C58 1990
523–dc 19 88 10290 CIP
ISBN 0 933 346 54 9 (Sky Publishing Corporation)
ISBN 0 521 36155 9 (Cambridge University Press)

For I can end as I began.
From our home on the earth we look out into the distances and
strive to imagine the sort of world into which we are born.
Today we have reached far out into space.
Our immediate neighborhood we know rather intimately.
But with increasing distance our knowledge fades...
until at the last dim horizon we search among ghostly errors
of observations for landmarks that are scarcely more substantial.
The search will continue.
The urge is older than history.
It is not satified and it will not be suppressed.

Edwin P. Hubble

Contents

Photograph Credits ix
Preface xi
Acknowledgements xiii

1. About this book   1

2. The human eye   4
Introduction   4
Rods and cones   4
Units of brightness   4
Dark adaptation   6
Contrast discrimination   7
Averted vision   12
Exposure time   14
Color   15
Vision and health   16
Summary   17

3. The eye and the telescope   19
Telescope basics   19
Aberrations   19
Types of telescopes   20
  The refractor   20
  The reflector   22
  Catadioptric telescopes   23
Telescope mountings   23
Eyepieces   24
Field of view   28
The role of a telescope's f/ratio   28
The exit pupil   29
Seeing and resolution   30
Filters   31
Using the telescope to find objects   38
  Celestial coooordinates   38
  Precession   40
Finders   44
Miscellaneous topics   46
  Caring for optics   46
  Dew   47
Summary   48

4. The faintest star visible in a telescope   49
Introduction   49
Magnification   49
Finding tonight's magnitude limit   50
Summary   53

5. Making drawings and keeping records   54
Introduction   54
Written descriptions   54
Drawings   55
  Drawing method 1: the initial-blind method   57
  Drawing method 2: filling in details   57
  Drawing method 3: the double-blind method   57
Summary   57

6. A case study: the Whirlpool Galaxy, Messier 51   59
Introduction   59
Brightness profile of M51   59
Visual detection   59
Discussion and summary   60

7. A visual atlas of deep-sky objects   64
The personal equation   64
Averted vs. direct vision   65
The observations and drawings   65
Definitions   67
M31, The Andromeda Galaxy (NGC 224), M32 (NGC 221), M110 (NGC 205), Galaxies in Andromeda   68
NGC 246, Planetary Nebula in Cetus   72
NGC 253, Galaxy in Sculptor   74
M33 (NGC 598), Galaxy in Triangulum   77
M74 (NGC 628), Galaxy in Pisces   80
M76 (NGC 650–651), Planetary Nebula in Perseus   82
NGC 891, Galaxy in Andromeda   83
M77 (NGC 1068), Seyfert Galaxy in Cetus   86
NGC 1365, Barred Spiral Galaxy in Fornax   87
M45, the Pleiades Open Cluster in Taurus   90
M1 (NGC 1952), the Crab Nebula: Supernova Remnant in Taurus   95
M42 (NGC 1976), M43 (NGC 1982), the Great Nebula in Orion   98
NGC 2023, NGC 2024, IC 434 (the Horsehead Nebula) Nebulae in Orion   106

Contents

M78 (NGC 2068) NGC 2071, Diffuse Nebulae in Orion — 110
NGC 2261, Hubble's Variable Nebula in Monoceros — 112
M46 (NGC 2437), NGC 2438, Open Cluster in Puppis with NGC 2438, a Planetary Nebula — 115
M67 (NGC 2682), Open Cluster in Cancer — 118
NGC 2903, Spiral Galaxy in Leo — 120
M81 (NGC 3031), Spiral Galaxy in Ursa Major — 123
M82 (NGC 3034), Peculiar Galaxy in Ursa Major — 126
M96 (NGC 3368), Galaxy in Leo — 128
M105 (NGC 3379), NGC 3384, NGC 3389, Galaxies in Leo — 130
M108 (NGC 3556), Galaxy in Ursa Major — 132
M97 (NGC 3587), the Owl Nebula: Planetary Nebula in Ursa Major — 133
M66 (NGC 3627), NGC 3628, M65 (NGC 3623), Galaxies in Leo — 135
M109 (NGC 3992), Galaxy in Ursa Major — 138
NGC 4038, NGC 4039, the Ringtail Galaxy in Corvus — 140
M99 (NGC 4254), Galaxy in Coma Berenices — 142
M106 (NGC 4258), Galaxy in Canes Venatici — 143
M100 (NGC 4321), Galaxy in Coma Berenices — 144
M84 (NGC 4373), M86 (NGC 4406), and 13 other galaxies in Virgo — 146
NGC 4449, Galaxy in Canes Venatici — 150
M87 (NGC 4486), NGC 4476, NGC 4478, Galaxies in Virgo — 153
NGC 4565, Edge-on Galaxy in Coma Berenices — 156
M90 (NGC 4569), Galaxy in Virgo — 158
M104 (NGC 4594), the Sombrero Galaxy in Virgo — 160
M94 (NGC 4736), Galaxy in Canes Venatici — 162
M64 (NGC 4826), the Black Eye Galaxy in Coma Berenices — 164
M63 (NGC 5055), Galaxy in Canes Venatici — 166
NGC 5128, Peculiar Galaxy in Centaurus — 169
NGC 5139 (Omega Centauri), the Great Globular Cluster in Centaurus — 172
M51 (NGC 5194), the Whirlpool Galaxy in Canes Venatici NGC 5195 — 176
M83 (NGC 5236), Galaxy In Hydra — 182
M4 (NGC 6121), Globular Star Cluster in Scorpius — 184
M13 (NGC 6205), Globular Cluster in Hercules — 186
M20 (NGC 6514), the Trifid Nebula in Sagittarius — 188
M8 (NGC 6523), the Lagoon Nebula in Sagittarius NGC 6530 — 192
M16 (NGC 6611), the Eagle Nebula in Serpens — 197
M17 (NGC 6618), the Omega Nebula and Open Cluster in Sagittarius — 201
M11 (NGC 6705), Open Cluster in Scutum — 205
M57 (NGC 6720), the Ring Nebula in Lyra — 208
M27 (NGC 6853), the Dumbbell Nebula in Vulpecula — 210
NGC 6888, the Crescent Nebula in Cygnus — 212
NGC 6946, Galaxy in Cepheus — 214
NGC 6960 and NGC 6992-5, the Veil Nebula in Cygus — 218
NGC 7000, the North America Nebula in Cygnus — 227
M15 (NGC 7078), Globular Cluster in Pegasus — 230
M2 (NGC 7089), Globular Star Cluster in Aquarius — 232
NGC 7293, the Helix Nebula in Aquarius — 235
NGC 7331, Spiral Galaxy in Pegasus — 239
NGC 7662, Planetary Nebula in Andromeda — 242
Conclusions — 244

Appendices
A. **Suggested reading**
  Recommended books — 245
  Beginning: learning the sky — 245
  Star atlases — 246
  Handbooks and observing guides — 247
B. **Star clusters for finding your limiting magnitude** — 249
C. **Air mass, atmospheric extinction, and other calculations**
  Calculating zenith angle and air mass — 268
  Other useful computations — 270
D. **Symbols and their definitions** — 279
E. **A catalog of deep-sky objects** — 283
F. **Optimum detection magnifications for deep-sky objects**
  How the computation is done — 318
  What the ODM means — 320
  The role of contrast — 321

Bibliography — 351

Index — 352

Photograph Credits

Chapter 3:
Figure 3.2b: Johnny Horne

Chapter 6:
Figure 6.1: Palomar Observatory

Chapter 7:
M31, M32, M110: Palomar Observatory
NGC 246: Jack B. Marling
NGC 253: Palomar Observatory
M33: National Optical Astronomy Observatories
M74: National Optical Astronomy Observatories
M76: Laird A. Thompson, Canada–France–Hawaii Telescope Corporation
NGC 891: Mount Wilson and Las Campanas Observatoires, Carnegie Institure of Washington
M77: Lick Observatories
NGC 1365: Wayne C. Annala, Copyright University of Hawaii, Institute for Astronomy
M45: Mount Wilson, and Las Campanas Observatories, Carnegie Institute of Washington
M1: Evered Kreimer, *The Messier Album*
M42, M43: Mount Wilson and Las Campanas Observatories, Carnegie Institute of Washington
Trapezium: Lick Observatories
NGC 2023, NGC 2024, IC 434: James E. Gunn
M78, NGC 2071: Evered Kreimer, *The Messier Album*
NGC 2261: National Optical Astronomy Observatories
NGC 2903: Palomar Observatory
M81: National Optical Astronomy Observatories
M82: Laird A. Thompson, Canada–France–Hawaii Telescope Corporation
M96: Evered Kreimer, *The Messier Album*
M97: Evered Kreimer, *The Messier Album*
M66, NGC 3628, M65: Laird A. Thompson, France–Canada–Hawaii Telescope Corporation
M109: Evered Kreimer, *The Messier Album*
NGC 4038, NGC 4039: Palomar Observatory
M99: Evered Kreimer, *The Messier Album*
M106: Palomar Observatory
M100: Evered Kreimer, *The Messier Album*

M84, M86, 13 other galaxies: Akita Fujii
NGC 4449: K.A. Brownlee, courtesy of *Deep Sky Magazine*
M87, NGC 4476, NGC 4478: Evered Kreimer, *The Messier Album*
NGC 4565: Palomar Observatory
M90: Evered Kreimer, *The Messier Album*
M104: National Optical Astronomy Observatories
M94: Palomar Observatory
M64: Martin Germano
M63: Evered Kreimer, *The Messier Album*
NGC 5128: David Malin, Anglo-Australian Telescope Board
NGC 5139: National Optical Astronomy Observatories
M51, NGC 5195: National Optical Astronomy Observatories
M83: Evered Kreimer, *The Messier Album*
M4: Martin Germano
M13: Palomar Observatory
M20: National Optical Astronomy Observatories
M8: National Optical Astronomy Observatories
M16: National Optical Astronomy Observatories
M11: Ben Meyer
M57: Laird A. Thompson, Canada–France–Hawaii Telescope Corporation
M27: National Optical Astronomy Observatories
NGC 6888: Martin Germano
NGC 6946: National Optical Astronomy Observatories
NGC 6960: Mount Wilson and Las Campanas Observatories, Carnegie Institute of Washington
NGC 6992-5: Mount Wilson and Las Campanas Observatories, Carnegie Institute of Washington
NGC 7000: Ron Pearson
M15: National Optical Astronomy Observatories
M2: Evered Kreimer, *The Messier Album*
NGC 7293: National Optical Astronomy Observatories
NGC 7331: Palomar Observatory
NGC 7662: Jack B. Marling

Preface

To stand beneath a dark, crystal-clear, moonless country sky is an awe-inspiring experience. Those thousands of stars, many larger than our own Sun, can make us feel small indeed. It seems possible to see to infinity, though we cannot reach beyond arm's length. The beauty of the universe defies description.

Turn a telescope on a seemingly empty part of sky and swarms of new stars come into view – and possibly a faint glow of fuzzy nebulosity. Yet the heavens are subtle. Imagine that the fuzzy patch at the threshold of visibility is really a trillion suns – a galaxy larger than our own, in which *our* Sun is but a tiny speck. Incomprehensible; yet somehow we try. Seeing that galaxy first-hand, even through a small telescope, is much more inspiring than the large, beautiful photograph in the astronomy book back indoors. Nothing can compare to viewing the universe directly.

The city dweller looks up at night and if lucky, sees a few stars and thinks "That's nice". But show that same person a very dark country sky and he or she will be awe-struck. Such a sky can be so spectacular that the Milky Way casts a shadow, and so many stars may be visible that even experienced observers have trouble finding constellations.

Even after many years as an amateur astronomer, I am still awe-struck on a dark moonless night. I was a very active amateur in the late 1960s and early 1970s, making both visual and photographic observations of everything within reach. By 1971, I had observed all the Messier and many NGC objects with an 8-inch telescope. I was especially interested in astrophotography, and taking pictures of the heavens consumed most of my spare time. Visual observing seemed just an enjoyment, something to do while getting warmed up for those long hours guiding the telescope for that prize astrophoto.

In the late 1970s, my amateur career came to a standstill while I worked on a Ph.D. in planetary science. By 1982 the amateur bug was biting hard again, and all my old, pleasurable observing memories brought me back to active status.

Now that I had seen "everything", I was willing to spend time making detailed drawings of what I saw. I began a literature search for material on how the eye performs in low-light-level conditions. After reading many of the handbooks for amateur astronomers available today, I found a surprising lack of information on observing deep-sky objects, or on what can actually be seen of them through telescopes.

The typical handbook devotes considerable space to observing the Moon and planets, but when the subject of galaxies and nebulae comes up, it has less to say. Usually it just recommends that because these objects are faint, low power should be used. Such works have inculcated the idea among amateur astronomers that one should have a "richest-field telescope" and low-power eyepieces for deep-sky work.

This concept seemed wrong to me. My impression at star parties in the dark skies of the Cascade Mountains of Washington state was that an 8-inch f/10 telescope gave a more detailed view of most objects than richest field telescopes of the same or even slightly larger aperture! This was one reason I built an 8-inch f/11.5 Cassegrain. Although I had started to build an 8-inch f/4.5, my opinions changed so much that I decided on the longer focal length partway through the project. I heard a few experienced amateur astronomers express the same thoughts–but none could give a good reason why high f/ratio telescopes seem to work better on faint objects.

During research for this book I found the answer. The impression that higher f/ratios by themselves give a better view is largely wrong. The true reason is partially what this book is about. The magnification used, not the f/ratio, determines what can be seen in a very faint object. A higher f/ratio telescope simply yields higher power with a given eyepiece, and therefore it is more likely to be used that way.

In August, 1982, I gave a talk to the Hawaiian Astronomical Society, an amateur club, in which I discussed how the human eye detects light at low levels and how various objects appear through telescopes. Many beginners see the beautiful pictures in astronomy books and expect the same views in their telescopes. This often results in disappointment. But I have found that if amateurs know more of the characteristics of the eye, and use the telescope a little differently, quite a lot of detail in galaxies, nebulae, and star clusters can be seen.

The response to that talk convinced me that this material should be presented to all amateurs. It was then that I decided to write this book.

Roger N. Clark
Denver, Colorado

Acknowledgements

As with any large work, there are many who helped in the creation of this book. I would like to thank Herman Dittmer, Ray Fabre, Ivan Giesler, Bob Gunnerson, Ruthi Moore, Mike Morrow, Milt Sher, and Bill Tittemore for their reviews of various drafts of the book. Their comments helped my ideas evolve to the present work.

I wish to extend a special thanks to Alan MacRobert of *Sky Publishing Corp.* for his extensive review and editing. Alan is the only editor I have known who can take a rough manuscript and change it into a flowing work without changing any of the original ideas. His knowledge of astronomy and attention to detail assured that every thought was correct and understandable.

Finally I wish to thank my wife, Susan, for reviewing parts of the book and sons Matthew, Christopher, and Tyler for putting up with my many hours spent at the computer, and the many nights observing far from home.

1
About this book

Before the late nineteenth century all astronomy – amateur and professional – was visual. Everything depended on skilled use of the eye. Today, however, the professional astronomer rarely looks through the large telescopes at his or her disposal. As photographic film became more sensitive, both professionals and amateurs devoted more time to photography. In the last few decades, the advent of sensitive electronic light detectors diverted professional and even some amateur work further from visual astronomy. Direct viewing – at least of deep-sky objects – is now mainly for those interested just in beauty and inspiration.

The typical amateur astronomer today has a telescope with a main mirror or lens 4 to 10 inches in diameter. A modern instrument of this size often provides a more detailed view of deep-sky objects than the larger telescopes used by the pioneering astronomers to make the great discoveries of the past. There are literally thousands of interesting objects, primarily galaxies but also star clusters and nebulae, within reach of the small telescope. Most are very faint, hundreds of times fainter than can be seen with the naked eye. However, a telescope with only 6 inches of aperture gathers about 400 times the light of the unaided eye, so these beautiful but dim objects can be brought into view.

The first-time user of a telescope is often disappointed that galaxies and nebulae do not look like the photographs in astronomy books. In some respects, the eye is no match for the camera. The beautiful photos are the result of very long exposures on sensitive film that builds up an image out of light too faint for the eye to see at all.

Nevertheless, the human eye is a very sophisticated light detector in its own way, and if used correctly, it can reveal considerable detail in galaxies, nebulae, and clusters.

Some beginning amateurs, inexperienced in the use of the eye and telescope, dismiss most deep-sky objects as faint smudges and merely search for a few easy ones before losing interest. Others, thrilled by the ability to see galaxies millions of light years away, diligently search out the faintest ones possible. Some amateurs turn to other sights such as the planets, where even small telescopes can match the largest ones because of limitations by the Earth's atmosphere. Other amateurs turn to photography. The difficulty of obtaining a spectacular astrophoto, however, soon becomes evident. Compensation for the Earth's rotation requires that the telescope be moved precisely to track the stars. This requires somewhat sophisticated equipment, lots of patience, and long spells of staring at a guide star while correcting the rate of a never-perfect telescope clock drive. Countless amateurs have gone through all of these stages and I am no exception.

With this book I hope to impart new knowledge that will rekindle enthusiasm for visual studies of galaxies, nebulae, and star clusters. Visual observing is not only highly enjoyable; it can even be a useful tool for research, such as in the search for supernovae in other galaxies.

Many books are available on the basics of astronomy, so such material is not presented here. The appendices will direct the reader to excellent sources of basic information. This book is, however, designed for the beginner as well as the most advanced observer. Some chapters contain complex diagrams and math, but the beginning astronomer with a limited math background will find a simple summary at the end of each chapter. It might be beneficial for all readers, during their first encounter with the book, to read the chapter summaries before the technical portions. Chapters 3, 5 and 7 contain little math and need not be skipped by the beginner.

The capabilities of the eye are presented in Chapter 2. We will see that the eye is more sensitive to faint, extended light sources if the light is spread out over about 5 to 10 degrees. Because the ability to see low-contrast features in faint objects depends on their angular size, the visual observer will detect different amounts of detail at different magnifications. The implications of this fact for observing faint astronomical objects are discussed in Chapter 3, which also presents an introduction to telescopes, mountings, eyepieces, and how to find celestial objects.

Chapter 4 analyzes the faintest star observable in a telescope and shows how this is a function of both the magnification and sky brightness, as well as the more commonly known factor of telescope aperture.

In Chapter 5 the techniques of making and keeping good records and drawings are discussed. We analyze the visibility of the galaxy Messier 51 in Chapter 6 to illustrate the detection of low-contrast features in deep-sky objects. In this instance, the visibility of a galaxy's spiral arms is addressed, and we see that there is an optimum magnification for viewing faint detail.

Chapter 7 describes the visual appearance of more than 90 deep-sky objects. A catalog of photographs and drawings is presented to illustrate what can be seen with a typical amateur telescope. This chapter is laid out so that a photograph and drawing are reproduced on facing pages at the same scale, to show exactly what in the photograph can be viewed in the telescope. Viewing distances from the drawing to the eye are given so the drawing can be seen at the same apparent size as at various magnifications in a telescope.

The appendices include charts of star clusters for use in determining the faintest star visible in a telescope, a catalog of more than 600 deep-sky objects and the optimum magnifications to use on each, and equations for determining rise and set times of an object, as well as its altitude and azimuth.

* * *

This book presents many new concepts, some of which may be difficult for the experienced amateur to accept. The most revolutionary is the "optimum magnified visual angle." This is the magnification at which an object should be viewed for best detection.

Research on how the eye detects faint objects contradicts several basic pieces of conventional wisdom. One is the belief that low magnification should be used to concentrate a faint nebulosity on a small area of the eye's retina. This would be true if the retina worked passively, like photographic film. But it doesn't. The visual system has a great deal of active computing power and combines the signals from many receptors to detect a faint extended object. Increasing the magnification spreads the light over more receptors, and the brain's processing power can then bring into view fainter objects having lower contrast.

Another interesting concept is how the optimum magnification for a given object varies with the size of the telescope. Since the surface brightness is less in a telescope of smaller aperture, the optimum power is *higher* than in a bigger telescope! This applies only to deep-sky objects, and is illustrated nicely in Appendix F. Brighter objects, such as details on planets, fall on a different part of the eye's detection curves. In that case, about the same magnification should be used on all telescopes. Planetary observing is not discussed in this book, but this conclusion is very interesting nonetheless.

Another new concept is the highest magnification that may ever be usefully employed on a telescope. It is normally accepted that the highest power is about 50 to 60 times the objective in inches. This limit is correct only for bright objects such as the Moon and planets. For fainter objects the eye has less resolution and needs to see things larger, so higher powers are called for. At the limit of the eye's detection ability, the highest useful magnification is on the order of 330 per inch of objective! These extremely high magnifications are useful in specialized cases such as detection of detail in a small planetary nebula. For the drawing of NGC 7662 in Chapter 7, a magnification of nearly 600 was needed with an 8-inch telescope.

These results are based on an elaborate study of the eye's performance carried out during World War II and published in 1946 by Blackwell (see the bibliography). This information has been around for quite some time and is occasionally presented in some forms (e.g. Roach and Jamnick, 1958). However, it has not been fully understood. It took considerable computer processing to convert Blackwell's original data into a form useful in an astronomical context. Thus, it is not surprising that these concepts have not been previously discovered.

2
The human eye

INTRODUCTION

The eye is a remarkably adaptable light detector, able to function in bright sunlight and faint starlight, an intensity range of more than ten million. This is a considerably greater range of brightness than can be handled by man-made instruments such as photographic film or television cameras.

The eye consists of the cornea, which provides most of the lens action; the lens, which provides the focusing ability; the iris, which restricts the aperture and the amount of light entering the eye; and the retina, which detects the light and sends the signal to the brain. The structure of the eye is shown in Figure 2.1.

In the retina, light is recorded by a photochemical reaction: when light impinges on a photochemical substance, a reaction sends an electrical signal to the brain. The sensitivity of the eye depends on the amount of chemicals present. It is primarily by varying the amount of photochemical material, not by opening the iris, that the eye can adapt to an amazing range of light levels.

RODS AND CONES

The light-detection devices in the retina are of two types: *rod* cells and *cone* cells. The cones are concentrated in the *fovea*, where visual acuity is greatest. This small area, appearing less than a degree in diameter, is the center of vision. We normally aim the fovea directly at an object to see maximum detail.

However, the cones are not as sensitive to light as the rods. At the center of the fovea (on the visual axis), there are no rods at all. The rods increase in density from zero at the center of the fovea to a maximum at about 18 degrees off-axis. The density of rods and cones is shown in Figure 2.2. There are no rods or cones at all on the spot where the optic nerve leaves the eye. This is called the *blind spot*.

The rods and cones respond best to different colors of light, with the rods slightly more blue-sensitive than the average of the cones. This is shown in Figure 2.3. The cones are entirely responsible for color vision, and since they require bright illumination one sees no color in dim light, when only the rods are working.

The rods and cones are connected to nerves via ganglion cells. In the fovea, a ganglion may serve a single cone, but in the periphery of the retina, where rods are prevalent, one ganglion may be connected to 100 rods. It is via the ganglion cells that the electrical signals from the rods and cones are transferred to the optic nerve and the brain. The detection of faint light apparently depends on how many ganglion cells are involved. Those in the periphery each serve rods covering an area about 20 minutes of arc in diameter. The detection and contrast-discrimination capabilities of the eye involve the summation effects of several ganglion cells. These are the capabilities that at low light levels are of interest to the visual astronomer.

UNITS OF BRIGHTNESS

Two factors govern the eye's detection of light. One is the total brightness of an object, and the other is its *surface brightness*. Surface brightness is the total amount of light divided by the area over which it spread.

A physicist might describe an object's total brightness by units of power (that is, energy flow), such as watts or number of photons per second. However, more specialized terms have been introduced for visual perceptions. The term *illuminance* describes the total light output of an object in the wavelengths seen by a typical human eye. One of the earliest units of illuminance measure was a candle. Requirements for precision led to the unit called the *candela*, the total visual light emitted in all directions by a standard candle made of a specific material and of a certain size. The *lumen (lm)* is another common unit of measure, and is equal to a candle divided by 4π (=12.5664).

Surface brightness, intensity per unit area, is described by another term, *luminance*. Note the subtle, but important, difference between luminance and illuminance. Common units of luminance are candelas per square meter, or lumens per square meter.

Astronomers use their own brightness unit: the stellar magnitude. And instead of linear measurements of distance on a surface, they use angular measurements of distance on the sky.

The magnitude was invented because the eye responds to light approximately logarithmically. One magnitude corresponds to a change in brightness by a factor of $100^{1/5}$, which is about 2.51. Five magnitudes is a factor of 100 in brightness and 10 magnitudes is a factor of 100 times 100 or 10 000.

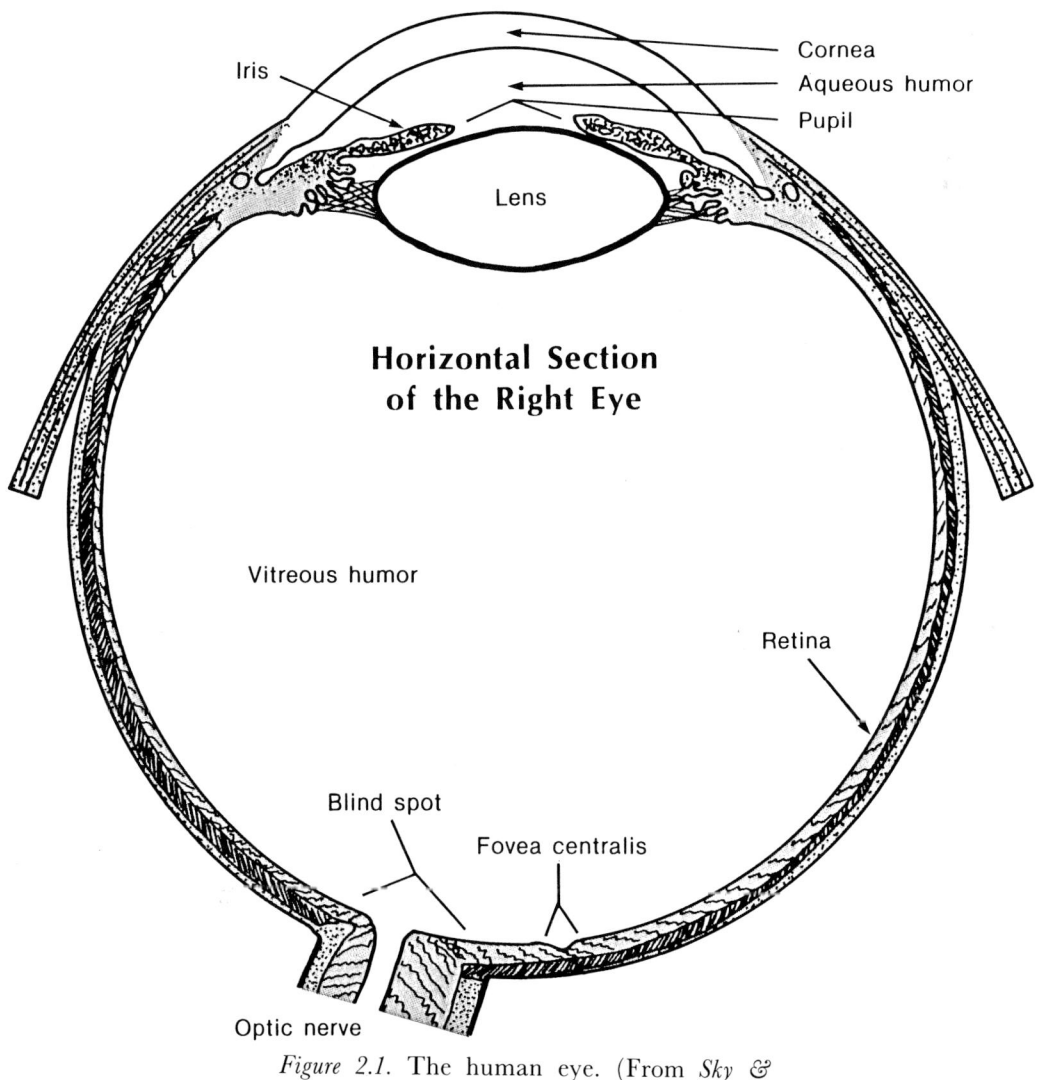

Figure 2.1. The human eye. (From *Sky & Telescope* April, 1984

Since astronomical objects cover an area of sky, their surface brightnesses are described in magnitudes per square arc-second. The full Moon, for example, is a half degree (1800 arc-seconds) in diameter, so it covers 2.5 million square arc-seconds of sky. Dividing its brightness by its area gives it a surface brightness of 3.6 magnitudes per square arc-second.

Astronomical objects differ vastly in both total brightness and surface brightness. The full Moon and the planet Mars have nearly the same surface brightness; their total amounts of light are so different only because the Moon covers a much larger area of sky. The Moon and the Sun, on the other hand, have nearly the same apparent size. Here it's a difference in surface brightness that causes such dissimilar amounts of light.

An object of a certain total brightness (such as the Moon) also illuminates the surface of the Earth with a certain number of lumens per square meter. Any astronomical object illuminates the Earth's surface in such a manner. Examples are in Table 2.1.

The illumination an object causes on the Earth's surface is directly relevant to astronomy. A telescope objective has a given area on which light from the object falls. The illumination per unit area times the area of the objective determines how many lumens are delivered to the eye.

Conversion between common units of surface brightness is shown in Table 2.2. The surface brightnesses of some familiar astronomical objects are shown in Table 2.3.

DARK ADAPTATION

When a person walks from daylight into a darkened room, the pupil of the eye opens to a maximum of about 7.5 millimeters (0.3 inch) in only a couple of seconds. However, the eye cannot yet see very well in the dim light. The effect of opening the iris changes the light entering the eye by no more than a factor of about 16. But as minutes go by, the eye's sensitivity increases by a factor of many thousands.

This *dark adaptation* is due to a chemical process. In darkness a chemical called rhodopsin, or visual purple, is manufactured and builds up in the rods and cones. The amount of visual purple governs the sensitivity of the eye. Dark adaptation is mostly complete after approximately 30 minutes, as seen in Figure 2.4a, though a slight buildup of visual purple continues for as long as two hours.

As can be seen from Figure 2.4b, the greater the angle an object subtends – that is, the

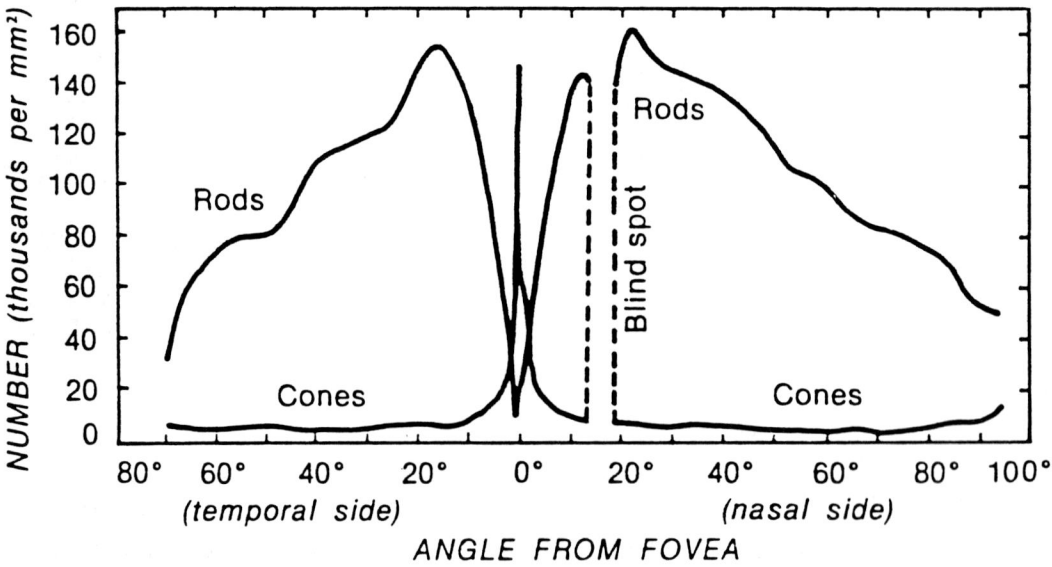

Figure 2.2. Distribution of rods and cones along a horizontal line across the retina. Parallel dotted lines represent the blind spot. (From *Sky & Telescope* April, 1984.)

Table 2.1. *Common illumination levels*

Source	Total stellar magnitude	Luminance caused at Earth's surface (lumens/square meter)
Sun	−26.7	130,000
Full daylight (not direct sun)	−24 to −25	10,000 to 25,000
Overcast day	−21	1,000
Very dark overcast day	−19	100
Twilight	−16	10
Deep twilight	−14	1
1 Candela at 1 meter	−13.9	1.00
Full Moon	−12.5	0.267
Total of all starlight	−6	0.001
Venus (at brightest)	−4.3	0.000139
Total of overcast starlight	−4	0.0001
Sirius	−1.4	0.0000098
0th-magnitude star	0.0	0.00000265
1st-magnitude star	1.0	0.00000105
6th-magnitude star	6.0	0.0000000105

Table 2.2. *Conversion of stellar magnitudes per square arc-second to candelas per square meter*

Magnitudes per sq. arc-sec	Candelas per sq. meter	Magnitudes per sq. arc-sec	Candelas per sq. meter
0	112700.0	14	0,283
1	44880.0	15	0.113
2	17870.0	16	0.0449
3	7114.0	17	0.0179
4	2832.0	18	0.00711
5	1128.0	19	0.00283
6	449.0	20	0.00113
7	179.0	21	0.000449
8	71.1	22	0.000179
9	28.3	23	0.0000711
10	11.3	24	0.0000283
11	4.49	25	0.0000113
12	1.79	26	0.00000449
13	0.71	27	0.00000113

larger its apparent size – the greater the sensitivity of the eye to it at all stages of dark adaptation, and especially after 30 minutes. This is an important concept to understand, for not only the brightness but the size of an object in the telescope will affect its visibility.

CONTRAST DISCRIMINATION

Visual astronomical observations depend not just on detecting faint light but also on contrast discrimination. Both abilities are involved in seeing such things as spiral arms of galaxies and dark rifts in nebulae, and in simply perceiving any object against the sky background. Contrast detection thresholds, as a function of background surface brightness for several object diameters, are plotted in Figure 2.5. This diagram shows that, for a given background (e.g. the night sky), less contrast is needed to see a larger object.

The data in Figure 2.5 were used to plot minimum detectable contrast versus angular size at constant values of background luminance to make Figure 2.6. Here we notice that for objects with small angular sizes, the smallest detectable contrast times the surface area is a constant. As an object becomes larger, this product is no longer constant. The angle at which the change occurs is called the *critical visual angle*. An object smaller than this angle is a point source as far as the eye is concerned. (A point can be considered the angular size smaller than which no detail can be seen.)

This critical angle is shown in Figure 2.7a plotted for various background luminances. Figure 2.7a shows that as the background becomes fainter, the size of a "point source"

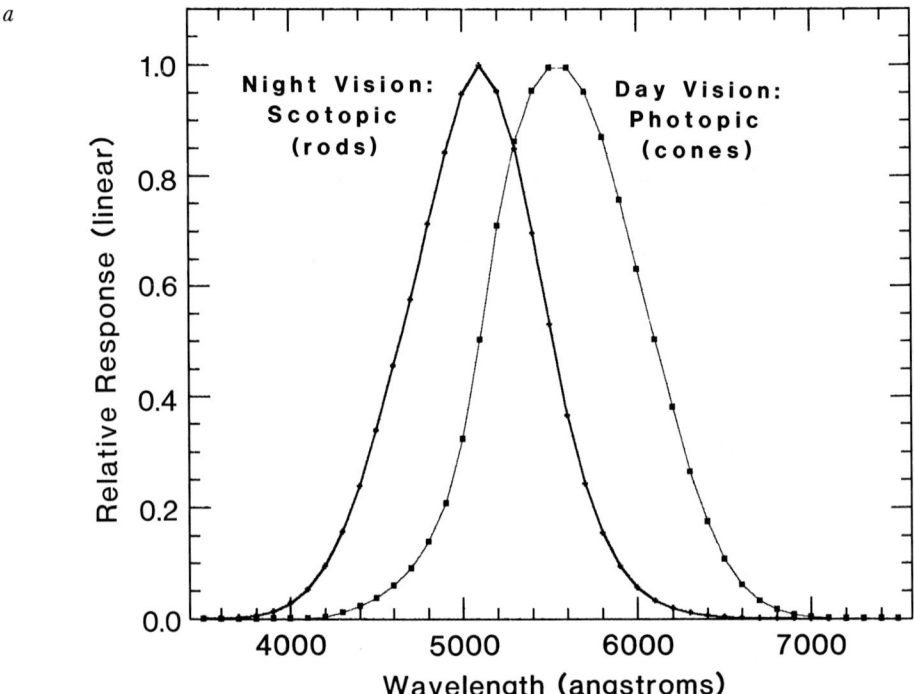

Figure 2.3. The relative response of the rods and cones as a function of wavelength (color) of light. Light of wavelength 5000 angstroms is green, 4000 angstroms is blue, and 6300 angstroms is red. Graph *a* shows the eye's response in linear units, while graph *b* shows the response in stellar magnitudes, a logarithmic scale. The data in both graphs *a* and *b* are scaled to the same value at their maximum to illustrate the color response of the rods and cones.

The cones are less sensitive than the rods, so the peak of the cone curve falls below that of the rod curve. How much below depends on the dark adaptation of the eye. When fully adapted, the rod peak is higher by slightly more than four magnitudes. This is shown in graph *c*. Here it is seen that the cone curve lies completely below the rod curve. Thus, the rods are more sensitive at all colors when dark adapted than the cones. Derived from data in Table II of Kingslake (1965). The cone relative to the rod sensitivity in graph *c* was derived from data in Crossier and Holway (1939).

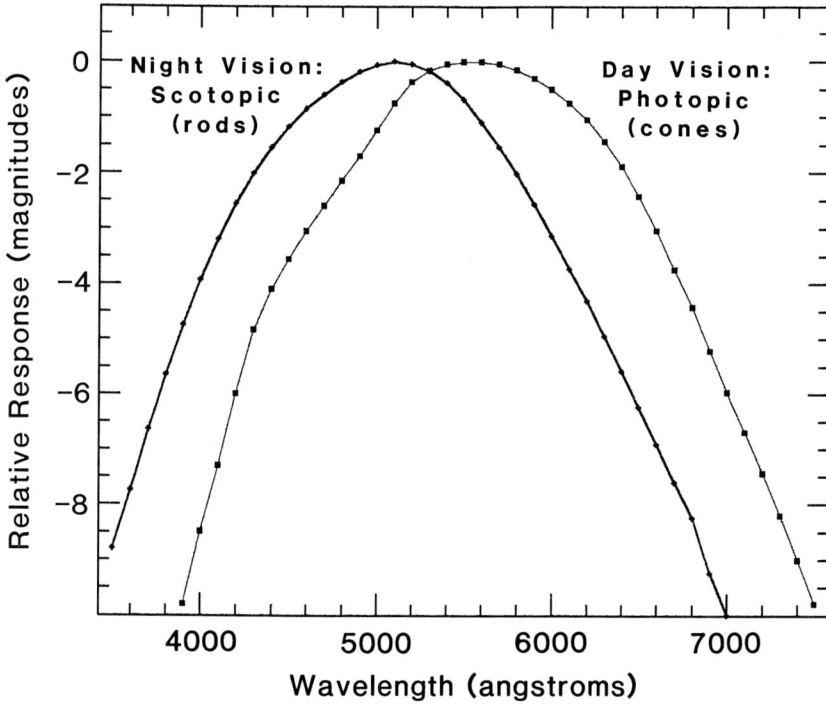

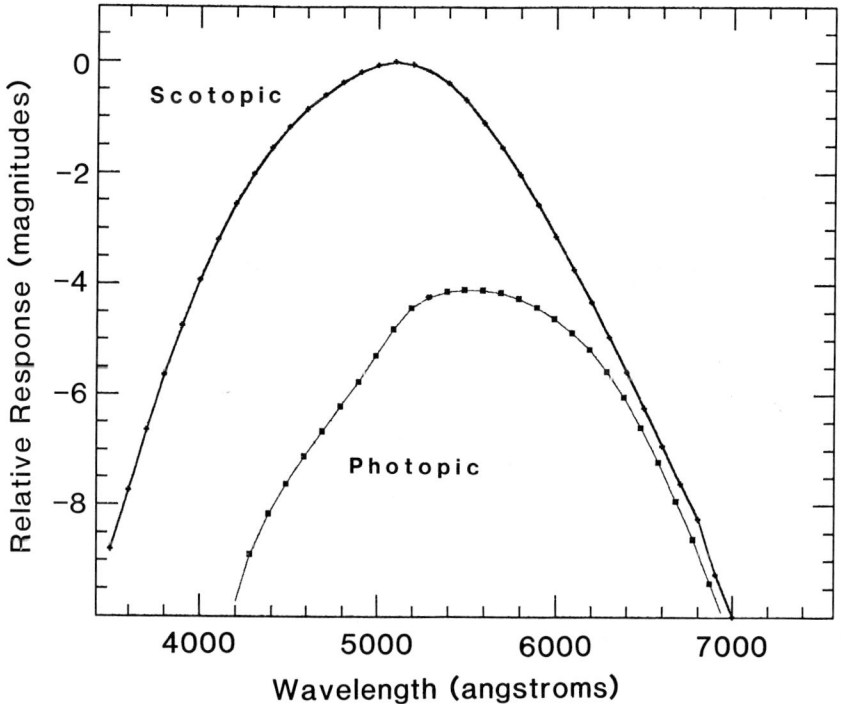

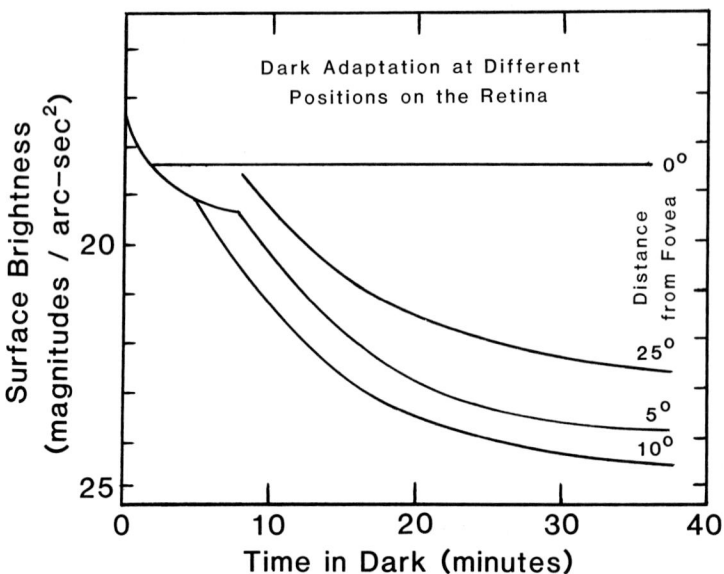

Figure 2.4. a) Dark adaptation measured with a 2° diameter test object placed at various angular distances from the fovea. Derived from data in Middleton, 1958.

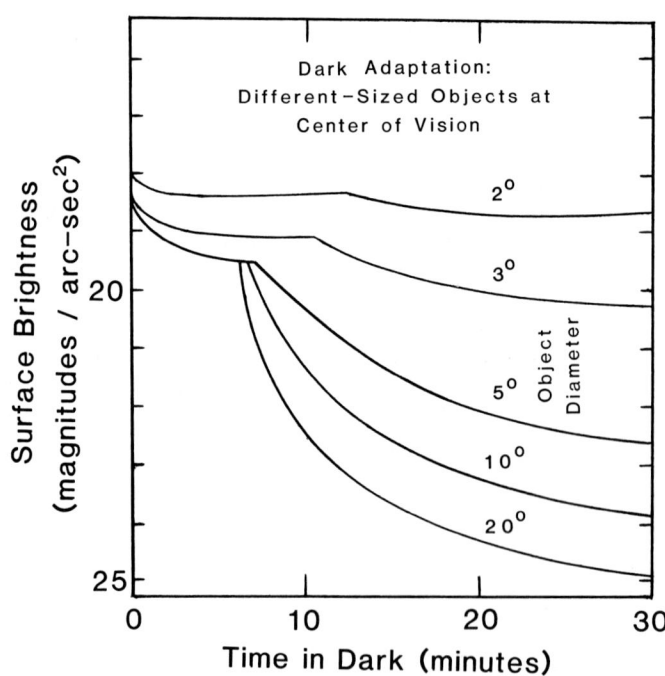

b) Dark adaptation as measured with test objects of different angular sizes located at the center of vision. The sizes are the diameter in degrees at which the objects could just be detected. Derived from data in Hecht et al. (1935) after Bartley (1951).

Table 2.3. *Approximate surface brightnesses of commonly observed objects*

Object	Candelas per sq. meter	Magnitudes per sq. arc-sec
Sun (magnitude −26.7, d = 1800″)	3 000 000 000	−10.7
Venus (at greatest elongation, −4.3, d = 24″, illum. = 50%)	15 000	1.9
Clear daytime sky (at horizon)	10 000	3
Full Moon (−12.5, d = 1800″)	6000	3.6
Mars at nearest opposition (−2.8, d = 25.1″)	4000	3.9
Overcast daytime sky (at horizon)	1000	5
Jupiter at opposition (−2.5, d = 49.8″)	800	5.7
Saturn at opposition (−0.4, d = 20.5″)	700	5.9
Heavy daytime overcast (at horizon)	100	8
Uranus (5.7, d = 4.2″)	60	8.6
Neptune (7.6, d = 2.5″)	30	9.3
Sunset at horizon, overcast	10	10
Clear sky 15 minutes after sunset (at horizon)	1	13
Clear sky 30 minutes after sunset (at horizon)	0.1	15
Fairly bright moonlight (at horizon)	0.01	18
Moonless, clear night sky (at horizon)	0.001	20
Moonless, overcast night sky (at horizon)	0.0001	23
Dark country sky between stars (at zenith)	0.00003	24

Note: Sky brightness values at the horizon can vary by a factor of ten and are adapted from Middleton (1958).

For the Sun, Moon, and planets, the total stellar magnitude and diameter (d) in arc-seconds (″) for which the computation was performed are listed in parentheses.

becomes larger for objects that are just detectable. In other words, the eye's resolution or ability to see detail is much coarser in the dark.

The eye's ability to see a point source, such as a star, increases as the background luminance gets dimmer. This is shown in Figure 8. Dark country skies are better than city skies for seeing faint stars because the background is fainter – *not* because the country skies are significantly more transparent. Astronomers often refer to sky darkness as "transparency," but this is a misnomer.

A low-contrast object is more easily detected if it is larger. For an extended object such as a galaxy viewed in a telescope, magnification does not change the contrast with the background, because both the sky's and the object's surface brightnesses are affected equally. Some visual observers have stated that a dim object's contrast with the sky background increases with higher magnification, but this is clearly wrong. The contrast merely *looks* greater because of the increased detection capabilities of the eye.

This facet of human vision has not been described in print to my knowledge, so I will coin a name for the maximum magnification that will help detection: the "optimum magnified visual angle". This angle is shown in Figure 2.6 and also Figure 2.7b. For those readers familiar with calculus and the slope of a curve, the optimum magnified visual angle occurs when the first derivative (the slope) of each curve in Figure 2.6 is equal to -1.

If an object is at the threshold of detection and smaller than the optimum angle, more magnification will make it easier to see. When the object is magnified beyond the optimum angle, its surface brightness decreases faster than the eye's contrast detection threshold, and the object will become harder to detect. Remember that even for an object somewhat above the detection threshold, higher magnification may bring out details *within* the object that are smaller than the optimum angle at a lower magnification. An example of what this means in practical terms for the observer is given in Chapter 6.

AVERTED VISION

We have seen that the rods are more sensitive to light than the cones, and that they are concentrated in the periphery, not in the central part of the eye where visual acuity is best. Most amateur astronomers quickly learn about averted vision: a faint object may be invisible when you look straight at it, but it pops into view if you look slightly to one side. As was shown in Figure 2.2, the density of the rods increases away from the fovea, reaching a maximum at about 18° to 20° off-axis. However, this generally does not correspond to the maximum sensitivity region. The sensitivity of the eye varies somewhat with the person, with the direction off-axis, and with the diameter of the object.

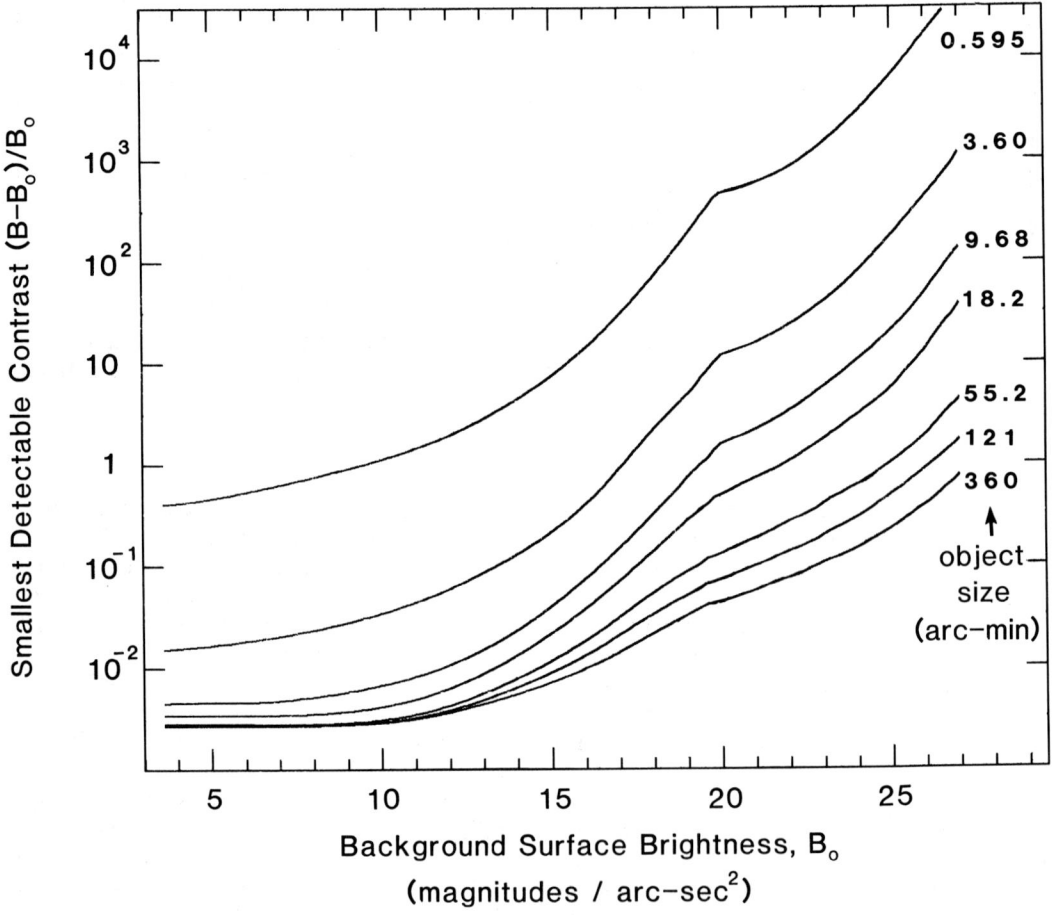

Figure 2.5. The minimum contrast needed to detect an object of a given angular size shown as a function of background surface brightness, B_o. The larger an object appears to the eye, the easier it is to detect. For small, bright objects on a bright background, a contrast less than 0.01 is enough for detection. But against the very dim night-sky background seen in a telescope (fainter than 25 magnitudes per square arc-sec), a large object must have a contrast of nearly 1.0, and a small object more than 100, to be detected. Derived from data in Table VIII of Blackwell (1946).

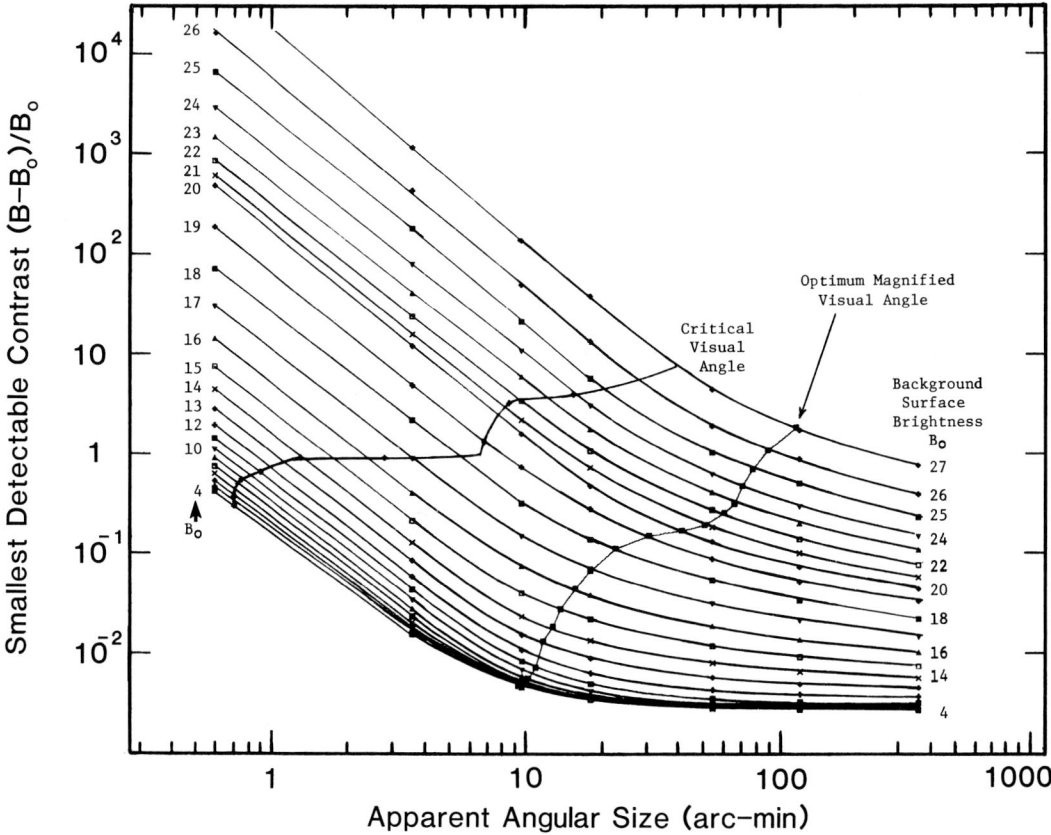

Figure 2.6. The smallest contrast needed to detect objects of various sizes on various backgrounds. This diagram is the most important one in the book, so it's worth taking the time to figure out its complexities. This is the same data as in Figure 2.5, except that contrast detection ability is plotted against angular size for various background surface brightnesses (magnitude per square arc-second).

When an object is magnified in a telescope, the contrast between object and background does not change since both are magnified equally. However, the object becomes larger as viewed by the eye. Therefore, moving horizontally across the chart corresponds to increasing the magnification.

As we start with low magnification on the left side, the contours of background surface brightness are diagonal straight lines. At a point called the critical visual angle, the lines begin to curve. Objects smaller than this value appear as point sources (the smallest detail that can be distinguished). As one moves to the right of the critical visual angle line, the faintest detectable surface brightness decreases faster than the background surface brightness. Thus, fainter objects – or detail within objects – can be seen as magnification is increased.

This is true only until the "optimum magnified visual angle" is reached. Thereafter, higher magnification decreases the detection threshold faster than surface brightness. A faint object is most visible when magnified to this angle. Chapter 6 is devoted to an example of the use and implications of this fact. Derived from data in Table VIII of Blackwell (1946).

An example of peripheral vision capabilities is shown in Figure 2.9. The periphery is about 40 times (4 magnitudes) more sensitive than the fovea. As Figure 2.10 shows, peripheral vision is best when the object appears 8° to 16° off-axis in the direction toward the nose. The areas up, down, and toward the ear (temporal side of the head) are not quite as sensitive. The blind spot lies 13° to 18° toward the ear.

EXPOSURE TIME

Contrary to what nearly every astronomer believes, the eye seems to have an integration capability similar to photographic film, though much more limited. For the detection of the faintest objects, the light must accumulate on the retina for around 6 seconds. When searching for faint objects in the telescope field, scanning the area too fast will considerably raise (worsen) the detection threshold. Fixating on a point in the field of view while using averted vision should work best. Practice is required to concentrate on a point while using averted vision, because the eye tends to jerk around slightly. Fatigue seems to compound the problem, an inherent problem for the casual astronomer trying to stay awake all night.

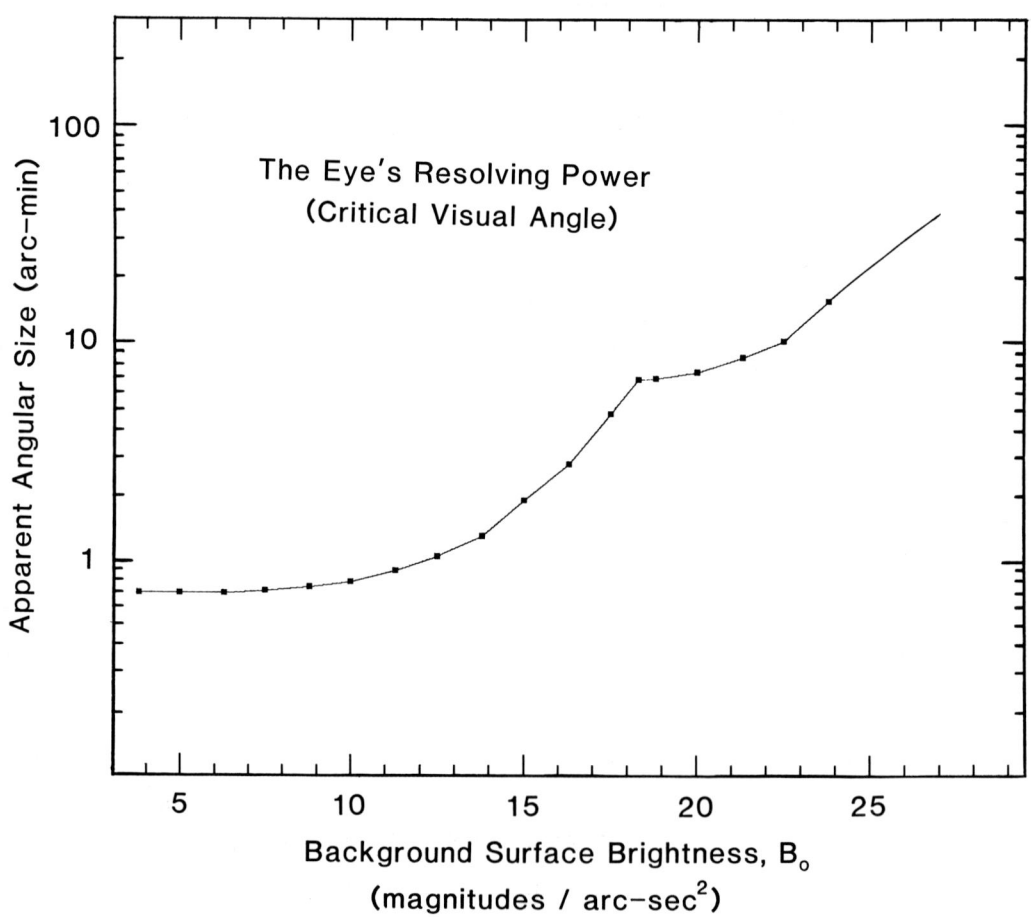

Figure 2.7a. The eye's resolving power, unlike a camera's depends on an object's surface brightness. The critical visual angle is the angle below which no detail can be seen and objects appear as point sources.

THE HUMAN EYE

COLOR

The human eye is a remarkable detector of color under bright, daytime conditions. But the color receptors, the cones, do not function at all in the low light levels of night, so no color is seen. The threshold of color is about 21.5 magnitudes per square arc-second (0.00003 candelas per square meter). At this intensity both rods and cones are working. Because rods and cones are most sensitive to different colors (Figure 2.3), the perceived brightness of an object near the transition light level can depend on what color it is. This *Purkinje effect* is well known to variable star observers, who often have to compare two stars of different colors.

In amateur telescopes, color can only be seen in the brightest portions of a few deep-sky objects, such as the Great Orion Nebula (M42) and bright planetary nebulae. Usually the only color seen is pale green, but if the contrast is high (as might be found in the mountains far from city lights), pastel reds sometimes appear in bright emission nebulae such as M42. However, color sensitivity seems to be among the greatest variables of the human eye. It is the author's experience that the sensitivity to seeing red in M42 varies considerably from person to person

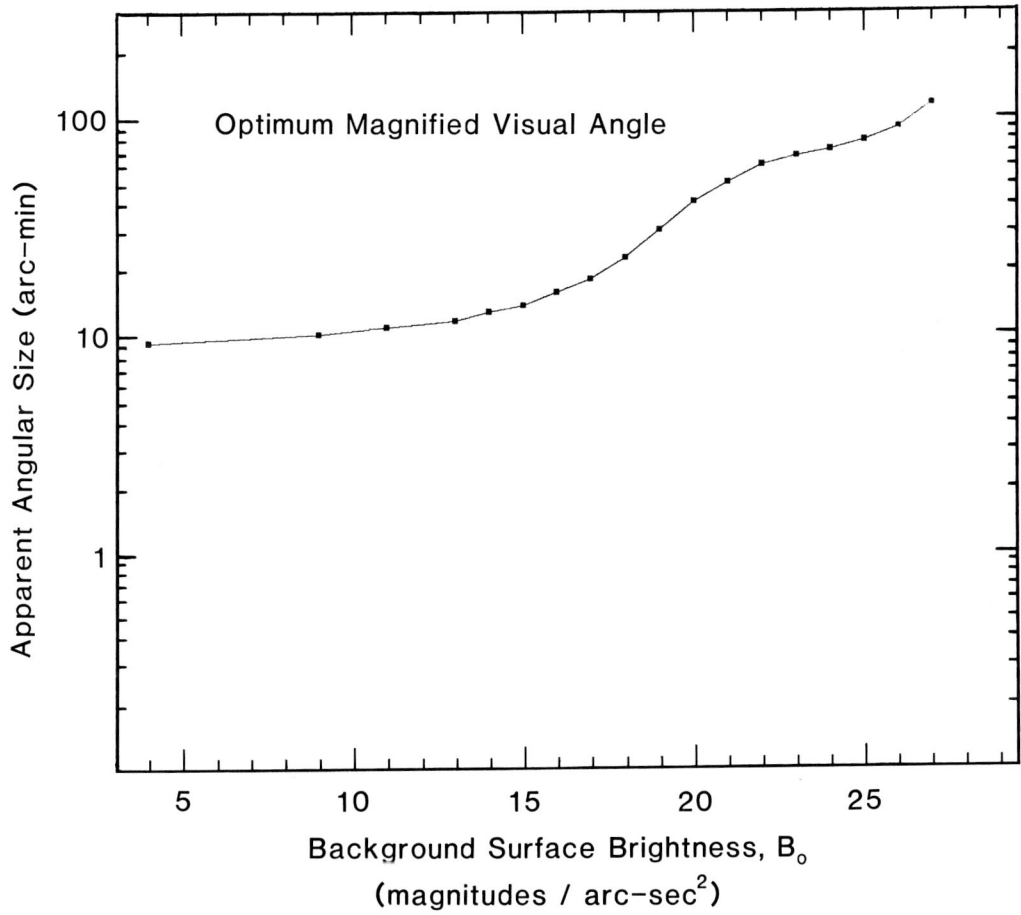

Figure 2.7b. The "optimum magnified visual angle" of an object depends on surface brightness. This angle is the size for which a faint object, or detail within an object, should be magnified in order to maximize the possibility of detection.

when all look through the same telescope. At the threshold of color vision, one should keep in mind, the perceived color may not be the true color of the object because of the Purkinje effect – and no deep-sky object is much above the threshold.

Note that the threshold of 21.5 magnitudes per square arc second is for the surface brightness at the eye, after the image is magnified by the telescope. Depending on the magnification, a telescope usually reduces the surface brightness by 0.6 to 7 magnitudes per square arc second. Appendices E and F give the mean surface brightnesses of various objects. These lists make it clear that few objects offer any possibility for detecting color.

(Many objects, however, have bright spots several magnitudes above the mean; see the entry for M42 in Appendices E and F and in Chapter 7).

VISION AND HEALTH

Many other factors affect the eye's sensitivity to faint light. For the eye to function properly there should be no vitamin deficiencies, and levels of blood sugar and oxygen should be adequate. At high altitudes, where the air is thin, some observers discover that heavy breathing (but not to the point of hyperventilating) enhances their night vision. Smoking will reduce the sensitivity of the eye. Contrast

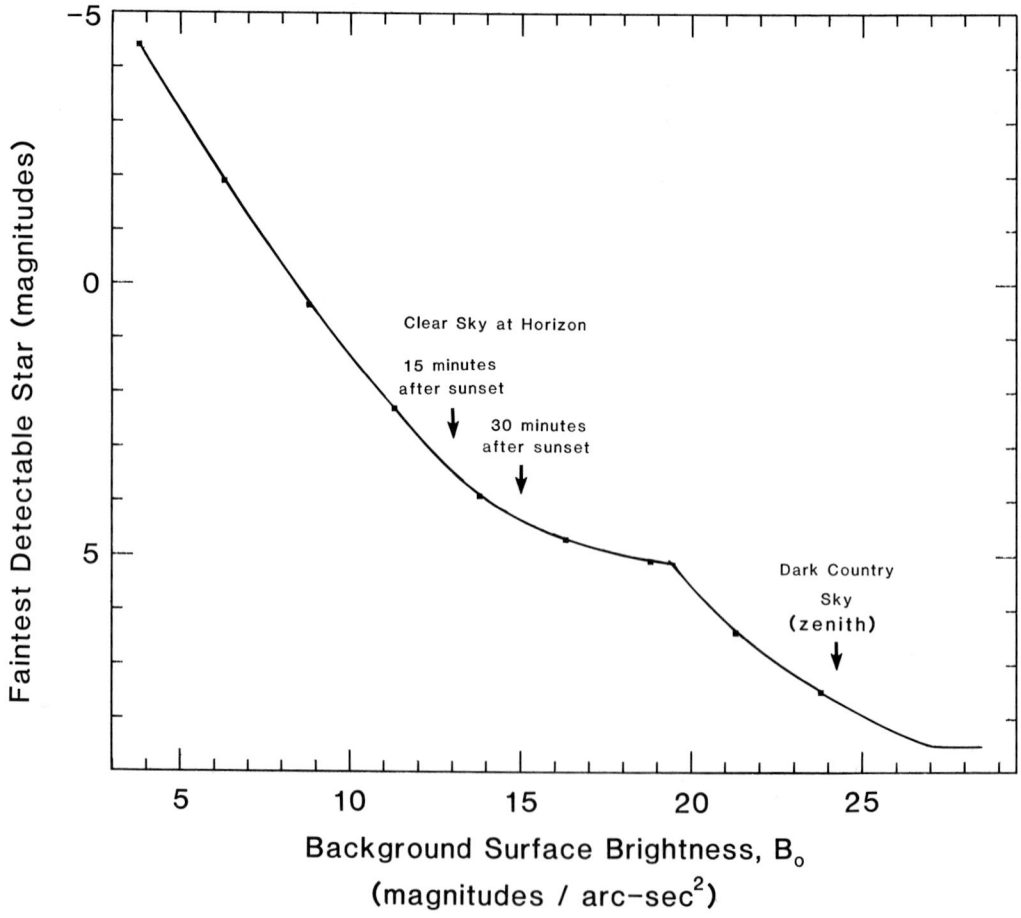

Figure 2.8. The faintest star that can be detected by the trained human eye, shown as a function of the background surface brightness. This curve was derived from the data in Figures 2.6 and 2.7a. The faintest point source, magnitude 8.5, was determined by Curtis (1901).

sensitivity decreases as blood-alcohol levels increase, so one should not drink alcoholic beverages before observing. Eating candy or other food gives a blood-sugar boost. Vitamin A and zinc are crucial for maintaining the eye's sensitivity, but taking supplements will not improve one's vision if the body already has enough.

Sunlight can be very detrimental to dark adaptation. When the eyes are exposed to bright sunlight for long periods, dark adaptation can take several days. Before an observing session, ultraviolet-filtering dark glasses should be worn outdoors for several days.

As a person ages the eye does too. The transmittance of the eye's lens declines and the pupil is not able to open as wide. Exposure to ultraviolet radiation reduces the lens transparency and also ages the retina, so unprotected exposure to bright sunlight (such as at the beach or in bright snow) should really be avoided at *all* times. Some sunglasses provide no protection at all against ultraviolet light. Ordinary eyeglasses, on the other hand, can be ordered with ultraviolet-filtering lenses. These are a fine idea for any astronomer who wears glasses.

The pupil will open to about 7 or 8 millimeters when a person is in his or her teens and early twenties, but by age 80, it may open to only about 3 to 5.5 millimeters. The age of the observer should be considered when choosing low power so that the telescope's exit pupil is not larger than what the eye can accept. The next chapter discusses exit pupils; see Bowen (1984) for further discussion.

SUMMARY

To detect faint objects use averted vision. The best direction to avert your view is so the object lies 8° to 16° toward the nose from the center of vision. Viewing 6° to 12° above or below the object is almost as good, but placing the object toward the ear should be avoided because of the blind spot.

A faint extended object (e.g. a galaxy, a bright spot within a galaxy or nebula) should

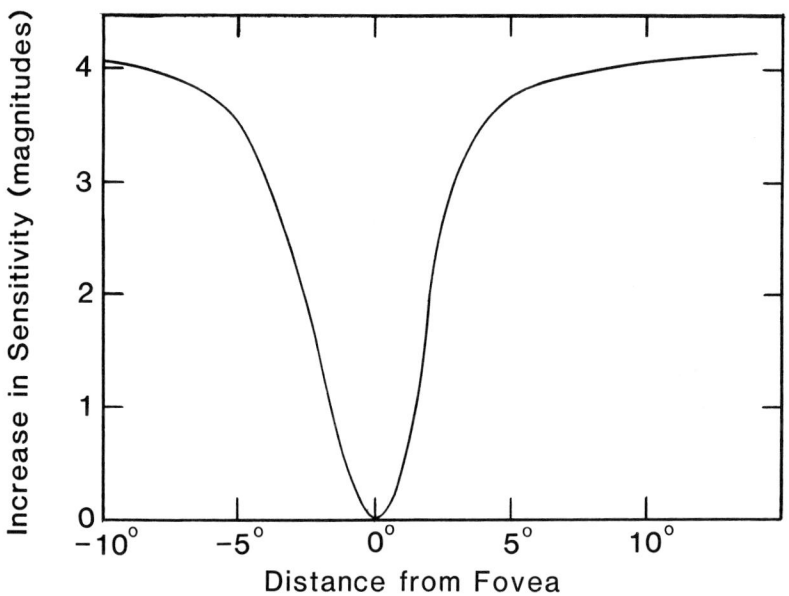

Figure 2.9. Peripheral vision. The average sensitivity of the eye is shown at different angular distances from the fovea (center of view). The eye can see about 4 magnitudes fainter when an object is placed a few degrees off-axis rather than stared at directly. The data were derived by averaging the response curves from three individuals. Derived from data in Crossier and Holway (1939).

be viewed with enough magnification so it appears several degrees across to the eye. To be detected, it must be surrounded by a darker or lighter background so the eye can distinguish contrast. Various magnifications should be tried to bring details into the range of best detection. At each magnification, considerable time must be spent examining for detail. Higher magnifications should be tried until the object is totally lost from view.

The eye should be dark adapted for at least 30 minutes so the photochemical visual purple is at full abundance. Bright stars and extraneous lights will tend to destroy dark adaptation.

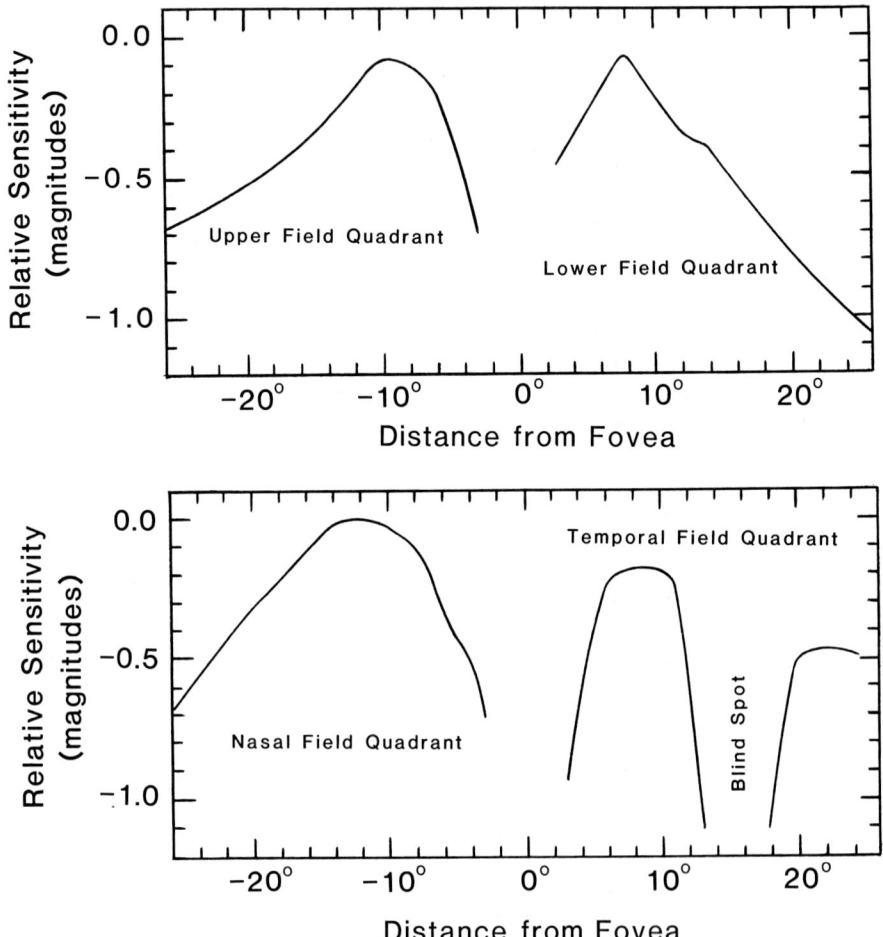

Figure 2.10. Peripheral vision further broken down. The detection threshold is shown for each field quadrant of the eye. The highest sensitivity occurs about 8° to 16° from the fovea where the image appears closer to the nose, although above and below the fovea are nearly as good. Placing the image further from the nose (in the temporal quadrant) should be avoided because of the blind spot. Derived from data in De Groot et al. (1952).

3
The eye and the telescope

TELESCOPE BASICS

In the last chapter, we saw that five factors influence the visibility of a faint astronomical object: its brightness, the brightness of the background, the object's apparent size as viewed by the eye, the eye's dark adaptation, and where the image falls on the retina.

The first three factors depend in part on the telescope. All that really influences them is the telescope's aperture and magnification. Other details of what goes on between the eye and the telescope objective are immaterial, as long as the optics do not significantly absorb or scatter light. This rule has interesting implications for the long-standing question of the best type of telescope for deep-sky objects. The answer is, it doesn't really matter.

The design of the telescope, however, may restrict the achievable magnifications and field of view, and some designs do absorb or scatter light less than others. Refractors are generally better in this regard than reflectors if the aperture is less than about 8 inches.

Telescope design is a compromise of many factors. The most important is the aperture. This is the size of the objective, the main lens or mirror which collects the light. The larger the objective, the more light it gathers and the brighter the image. A bigger objective can also resolve finer details, but in practice the Earth's turbulent atmosphere often blurs the image so that a large telescope can see no more detail than a small one.

A large-aperture telescope certainly costs more than a small instrument of similar design and is also bulkier and harder to store and transport. For the amateur astronomer, portability is often crucial because the telescope will be carried by car far from city lights. If the telescope is kept indoors, it should not be too heavy or complex to move outdoors and set up. A troublesome giant of an instrument will be used only infrequently. There are many telescopes designed to optimize performance while minimizing the inconveniences.

A telescope acts as both a light-gathering and magnifying device, whether used visually or for taking photographs. The objective gathers light and focuses it to form an image, which floats in the air like a tiny upside-down picture of the scene. The distance from the objective to the image is called the *focal length (F)*. (Some complex telescope designs have *effective focal-lengths* that differ from this distance.)

A key characteristic of any telescope is its focal ratio or f/ratio. This is simply its focal length (F) divided by its aperture diameter (D):

$$\text{f/ratio} = F/D. \qquad \text{(equation 3.1)}$$

A telescope with a 4-inch diameter objective and a focal length of 40 inches, for instance, is called f/10.

In order to see through the telescope, an eyepiece must be used. The eyepiece magnifies the image the same way a hand lens magnifies newspaper print. The telescope's power, or magnification (m) equals the effective focal length of the telescope divided by the focal length of the eyepiece (f):

$$m = F/f. \qquad \text{(equation 3.2)}$$

In general, the higher the magnification, the less sky can be seen at once.

ABERRATIONS

There are many telescope designs, and every one suffers to some degree from optical defects known as aberrations. These result when rays of light from a point on a distant object fail to reach the correct point on the image (also called the focal surface). Some such errors are inherent in the system design; others are added by imperfect construction. The six main ones are: *chromatic aberration, coma, astigmatism, spherical aberration, distortion,* and *Petzval curvature.*

Chromatic aberration arises because different colors of light are bent by different amounts when they pass through a lens. Unfortunately, no kind of glass known has the same index of refraction (light-bending strength) for all colors. Simple lenses suffer from chromatic aberration the most. Images formed by a simple lens are surrounded by obvious blue and red fringes. When two or more lenses with different index-of-refraction properties are combined, the chromatic aberration can be controlled to some extent because two or more different colors can be made to focus at the same place.

Coma is named for the comet-like images of stars seen off-axis (away from the center of the field). The farther the star is from the objective's optical axis, the worse the coma will be. Coma also depends on the f/ratio of the system, and for a given f/ratio and angle off-axis, it increases with the aperture.

Astigmatism may be caused by errors in the manufacture of an optical component or may be inherent in the design. An example might be a "spherical" lens surface that is not perfectly spherical but slightly cylindrical, for instance if the light from the left and right edges of the lens is focused at a different point than light from the top and bottom edges. Astigmatism can also be caused by improper alignment of an optical system.

Spherical aberration is a telescope's inability to focus all the light at the same distance from the objective. For example, a beam of parallel light that encounters a spherical mirror will not all be focused at the same distance from the mirror. The light that encounters the mirror's edge will be focused closer than light that strikes the center. Parabolic mirrors do not suffer from spherical aberration when focused on objects very far away. *Zonal Aberration* is spherical aberration at a particular radius from the center of a lens or mirror.

Distortion is the squeezing or stretching of a part of an image toward or away from the center. A common test of distortion uses a square grid; the lines may bend toward or away from the center, giving the appearance of a pincushion or a barrel. Hence the common names of *pincushion distortion* and *barrel distortion.*

Petzval curvature (or field curvature) results from the image being not flat but slightly concave or convex. Typically, the best focus at the edge of the field lies closer to the objective than the best focus at the center. (Astigmatism also affects the surface of best focus.) Eyepieces too can have field curvature, so stars at the edge of the view are rarely as sharp as on-axis – unless the eyepiece curvature happens to compensate for that of the objective.

TYPES OF TELESCOPES

Telescope design is a compromise between aberrations, cost, portability, focal length and field of view, among other things. The different types of telescopes described below are illustrated in Figure 3.1.

The refractor

A refractor is a telescope whose objective is a lens. To reduce chromatic aberration, the lens is almost always *compound,* or made of two or more disks of glass (lens elements). Usually the objective is *achromatic* (made of two elements), though a few expensive ones are *apochromatic* (usually having three elements).

The main defect of refractors is chromatic aberration, but astigmatism and spherical aberration can also be problems depending on field size, focal ratio, and design. A high focal ratio is required in most achromatic lens designs in order to minimize the aberrations. The commonly accepted f/ratio for refractors is f/15, though if they are small and not intended for high magnifications, f/8 or even f/6 will give acceptable definition. Binoculars are refracting telescopes with an image-erecting

THE EYE AND THE TELESCOPE

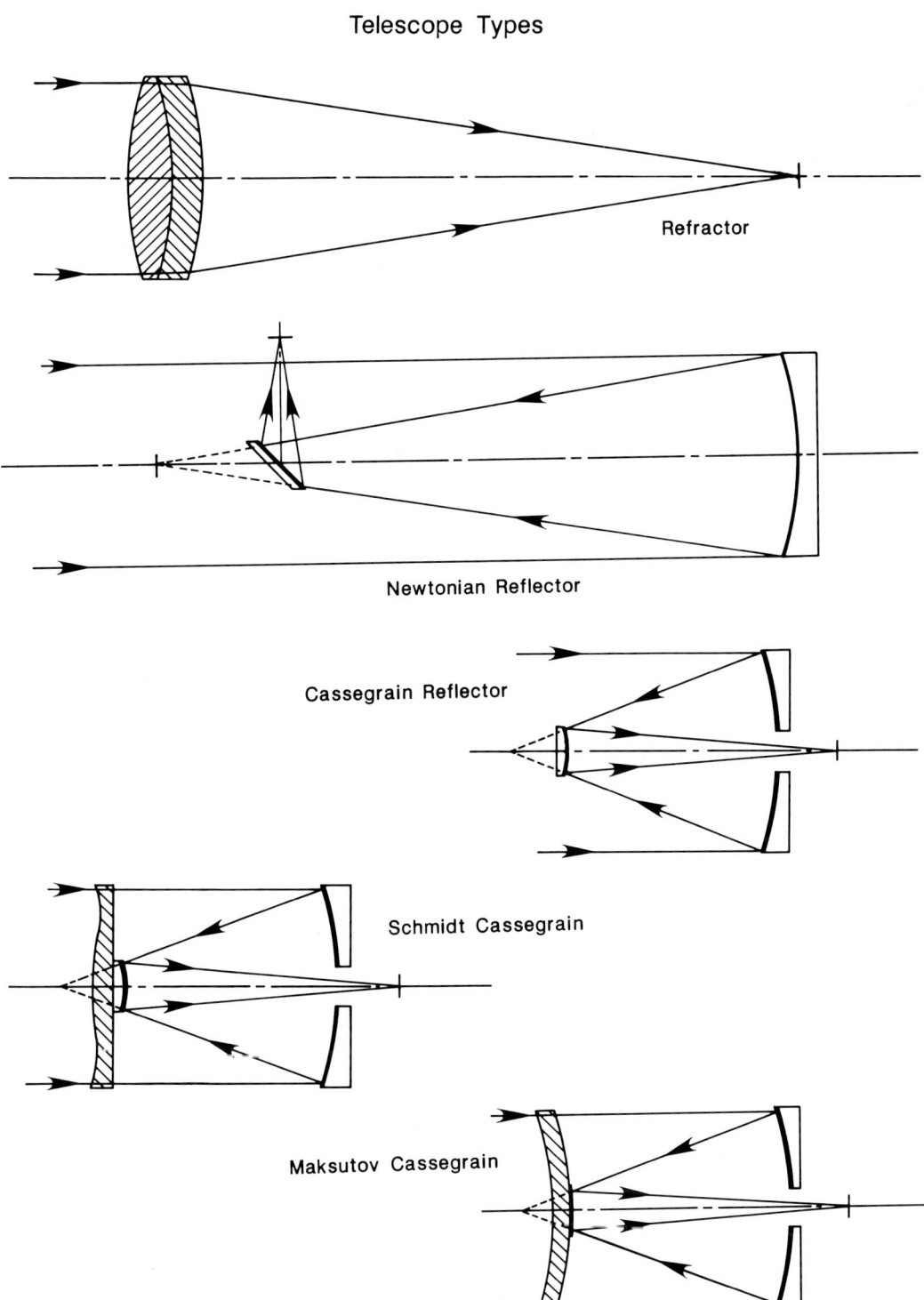

Figure 3.1. Basic telescope designs. In each case, starlight arrives from the left and is focused to a point on the image plane, which is seen in profile as a short line.

system and a low-power eyepiece for each eye.

Advantages of the refractor:

  The enclosed tube helps keep the optics clean.
  Refractors are relatively free of coma.
  There is no central obstruction (see reflecting telescopes below).
  The construction is rugged and stays in optical alignment.
  A lens scatters less light than a mirror.

Disadvantages:

  The cost is often many times that of the same size reflecting telescope.
  The long focal length makes it difficult to achieve a large field of view.
  The long focal length means a long tube, which requires a larger mounting, reduces portability and increases susceptibility to being shaken by wind.
  Astigmatism is present off-axis.

The reflector

A reflector uses a mirror for its objective, so it is free from chromatic aberration. (All eyepieces, however, employ lenses, so the eyepiece can add noticeable false color to a reflector.) Only the surface of the mirror needs to be optically perfect, whereas in a refractor the entire volume of glass must be of high optical quality. An objective mirror has only one optical surface, making it less expensive to manufacture than an objective lens, which has at least four.

Reflectors come in two basic designs: Newtonian and Cassegrain.

The Newtonian reflector

The Newtonian telescope, named for its inventor Isaac Newton, usually has a parabolic objective mirror. The focal point is between the mirror and the object to be viewed. In order to see the image, a small flat mirror is usually placed just inside the focus to reflect the converging light to the side of the tube where the observer can place an eyepiece or other equipment such as a camera.

Advantages:

  Low cost.
  Freedom from chromatic aberration.
  Focal ratio can be short (as low as about f/4), giving a large field of view and good portability.

Disadvantages:

  Substantial coma at low f/ratios.
  Alignment may often need some adjustments
  The reflective coatings are easily scratched and more difficult to clean than a lens.
  The small diagonal mirror blocks some light, and also slightly reduces resolution and contrast.

The Cassegrain reflector

The classical Cassegrain telescope has two curved mirrors: a paraboloid just like the Newtonian, and a small convex hyperboloid instead of the Newtonian's flat secondary. The hyperboloid secondary mirror magnifies the image formed by the primary. In the most common configuration, the image is directed back through a hole in the primary to a convenient location behind, where eyepieces and other equipment can be placed.

The coma in a Cassegrain is the same as in an equivalent focal length Newtonian. In effect, the Cassegrain is a compact high f/ratio Newtonian. In addition, a flat diagonal mirror may be temporarily substituted for the hyperboloid secondary, creating a short-focus Newtonian telescope.

Advantages:

  Freedom from chromatic aberration.
  The f/ratio can be fairly low, giving a relatively large field of view, though not as large as in a Newtonian.
  A Cassegrain is more compact than an equivalent Newtonian.

Disadvantages:

  Substantial coma.
  Alignment is more difficult than for Newtonians.
  The reflective coatings are easily scratched and more difficult to clean than a lens.
  The hyperbolic secondary mirror is larger than the flat secondary of a Newtonian,

blocking more light and further degrading image sharpness and contrast.

Variations on the Cassegrain include the Gregorian, which has a parabolic primary and a concave ellipsoidal secondary, and the Ritchey-Chrétien and Dall-Kirkham, which look like ordinary Cassegrains but have slightly different mirror figures.

Catadioptric telescopes

Catadioptric or compound telescopes use a combination of lenses and mirrors. The lenses do not refract light enough to cause chromatic aberration, but merely correct for aberrations in the mirrors.

The Schmidt-Cassegrain

The Schmidt-Cassegrain has a spherical primary mirror and a convex, usually spherical secondary mirror. A thin lens or corrector plate in front corrects the spherical aberration. This lens has an unusual aspheric figure.

Advantages:

All the advantages of a Cassegrain.
A sealed tube keeps dust off the mirrors.
Better freedom from aberrations than a classical Cassegrain.

Disadvantages:

Same as those for Cassegrains.
More expensive than Newtonians or classical Cassegrains.
Curved focal surface.

The Maksutov Cassegrain

A Maksutov telescope has a spherical primary and secondary much like the Schmidt-Cassegrain, but the corrector lens is a deeply curved meniscus.

Advantages:

Similar to those of the Schmidt-Cassegrain except the tube is even shorter.
No aspheric surfaces. (Minor spherical aberration is sometimes corrected by aspherizing the meniscus.)

Disadvantages:

Similar to those of the Schmidt-Cassegrain. Curved focal surface, but better than for similar Schmidt-Cassegrains.

TELESCOPE MOUNTINGS

A telescope's optics are only half the story. The best optical system will be useless if it is on a wobbly, quivery mounting.

A telescope mounting must not only hold the instrument solidly, but allow it to be pointed anywhere in the sky smoothly and easily. Three basic types of mountings are in common use: 1) the German equatorial, 2) the fork, and 3) the Dobsonian. They are illustrated in Figure 3.2. There are many variations on these basic designs.

All mountings can be divided into two groups: *altazimuth* and *equatorial*. If one of the mount's axes is made parallel to the earth's axis, then the telescope is called an equatorial. It swings in the direction of celestial north-south and east-west, and can be made to follow the stars as the earth rotates by turning only one axis. This tracking of the stars can be done with a motor known as a clock drive.

An altazimuth mount, on the other hand, swings up and down (in altitude) and side to side (in azimuth). The German and fork mountings are equatorial, whereas the Dobsonian is altazimuth.

The German equatorial mounting (Figure 3.2a) is good for long telescopes such as refractors and long-focus Newtonians. But it requires a heavy counterweight, and when it follows an object crossing the meridian, the telescope must be taken off the object, swung around to the other side of the mount, and re-aimed.

The fork mount (Figure 3.2b) is probably the best for telescopes with short tubes. The eyepiece does not move much as the telescope is pointed to different parts of the sky, and there is no counterweight. A disadvantage is that the sky around the celestial pole sometimes cannot be viewed.

A variation of the fork mounting is the yoke, which is supported by two piers, one on either end of a framework that holds the telescope. The yoke mounting is very steady but not portable and not commonly found in commercial instruments for amateurs.

The Dobsonian mount (Figure 3.2c) is the most stable design commonly used by amateurs. It is sort of a fork mount, but because it is an altizmuth the design is simple while mechanical stresses are minimized. Commonly built of plywood with teflon pad bearings, the Dobsonian mount can be very compact, and it has no counterweights. Its main disadvantage is difficulty in tracking stars, especially near the zenith. It is often used to mount short-focus Newtonian telescopes of very large aperture. Such a mount costs very little to build – basically the price of the wood – yet it can be as stable as a massive equatorial.

EYEPIECES

Eyepieces are short-focal-length lens systems that magnify the image formed by the telescope objective. Because the eye is most comfortable when focused on objects at infinity (10 meters or more in practice), the eyepiece is designed to convert the light from a telescope image into a system of parallel rays. The eye lens and cornea focus these rays onto the retina just as a camera lens focuses light from a distant object onto film.

Eyepiece designers try to minimize aberrations while maximizing the field of view. Most eyepieces have two or more lenses (Figure 3.3). The one closest to the eye is called

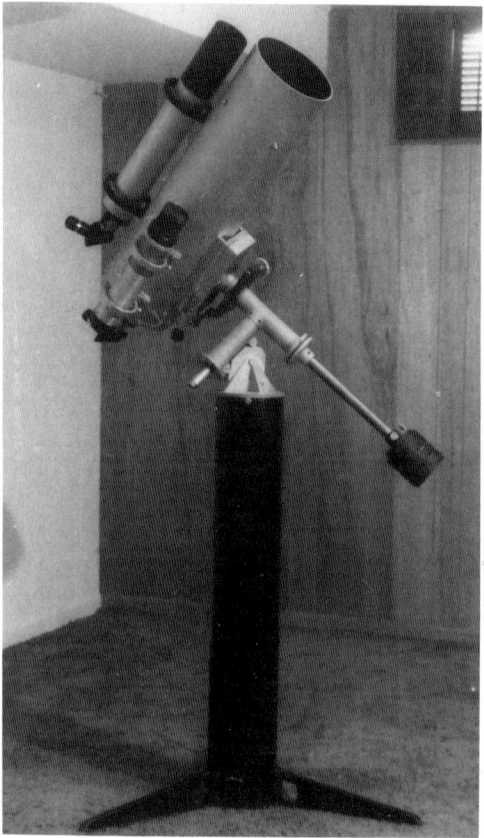

Figure 3.2a. The author's homemade telescope, an 8-inch classical Cassegrain on a German equatorial mounting. The primary mirror is f/3.5, and the secondary mirror magnifies the focal length to 2336 mm (f/11.5). The smaller finder has a 60-mm objective with a 300-mm focal length and a 38-mm Erfle eyepiece, giving 7.9× and an 8° field of view. The larger finder is a 3.1-inch f/8 refractor normally used with a 20-mm Erfle eyepiece giving a 2° field of view and 31×.

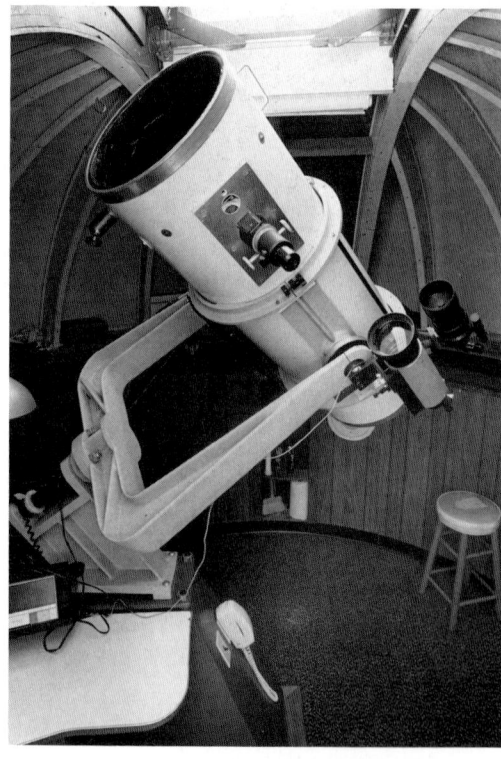

Figure 3.2b. Johnny Horne's 12.5-inch combination Newtonian-Cassegrain telescope is shown on a fork mount.

the eye lens, and the one farthest is called the field lens. The focal plane of the eyepiece, commonly called the image plane, should be positioned to correspond to the image plane of the objective. This is what you do when you focus a telescope.

The light that leaves the eyepiece forms a cone of parallel ray bundles. The angle of this cone is called the apparent field of view, because this is the angular size the eye sees as the round "window" of the telescope's view.

Each bundle of rays from a point on the image has the same diameter as it leaves the eyepiece. This diameter depends on the magnification of the telescope and the size of the objective. These ray bundles come together behind the eyepiece to form a disk called the "exit pupil." It can be seen as an actual disk of light floating just behind the eyepiece. The iris of the eye is placed here for optimum viewing. If the exit pupil is too close to the eyepiece, the observer will not be able to get his or her eye close enough to see the whole field of view. The distance from the eye lens to the exit pupil is called, appropriately enough, the "eye relief".

Eyepieces used for astronomy should have antireflection coatings on all air-to-glass surfaces, in order to reduce loss of light, loss of contrast, and spurious images caused by reflections from glass surfaces. "Multicoatings" are more effective than single-layer coatings.

Many eyepiece designs have been invented. Here are some of the more common ones.

Huygens. The Huygens eyepiece has two plano-convex lenses and overall low performance. The curved sides of the lenses face toward the objective, and the image plane lies between them. The apparent field ranges from 25° to 40°. This eyepiece works satisfactorily on telescopes with f/ratios of f/8 and higher. The eye relief is only about 25 percent of the focal length.

Ramsden. The Ramsden eyepiece also consists of two simple plano-convex lenses, but the curved surfaces face each other. The image plane is just outside the field lens. The Ramsden is a very low-cost, low performance eyepiece, but slightly better than the Huygens. The apparent field ranges from 30° to 40°. It works satisfactorily on telescopes with f/ratios of f/7 and higher. The eye relief is about 30 percent of the focal length.

Kellner. The Kellner comes in three types, I, II, and III. Types I and II consist of a simple and an achromatic lens. Type I has a single plano-convex field lens and type II has a single double-convex eye lens. Type III is also known as a Plössl and is discussed below. In all cases the image plane is just outside the field lens, and the apparent field is 35° to 50°. Types I and II have an eye relief of about 30 percent of the focal length, and all three work well with f/ratios of f/6 and higher. The Kellner provides better color correction than either the Ramsden or Huygens.

Orthoscopic. This is one of the finest telescope eyepieces. It consists of a triplet field

Figure 3.2c. A telescope in a typical Dobsonian mounting. This 6-inch f/5 Newtonian with a square tube was built by the author. The Dobsonian mount is lightweight and very portable.

lens and a plano-convex eye lens. A well-designed orthoscopic can be fully corrected for distortion, and has very good color correction. It works well with telescopes having f/ratios as short as f/4.5. The apparent field is 30° to 50°, and the eye relief is as much as 80 percent of the focal length.

Plössl. The Plössl eyepiece consists of two achromats in a design similar to the type I and II Kellners. It rates as one of the finest telescope eyepieces. The apparent field is 35° to 50°, and the eye relief is about 75 percent of the focal length. The Plössl works well with f/ratios as short as f/4.5.

Erfle. This eyepiece has a very wide apparent field, about 50° to 70°, with 65° common. It has three achromats and provides good definition in the center, but aberrations become obvious toward the edges. The eye relief is about 30 to 40 percent of the focal length. The Erfle works well with f/ratios of f/4.5 and up.

Nagler. The Nagler is a proprietary design having seven elements and an astonishing 82° apparent field. It is designed for use with f/4.5 Newtonian telescopes, though it will work well with higher f/ratios. The field of view can be breathtaking, though it is sometimes difficult to see the entire field at once because it appears so large. This is a very expensive eyepiece, often costing four to five times as much as a good quality Erfle, Plössl, or orthoscopic.

In recent years eyepiece sales have become a very competitive business, and new (and expensive) proprietary designs are coming to the market.

The Barlow lens. A Barlow is a negative lens that enlarges the image plane in a telescope. It is not an eyepiece but an eyepiece accessory. Barlow lenses are usually used to increase the telescope's effective focal length two to three times so an eyepiece will give a correspondingly higher magnification. A Barlow lens is a necessity for short-focus telescopes if high magnifications are desired.

For example, say your telescope has a focal length of 1200 mm (47 inches), and your shortest focal length eyepiece is 6mm. That eyepiece on that telescope gives a magnification of 200×. Shorter focal length eyepieces are impractical because their eye relief is too small. But using a 2× Barlow lens would give 400× and a 3× Barlow 600× with the same 6mm eyepiece. With three eyepieces and a Barlow lens that is continuously variable, almost any reasonable magnification can be obtained. The Barlow does add more optical surfaces in the light path, reducing the light transmission of the telescope. But the nega-

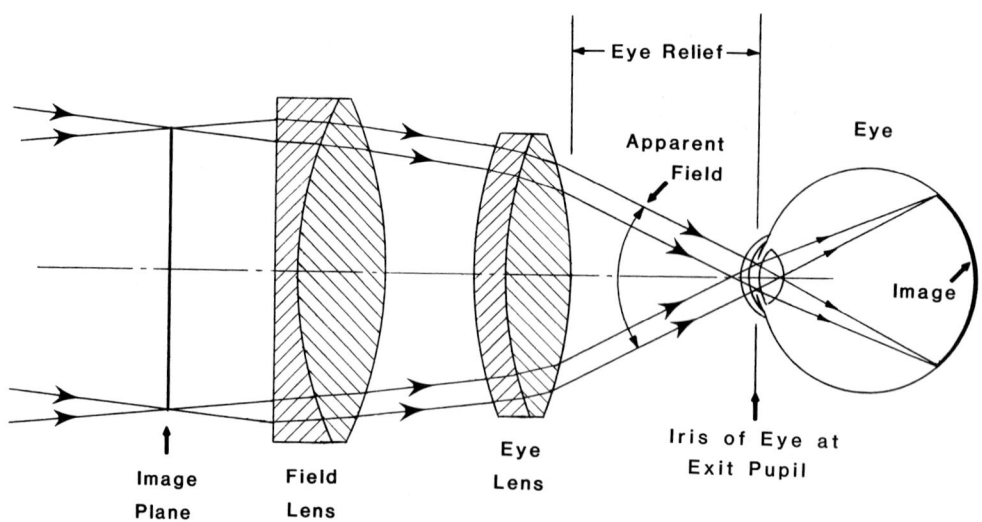

Figure 3.3a. How an eyepiece focuses light for the eye.

THE EYE AND THE TELESCOPE

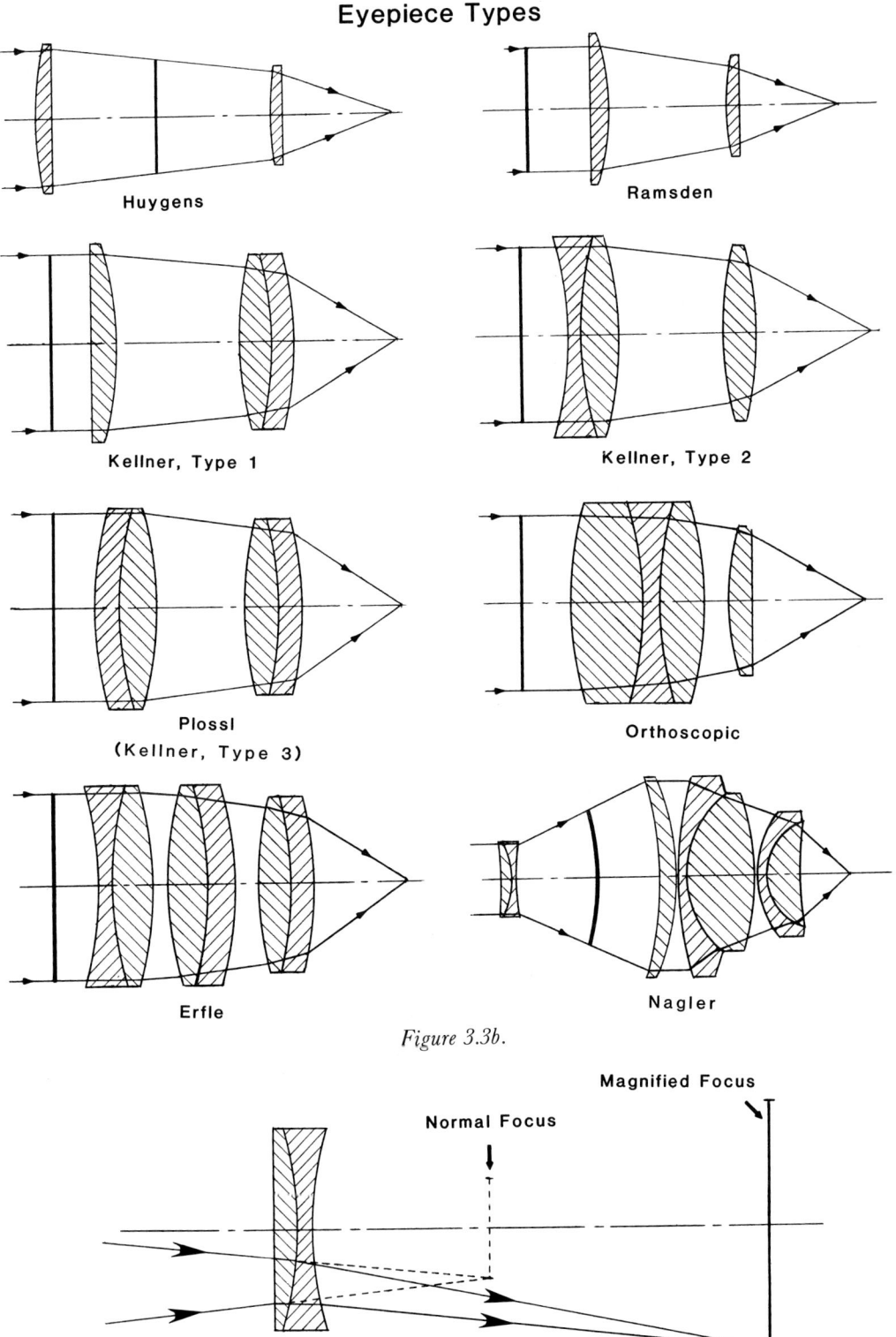

Figure 3.3b.

Figure 3.3c. The Barlow lens.

tive Barlow lens actually flattens the Petzval curvature of most visual telescopes and thus makes them perform better.

FIELD OF VIEW

When telescopes first came into common use in the 18th and early 19th centuries, eyepieces were unsophisticated – at first a single simple lens, then later the Ramsden and Huygens. These eyepieces had small apparent fields of view. Looking through them was almost like looking through a peashooter. To obtain a large true field of view on the sky, long focal length eyepieces had to be used, resulting in low power. As better eyepieces were invented that offered wider views, visual discoveries of wide-field objects became possible.

For a given telescope focal length, it is the choice of eyepiece that determines both the magnification and the true field of view. The eyepiece magnifies the true field of view to a large circle projected in front of the eye. The size of this circle is the apparent field of view. The magnification (m) also determines the apparent diameter (a_p) of an object with a true angular diameter (a_t) in the sky:

$$a_p = 2\arctan[\,m\tan(0.5\,a_t)], \quad \text{(equation 3.3)}$$

which can be simplified to

$$a_p \simeq m\,a_t. \quad \text{(equation 3.4)}$$

Equation 3.4 is accurate to within a couple percent when the apparent diameter, a_p, is less than about 30° and m is greater than about 10.

An eyepiece for a given magnification should be able to view as much as possible of the image formed by the telescope's objective. The linear size (s) of an image in the focal plane is given by

$$s = 2\,F\tan(0.5\,a_t), \quad \text{(equation 3.5)}$$

where F is again the telescope focal length. An eyepiece of focal length f will magnify the image so that, to the eye, it will subtend an apparent angle

$$a_p = 2\arctan(0.5\,s/f). \quad \text{(equation 3.6)}$$

If the image's linear size, s, just fills the field of the eyepiece, we find the maximum apparent angle that can be viewed by the eyepiece. This is the apparent field of the eyepiece, a_e.

Different types of eyepieces have different apparent fields. A simple lens may have an apparent field of only about 10°. Table 3.1 lists the characteristics of several eyepiece types.

At low powers and wide fields, we run into a practical restriction. The eyepiece should not allow you to see a larger image than the rest of the telescope provides. Otherwise, you'll just be viewing the inside walls of the eyepiece holder or other telescope parts, such as light baffles.

Modern telescopes commonly have eyepiece holders of either 1.25 or 2.00 inches (31.7 or 50.8 millimeters) inside diameter. Since the eyepiece barrel must fit within this, the maximum achievable image size is about 1.14 and 1.90 inches (29 and 48 mm) for 1.25- and 2.00-inch holders, respectively. For a given eyepiece apparent field, the maximum usable focal length (f_m) can be computed by rearranging equation 3.6:

$$f_m = s\,/\,[2\tan(0.5\,a_e)]. \quad \text{(equation 3.7)}$$

where s is 29 or 48mm, for the two sizes of eyepiece holder. The maximum useful focal lengths of eyepieces with various apparent fields are shown in Table 3.1.

Erfle and Nagler-type eyepieces have the largest apparent field. A focal length of about 23 millimeters is the maximum for an Erfle in a 1.25-inch eyepiece holder. Buying a 40-mm Erfle in a 1.25-inch tube defeats its purpose, since the apparent field would be restricted to 40°.

THE ROLE OF A TELESCOPE'S F/RATIO

The angle of the converging light cone from the objective to the image plane is determined by the f/ratio. Light cones are illustrated in Figure 3.1. The refractor and reflector diagrams have f/ratios of f/3.3 and the Cassegrains f/6.5. Even shorter f/ratios are illustrated by the primary mirrors of the Cassegrains: f/1.3. For narrower light cones, the f/ratio is larger, the telescope's aberrations are smaller, and eyepieces work better.

As the f/ratio decreases, the surface brightness in the image plane increases. Given two six-inch telescopes with focal lengths of 30 and 60 inches (f/5 and f/10, respectively), the f/5 telescope will produce images with higher surface brightness than the f/10 telescope

Table 3.1. *Eyepiece characteristics*

Type	Apparent field degrees	Focal length (mm) that views an image of diameter:	
		29.0 mm	48.0 mm
Nagler	82	16.7	27.6
good Erfle	65	22.8	37.7
average Erfle	60	25.1	41.6
	55	27.9	46.1
Kellner, some orthoscopics	50	31.1	51.5
Plössl	45	35.0	57.9
limit for Huygens	40	39.8	65.9
limit for Ramsden, some orthoscopics	35	46.0	76.1
	30	54.1	89.6
	25	65.4	108.3
single achromat	20	82.2	136.1
single simple lens	10	165.7	274.3

when viewed with the same eyepiece. A 12-inch f/5 telescope will produce images with the same surface brightness as the six-inch f/5. However, in the 12-inch the image would be twice as large.

The f/ratio determines the useful range of magnifications that may be obtained with standard equipment. With long focus (high f/ratio) instruments, low powers and wide fields are difficult to obtain. For short focus (low f/ratio) telescopes, it is difficult to obtain high power. You can always use a barlow lens to boost magnification, but another optical element is introduced and some low f/ratio systems may begin to show aberrations. Choosing the telescope's f/ratio is important because you want to be able to use low power to get wide fields of view to see large objects against the sky background, yet you also want high power to see fine detail.

The eye, however, is not affected by the f/ratio of the telescope. Only the size of the object in view (the magnification) and its brightness (telescope aperture) are ultimately important. Consider two 8-inch (203 millimeter) telescopes, one an f/4 Newtonian, the other an f/12 Cassegrain. Suppose two eyepieces with the same apparent fields of view are available, one of 21 millimeters focal length, and one of 7 mm. The 7-mm eyepiece on the f/4 telescope gives a magnification of 116 (116×), and the 21-mm eyepiece on the f/12 Cassegrain also gives 116×. Each telescope-eyepiece combination provides the same true field of view on the sky. In this case, all deep-sky objects viewed through the two telescopes will appear identical. Neither telescope has an advantage. We can state a general rule:

> Telescopes of equal aperture and good, clean optics that do not scatter light will show identical views of deep-sky objects at the same magnification, regardless of their f/ratios.

THE EXIT PUPIL

As mentioned earlier, a telescope's exit pupil is the little disk of light projected behind the eyepiece. The eye should be placed here to see the field of view best.

The exit pupil is actually a small image of the telescope objective. In a reflector, it will show a silhouette of the central obstruction and spider (the support vanes to the secondary mirror) and may become distractingly visible if the eye is kept too far back from the eyepiece.

The diameter of the exit pupil, e_p, is the objective diameter, D, divided by the magnification, m:

$$e_p = D / m. \qquad \text{(equation 3.8)}$$

Table 3.2. *Minimum useful telescope magnifications*

Aperture		
inches	cm	Magnification
1	2.54	3.4
2	5.08	6.8
3	7.62	10.2
4	10.1	13.6
5	12.7	17
6	15.2	20
7	17.8	24
8	20.3	27
10	25.4	34
12	30.5	41
14	35.6	47
16	40.6	54
18	45.7	61
20	50.8	68
22	55.9	75
24	61.0	81
30	76.2	101
36	91.4	121

The magnifications are for a 7.5 mm exit pupil.

If all the light collected by the telescope is to enter the eye, then the exit pupil must be smaller than the eye's pupil. In Chapter 2 we saw that the iris of a young adult will open to about 0.3 inch (7.5 millimeters) when fully dark-adapted. This value sets a lower limit to the magnification that can be used on a given telescope. By setting D to 25.4mm (1.0 inch) and e_p to 7.5mm, we find that the minimum useful magnification is 3.4× per inch of objective diameter.

Table 3.2 gives the minimum magnification for common telescopes based on this relation. However, it is usually wise to be conservative since individual observers' eyes may vary, and the maximum size of the pupil shrinks as a person ages. Thus a value of 4× or even 5× per inch of objective may be more appropriate. If lower magnifications are used, the iris of the eye may block some light, reducing the effective aperture of the telescope.

SEEING AND RESOLUTION

The size of the smallest detail that a telescope can reveal (ignoring atmospheric turbulence) is called *resolution*. This depends on the size of the objective, since that is what determines the size of the telescope's *diffraction pattern*.

The diffraction pattern of a point source, such as a star, is a circular disk of light surrounded by bright rings. At visual wavelengths, the angular diameter, a, of the disk is

$$a = 5.45 / D \qquad \text{(equation 3.9)}$$

where the aperture of the telescope, D, is in inches and a is in arc-seconds. This size represents the diameter of the diffraction pattern's first dark ring, just outside the central spot (Figure 3.4). (The diameter of the second dark ring is twice that of the first.) The angular size a of the disk is called the Rayleigh resolution limit.

If a bright image in the telescope is magnified too much, it appears fuzzy due to the telescope's limited resolution. The generally accepted magnification limit for bright objects is about 60 times the diameter of the objective in inches. As an example, consider a 3-inch telescope; the central disk of the diffraction pattern is nearly 2 arc-seconds wide. The eye is able to resolve about 1 arc-minute for bright objects, so only about 30 power is needed to enlarge the diffraction disc up to the resolving power of the eye. This is only 10× per inch of telescope objective diameter. At 90 power (30× per inch of objective), the eye can easily see the diffraction disk and rings if there is little atmospheric turbulence, and at 180 power (60× per inch of objective) the disk is obvious. On extended sources such as planets or scenes on the ground, the image is made up of multiple overlapping diffraction patterns, so it appears fuzzy.

With fainter subjects it's a different story. As we saw in Chapter 2, the eye has less resolution in dim light. The magnification limit is found by dividing the eye resolution (1800 arc-seconds when the light is extremely dim) by the telescope resolution, using equation 3.9. The limit is 330× per inch of objective. And so, for faint stars and most deep-sky objects, there is essentially no limit to the useful magnification at all. For example, many planetary nebulae have small angular sizes but relatively high surface brightnesses. Very high magnifications may be employed to bring detail within such a nebula into view.

High magnifications, however, cause more problems than limiting the field of view. Any vibration or looseness in the telescope mount becomes intolerable. Irregularities in the atmosphere – the constant quivering, churning and boiling of an image, which astronomers call the seeing – usually make an image fuzzier than the telescope's theoretical resolution limit, and high power only magnifies this problem. Furthermore, the 330× per inch of objective mentioned above is useful only at the very limit of detection. Any object brighter than the threshold of detectability will appear fuzzy at such powers. Because most fields will contain some brighter subjects, the overall impression will be that everything looks fuzzy (and it is, all except the one thing at the extreme limit of detectability).

Magnification also reduces the surface brightness of everything in view. It must not reduce an object's surface brightness below the eye's detection limit, of course, or the object will disappear. Therefore, the only case where 330× per inch might be justified is the detection of a star against a bright background, as discussed in Chapter 4.

FILTERS

Wouldn't it be wonderful if there were a device that darkened the sky background while leaving stars and nebulae bright? We saw in Chapter 2 that the detection capability of the eye depends on an object's contrast with its background as well as on its size and surface brightness. One way to reduce the sky background is to get far away from city lights. But even where there is no artificial pollution, the Earth's sky glows from scattered starlight and from airglow. Airglow is like a permanent, low-grade aurora. It is caused by charged particles from the Sun encountering our atmosphere at very high altitude.

Airglow occurs at only a few wavelengths (isolated colors), and if a filter could be made to remove these wavelengths while passing the rest of starlight and nebular light, contrast would be increased and the view would

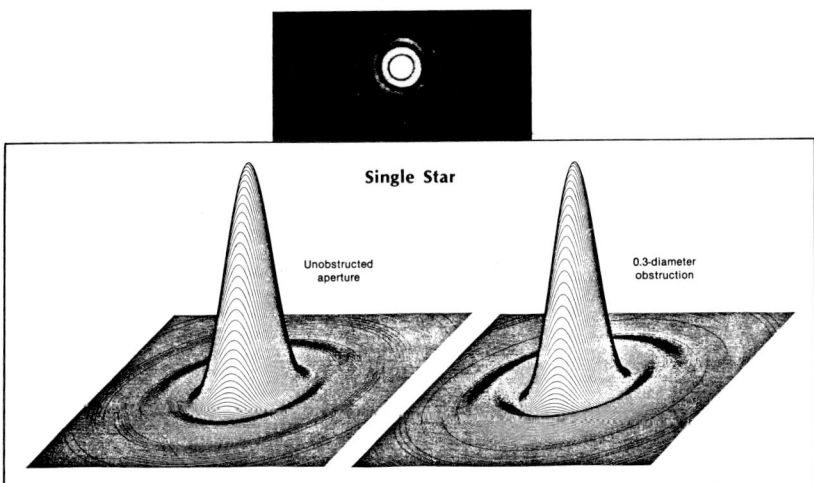

Figure 3.4. A typical diffraction disk of a star image in a telescope. At top center is a photograph showing the diffraction rings, taken by Allyn J. Thompson. The central spot was intentionally overexposed to show the rings. The 3-dimensional plots of diffraction patterns indicate brightness by height above a square base, viewed obliquely. Faint diffraction rings appear as concentric ripples around the center of a star's Airy disk. At left is the pattern of a star in a telescope having no central obstruction, while at right is the pattern for a telescope having a round central obstruction 30% of the objective's diameter. The central obstruction removes light from the central spot and redistributes it into the rings. From Stoltzmann (1983).

Table 3.3. *Typical emission lines of nebulae*

Color	Wavelength (angstroms)	Atom or Ion	Notes
Strongest lines:			
Violet	3727	Oxygen II	Forbidden line, often strong
	3869	Neon III	Forbidden line
	4340	Hydrogen I	H gamma, 40% as strong as H beta line
Blue-Green	4861	Hydrogen I	H beta, 30% as strong as H alpha line
Green	4959	Oxygen III	Forbidden, 30% as strong as 5007 line
Green	5007	Oxygen III	Forbidden, usually strongest of all
Red	6548	Nitrogen II	Forbidden, 30% as strong as 6584 line
	6563	Hydrogen I	H alpha
Red	6584	Nitrogen II	Forbidden line
Weaker lines:			
Violet	3798	Hydrogen I	
	3835	Hydrogen I	
	3888	Hydrogen and Helium I	
	3969	Hydrogen I and Neon III Forbidden	
	4102	Hydrogen I	
Blue	4471	Hydrogen I	
Blue	4686	Helium II	
Yellow	5876	Helium I	
Red	6300	Oxygen I	Forbidden line
	6364	Oxygen I	Forbidden line
	6717	Sulfur II	Forbidden line
Red	6731	Sulfur II	Forbidden line

appear better. If city lights emitted only a few wavelengths, they too might be filtered from the observer's view. Fortunately, this can be done – at least partially. To understand how, we'll need to examine the nature of light from celestial objects.

The light from a star consists of many wavelengths or colors. Such light is called *continuum* radiation, since its spectrum appears nearly continuous.

Some nebulae also give off continuum radiation. But many of the best and brightest emit light at only a few, specific wavelengths. These two types are *reflection* and *emission* nebulae, respectively.

Reflection nebulae shine by reflected starlight, similar to the way a cloud in the Earth's sky reflects sunlight or the way the molecules in the Earth's atmosphere scatter it. Because the molecules and some dust particles in our atmosphere are smaller than the wavelengths of visible light, blue light is scattered more efficiently than red, and thus the sky appears blue. This type of scattering is called *Rayleigh scattering* after the scientist who first described the effect. Reflection nebulae "shine" in part by Rayleigh scattering of starlight, so the spectrum is continuous, much like that of a star but usually bluer.

Emission nebulae, on the other hand, shine in only a few colors. They come in two types: planetary and diffuse nebulae. Both shine because starlight excites specific types of atoms, which re-emit this energy only at certain wavelengths.

The brightest spectral lines (colors) of emission nebulae are listed in Table 3.3. The atom that emits the radiation is listed, followed by a Roman numeral that indicates its ionization state. (I is un-ionized, II means the atom is missing one electron, III means two electrons are missing, and so on.)

A *forbidden* line is not really forbidden to happen; it just cannot be observed in labor-

Table 3.4. *Natural and manmade light pollution*

Source	Wavelength (angstroms)	Comments
Manmade:		
Mercury Vapor	3660	Strongest mercury line
Mercury Vapor	4050	40% as strong as 3660 line
Mercury Vapor	4360	75% as strong as 3660 line
Mercury Vapor	5460	98% as strong as 3660 line
Mercury Vapor	5750	96% as strong as 3660 line
Mercury Vapor	3200–7300	Continuum; 3% as strong as 3660 line
Lucalox Mercury Vapor:	5500–7000	Peak is 5700–6200 angstroms
Lucalox Mercury Vapor:	5000	15% of Peak
Lucalox Mercury Vapor:	4000–7500	Continuum; 5% of Peak
Low Pressure Sodium	5893	Nearly monochromatic
High Pressure Sodium	3500–7000	Partly continuum; brighter in red
Incandescent	4000–7000	Continuum; brighter in red
Natural:		
Airglow	5577	Oxygen
Airglow	5893	Sodium
Airglow	6300	Oxygen
Moon	3500–7500	Continuum; similar to sunlight

atories because the gas density must be lower than in the best artificial vacuums. Normally when an atom is excited, it emits a photon within about 10^{-8} second. But the excited states that result in forbidden lines can last minutes to hours. In the laboratory, atoms cannot remain undisturbed that long because they collide with each other or the walls of the container.

Because less energy is needed to excite the forbidden lines, they are much stronger than ordinary lines when they can occur at all. Hydrogen is the most abundant element in emission nebulae, but the forbidden lines of oxygen emit the most light.

Just as light from deep-sky objects comes in two types, continuum and discrete wavelengths, so does light pollution. For example, moonlight and light from incandescent bulbs is of the continuum type. Fortunately, the light from airglow and most streetlights is at discrete wavelengths in different parts of the spectrum than most of the light from nebulae. So the two can be separated with an appropriate filter.

Such filters are known as *nebula filters* or *light-pollution* filters. The term "nebula filter" is more appropriate, since views of nebulae are improved most.

The usefulness of such a filter depends on the nature of the interfering light. Table 3.4 lists common sources.

In examining Tables 3.3 and 3.4, we see that most light pollution is at wavelengths different from the light of nebulae. This is illustrated graphically at the tops of Figures 3.5 to 3.8. Referring back to Figure 2.3, recall that the peak response of the human eye's night vision is near 5000 angstroms (at the color green), or the same wavelength as the strongest nebular line at 5007 angstroms. But there are strong light pollution lines close by, near 4400 and 5400 angstroms. A filter must cut the spectrum pretty finely to reject these wavelengths and still have a high transmission at 5000 angstroms. This can only be achieved with modern interference-filter technology.

Interference filters have very thin layers of partially reflecting material separated by thin transparent layers. Depending on a layer's thicknesses, different wavelengths of light will undergo constructive or destructive interference. The number of wavelengths that can be controlled depends on how many layers are deposited. Ordinary colored filters are added to block wavelengths far from the primary transmittance region.

Interference filters can have very high transmittance at very narrow wavelengths. They are ideal for isolating the light of nebulae and have uses in many other areas of science. In the early 1970's the technology to make them was quite expensive, but now they can be mass produced and cost no more than a high quality eyepiece.

What about continuum light pollution? This is harder to deal with. It can be rejected only by making the filter absorb *all* light not at the wavelengths of the nebula. The wavelength bandpass of an interference filter can indeed be made extremely narrow. For example, filters used by amateurs for observing solar prominences have a bandpass of only 0.7 angstrom at the wavelength of hydrogen alpha, 6563 angstroms. At wavelengths as far away as 6500 angstroms, the filter may transmit less than one thousandth as much as

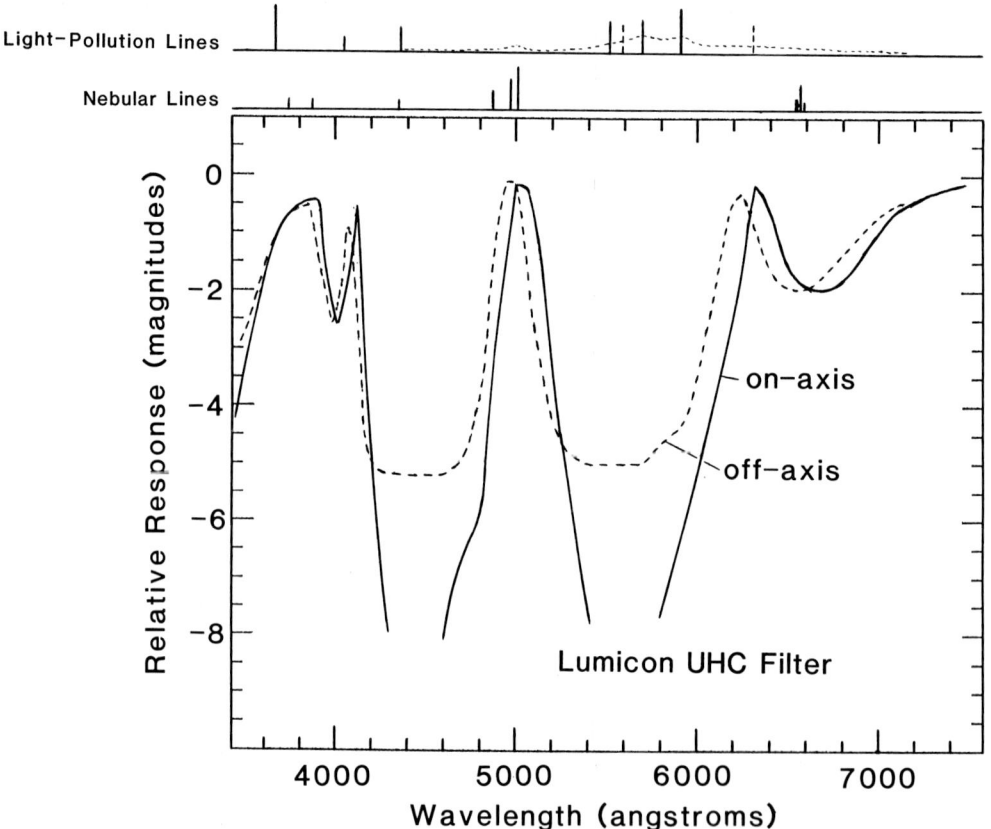

Figure 3.5. The spectral transmission of the Lumicon UHC filter is shown for light arriving face-on (solid line) and at an angle of about 20° (dashed line). Note how the bandpass (the central peak) shifts to shorter wavelengths as the filter is tilted, and how the blocking of the light outside the bandpass becomes less effective. A relative response of -2.0 magnitudes is a transmittance of 16%, -5.0 magnitudes is 1%, and -8.0 magnitudes is 0.06%. The positions of major nebular lines from Table 3.3 and the major light-pollution lines from Table 3.4 are shown above the graph. The dashed pollution lines are natural airglow lines of oxygen. The broad dashed line that peaks around 5800 to 6000 angstroms is the Lucalox streetlamp. Continuum pollution sources are not shown because they affect all wavelengths to some degree.

at its peak. Similarly, a nebula filter could be constructed that transmitted only the light at 5007 angstroms.

However, interference filters have some undesirable characteristics. In particular, they work at the correct wavelength only if the layers making up the filter are exactly the correct thickness. Two things disturb the thickness of these layers: temperature and tilting of the filter.

As the temperature changes, the layers will expand or contract, and the transmitted wavelength changes accordingly. The narrow-band solar filters are mounted in a small oven to precisely control their temperature. A solar filter's wavelength can actually be scanned slightly by changing the oven temperature.

When a filter is tilted, the transmitted wavelength becomes shorter, the bandpass broadens, and more unwanted light is transmitted. If you have access to a nebula filter, try tilting it and you will see its color change from greenish to purple.

These effects limit the design of nebula filters for practical purposes. All nebula filters are made with bandpasses broad enough so temperature changes have no effect. Tilting the filter, however, does alter its characteristics.

Because stars emit continuum light, when any wavelengths at all are blocked, stars will appear fainter. The narrower the bandpass, the more starlight will be lost. Furthermore, many stars are bright enough to stimulate color perception, so if the filter transmitted only the nebular line at 5007 angstroms (green), all stars would appear green. Thus some manufacturers have made filters that transmit some blue and red light so the color balance is closer to normal. Light pollution at these wavelengths is not very great.

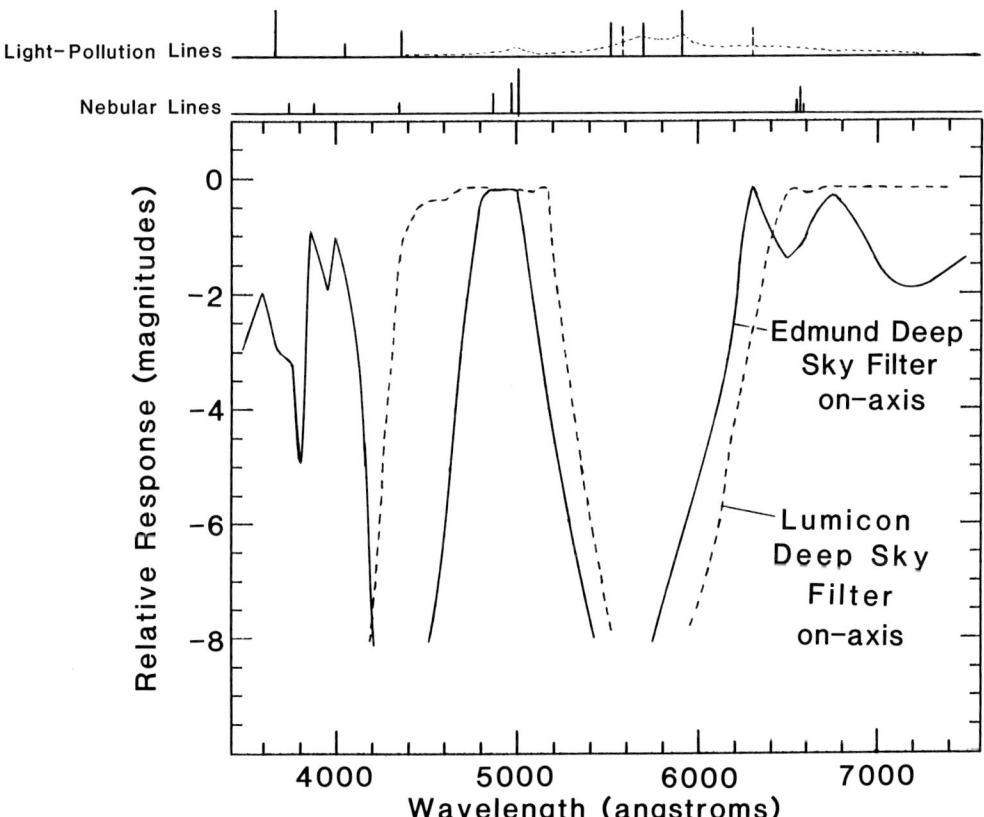

Figure 3.6. Transmittance curves of the Edmund Deep-Sky Filter (solid line) and Lumicon Deep Sky Filter (dashed line).

Light from nebulae is also reduced because no filter transmits 100% at any wavelength. At its peak response, it may transmit 80% to 90%, which would reduce the light from the nebula by 0.1 to 0.2 magnitude even if all the light were emitted in the filter bandpass. Nebula filters work because they *increase the contrast* between the object and background more than they *reduce the light* from the object. A nebula filter improves visibility only when the improved contrast outweighs the loss of light. This depends on the eye's characteristics in the particular situation, the filter design, and the type of the light pollution.

Let's examine some filters and the implications for their use. The spectral response of one of the better nebula filters in common use, the Lumicon UHC Filter, is shown in Figure 3.5. Note that the main bandpass (the central peak) is centered at a wavelength a little longer than the nebular lines at 5007 and 4959 angstroms. This may seem like a mistake, but is a very clever design. When the filter is tilted 20° (dashed line), the bandpass shifts to a shorter wavelength and the 5007 line still falls within the long-wavelength side of the bandpass.

Light from a telescope is always a converging or diverging cone, so the filter must be able to transmit the strong nebular line over a range of angles. For example, if the apparent field of an eyepiece is 40°, then the light at the edge of the field is tilted 20° from the optical axis. If the Lumicon UHC filter is placed between the eye and eyepiece, then all the light from the 5007 line reaches the eye. However, if an Erfle eyepiece is used, any nebula near the edge of the field of view (30°

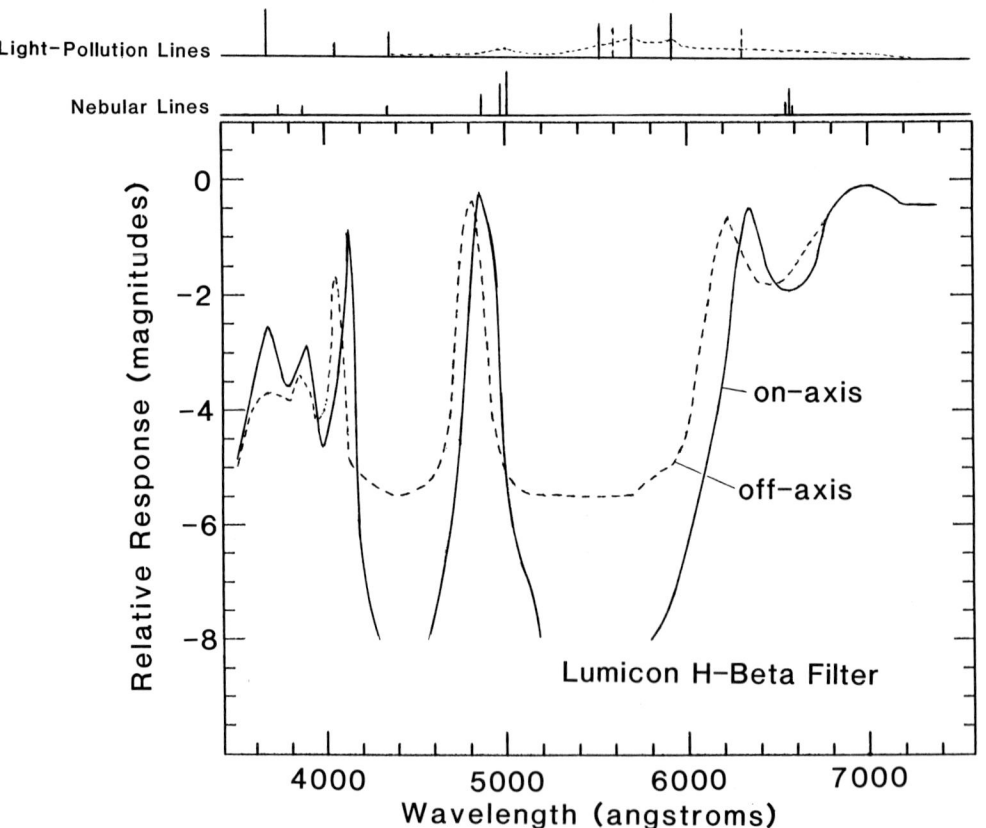

Figure 3.7. The Lumicon H-Beta Filter spectral response is shown for light on-axis (solid line), and off-axis (dashed line). Note that a small tilt of the filter shifts the bandpass so the H-beta line at 4860 angstroms is no longer transmitted.

off-axis) may be completely invisible because the filter bandpass shifts enough to cut off the nebular line.

On the other hand, the light from the edge of the objective even in a fast telescope of f/4.5 is only 6.3° from the optical axis. So, if a wide-field eyepiece is used, the filter should be placed between the eyepiece and the objective, not between the eyepiece and the eye.

One observer's trick takes advantage of a filter's off-axis problem. The filter can be used to help detect a faint object by tilting it back and forth between the eye and eyepiece. The nebula will blink in and out of view.

If we examine other manufacturers' filters, we see some interesting results. The Edmund Scientific Deep-Sky Filter (Figure 3.6, solid line) has a slightly wider bandpass than the Lumicon UHC so that the 4861-angstrom H-beta line is also transmitted when the light arrives on-axis. However, when the Edmund filter is tilted only slightly, the 5007-angstrom line is lost. Some objects do have their strongest emission in H-beta. The nebula IC 434 surrounding the Horsehead (B33) is a good example. Thus the Edmund filter appears to be a good H-beta filter. The Meade Instruments Deep-Sky Filter has a spectral response very similar to the Edmund filter, but does not transmit any blue or ultraviolet light.

The Lumicon Deep-Sky Filter (Figure 3.6, dashed line) follows a different strategy. It has a very wide bandpass that transmits light from both the 4861- and 5007-angstrom nebula lines even when the light is substantially off-axis. Thus it is good for use with wide-field eyepieces and fast f/ratio telescopes. Such a wide-bandpass filter will transmit more light from galaxies and star clusters but also more light pollution.

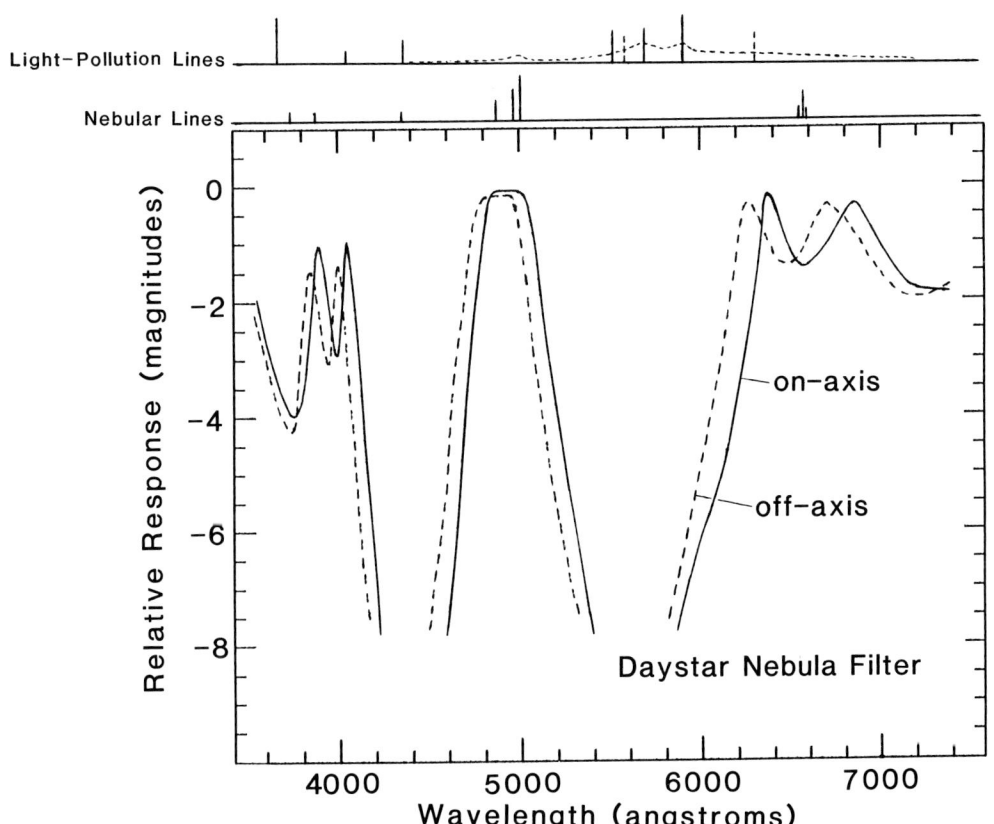

Figure 3.8. The Daystar Nebula Filter transmission on-axis (solid line) and about 20° off-axis (dashed line).

The Lumicon H-Beta Filter (Figure 3.7) has a narrow bandpass designed for use on objects that emit most of their light at 4861 angstroms. But when light passes through this filter only slightly off-axis, the bandpass wavelength becomes too short to transmit much of this light (dashed line). This filter is only good for use between the eyepiece and the objective and should not be used even for casual observations between the eyepiece and eye.

The Daystar Nebular Filter (Figure 3.8) acts similarly to the Edmund filter, but the bandpass is at a slightly longer wavelength, the transmittance is higher at the center of the bandpass, and the off-axis blocking is as good as on-axis. These factors are probably why some observers declare this filter better than the Lumicon UHC filter in actual observing practice, though the author believes them to be very close.

The manufacture of nebula filters is so new that different ones may be available by the time this book is published. The factors discussed here, however, will help in evaluating any of them. Finally, observers about to make a purchase should remember that the difference between even the best nebula filter and none at all is usually rather subtle.

USING THE TELESCOPE TO FIND OBJECTS

So far this book has discussed observing techniques. But the best deep-sky techniques are useless unless the objects can first be found.

In order to find anything in the sky, you must become familiar with the bright stars and constellations – just as when you drive into a new city you get oriented by major landmarks. Once you're in the right celestial neighborhood, a detailed map is required for locating small, faint objects. Constellation guides and star atlases are listed in Appendix A.

Celestial coordinates

The stars appear fixed on an imaginary *celestial sphere*, which is always centered on the observer. The celestial sphere appears to rotate once every day because we observe it from the rotating Earth. Lines projected from the center of the earth through the north and south poles extend to the north and south poles of the celestial sphere. The stars appear to rotate around these poles, as was known in ancient times.

Astronomers have set up a coordinate system for describing positions on the celestial sphere similar to latitude and longitude on Earth. The Earth's latitude lines are projected outward from our globe onto the sky to become the lines of declination. The celestial equator, straight overhead at the Earth's equator, defines 0° *declination* on the celestial sphere. The north celestial pole is at +90° declination, the south celestial pole at -90° declination.

"Longitude" on the celestial sphere is called *Right Ascension*. Longitude lines are a little more difficult than latitude to project onto the sky because the stars are constantly rotating across the longitude lines on the Earth. For the 0° point astronomers have chosen the place at which the Sun crosses the celestial equator from south to north on its yearly journey around the sky. This point is called the First Point of Aries (though it is now in Pisces because of a slow movement of the coordinate system over the centuries, known as precession).

Imagine aiming a telescope at the First Point of Aries and keeping the telescope fixed with respect to the earth (on a motionless mount). As the Earth turns, the right ascension of stars crossing the view increases. Right ascension is commonly measured in hours, minutes, and seconds of time instead of degrees, minutes and seconds of arc. There are 24 hours of right ascension just as there are 360 degrees of longitude. (Therefore, a minute or second of right ascension is not the same as an angular minute or second of arc. Each is 360/24 or 15 times greater than the corresponding unit of longitude.)

Note that as declination increases (positive) or decreases (negative) from the celestial equator, the right ascension lines become closer together. They converge at the poles.

Since the Earth's axis is tilted with respect to its path around the Sun, the apparent track of the Sun on the sky is north of the celestial equator for half the year and south for half. The sun's apparent track on the celestial sphere is called the ecliptic, and is tilted 23.5° to the celestial equator. The planets generally follow the ecliptic too because the orbits of all except Pluto are inclined only a few degrees from the plane of the Earth's orbit.

The line on the celestial sphere that passes through the pole and directly overhead, and extends to the north and south points on the horizon, is called the meridian. The point directly overhead is called the zenith.

The Sun is at 6 hours right ascension at the beginning of summer in the Northern Hemisphere, so right ascension 18 hours (the opposite side of the sky) will be on the meridian at midnight.

At the beginning of autumn in the Northern Hemisphere, right ascension 0 hours is on the meridian at midnight. At the beginning of winter, 6 hours R.A. is on the meridian, and at the beginning of spring, 12 hours.

With these facts, and a little study of the approximate right ascensions of the constellations, you should be able to figure out what constellations are up at night any time of the year.

For example, say it is 9 p.m. on August 20; what constellations are on the meridian? Consider that one month later is the beginning of fall, when 0 hours R.A. is on the

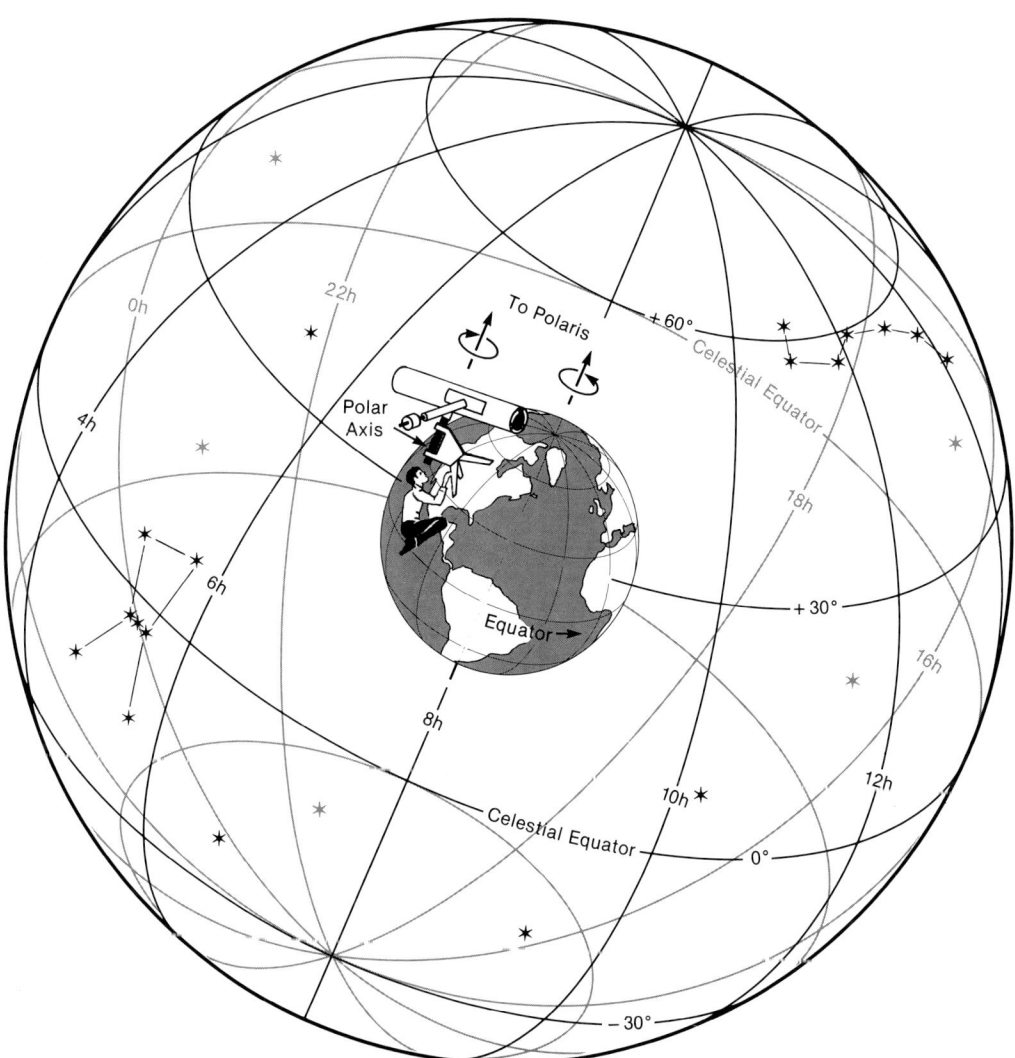

Figure 3.9. The celestial sphere. The observer is aligning the equatorial mount's polar axis by making it parallel to the Earth's axis. (Courtesy of *Sky & Telescope*.)

meridian at midnight. Since there are twelve months in a year and 24 hours of right ascension, the constellations move 2 hours per month. Counting 2 hours backwards from 0 hours (the same as 24 hours), we find that on August 20, right ascension 22 hours should be approximately on the meridian at midnight. Three hours earlier the right ascension would be three hours less, so at 9 p.m., right ascension 19 hours should be on the meridian. A star map shows that constellations near R.A. 19 hours include Lyra, Aquila, Sagittarius, Corona Australis, Telescopium, and Pavo.

Precession

The Earth is spinning like an off-balance top in that the direction of its axis is very gradually changing. Thus the stars are gradually changing their declination and right ascension. The direction of the Earth's axis follows a slow cyclical motion, with a 26,000 year period, called *precession*. Thus, star charts need to be updated with a new coordinate grid about once every 50 years. Since precession is very predictable, the coordinates for a given year can be changed to those for another year by putting them through mathematical formulae.

Whenever a coordinate is given for an object, the year is specified when that coordinate is correct. This time is referred to as the *epoch* or, more precisely, the *equinox*. Most coordinates are currently given for equinox 2000 if the object is "fixed" like a star, or for the actual date (e.g. 1988.5) if the object is a short-term phenomenon like a comet.

The following simplified formulae for precession are reasonably accurate. If RA and Dec are an object's right ascension and declination (RA expressed in decimal hours and Dec in decimal degrees), the yearly changes ΔRA and ΔDec in these coordinates are given by:

$$\Delta RA = 0.051 + 0.022 \sin(15\,R.A.) \quad \text{(equation 3.10a)}$$

$$\Delta Dec = 0.33 \cos(15\,R.A.) \quad \text{(equation 3.10b)}$$

where ΔRA is expressed in minutes of time and ΔDec in minutes of arc. Multiply by the total number of years desired, negative if going back in time, and apply the corrections to RA and Dec. More complex formulas must be used for stars near the poles, but the ones given here work well in most of the sky.

Many modern observatories have very precise positioning capabilities, and change coordinates to the epoch of the date and time the observation is made. By adding further corrections such as for the flexure in the telescope mount and atmospheric refraction, the telescope can be aimed to within a couple of arc-seconds of an object.

Such modern technology has led to a generation of professional astronomers who do not need to know where something is in the sky. If they know its coordinates, a computer tells when it is observable and points the telescope right at it. Amateur astronomers are now beginning to use such technology too. But even a totally automatic telescope must first be accurately aligned on the celestial pole for the system to work, and that can only be done with a knowledge of where things are in the sky.

Furthermore, "flying on instruments" takes much of the fun out of skywatching, and leaves the viewer helpless in the face of problems. Most amateurs recommend against it for beginners, who need to learn the sky for themselves. Once the brighter stars and major constellations can be identified, the fainter objects that cannot be seen with the unaided eye can be found by using one of the following methods.

1. Star hopping

Star hopping should be learned by all amateur astronomers. It is the method of starting at a known naked-eye star and using a chart to move carefully from star to star until the desired object is reached.

The best way to plan such a search is to draw a circle on a piece of clear plastic representing the field of view of the finder telescope at the correct scale for the star chart. A wire ring the right size works just as well. The search route can be tested by sliding the circle across the chart. The route is then duplicated on the sky. Examples of star hopping are illustrated in Figures 3.10a through e.

2. Using setting circles

Setting circles are graduated scales on a telescope's equatorial mount. They allow the telescope to be positioned on an astronomical object by "dialing in" the object's coordinates. This is not as simple as it sounds, and it should not be considered a substitute for learning to star-hop.

The first step in using setting circles is to align the mount's right-ascension axis (also called the polar axis) so it points to the celestial pole. One method is to insert a special finder telescope right into the mount's polar axis. Such finders have a special reticle that shows the star pattern at the celestial pole. It is a simple matter to align the star pattern on the reticle with that on the sky. A few manufacturers offer such finders inside their mountings.

A second method of alignment is to chose two stars at the same declination about 6 hours apart, with one on the meridian. The telescope is aimed at one star and then swept to the other; the mount is adjusted until the telescope can be made to point to both stars by only moving in right ascension (that is, with the declination axis locked).

For example, consider one star on the meridian and one 6 hours further west. If you center on the first and try to move to the second, any error will be largely due to an alignment error in azimuth (the polar axis is pointed to one side of the pole). If the telescope ends up pointing south of the second star, then turn the entire mount until the star is centered. Now move back to the star near the meridian. This move is sensitive to an alignment error in altitude (the polar axis points too high or low). If the star is not centered, then move the mount in altitude to correct the position. Theoretically the mount should now be aligned, but repeat the procedure to check.

The accuracy of polar alignment you need depends on how accurate your setting circles are, what your field of view is when searching for objects, and how closely you want them centered. The accuracy requirement for visual observing is much less severe than for astrophotography. If your field of view is about 1°, and the setting circles are graduated

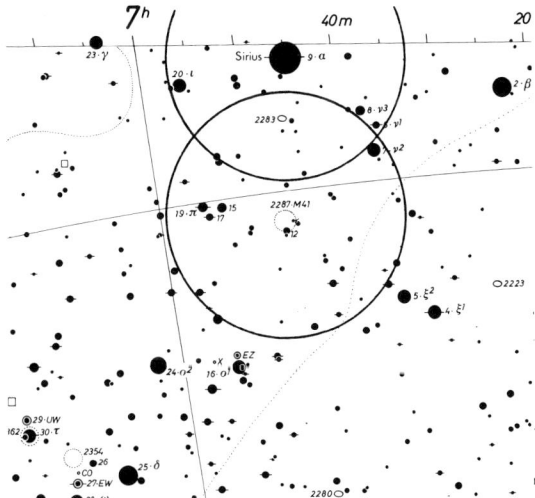

Figure 3.10a. How to find the open cluster M41 by star hopping from Sirius in the constellation Canis Major. The circles show a typical finder telescope's 6° field of view. First center Sirius in the finder, then sweep 2/3 of a finder field south and a bit east, keeping track of star patterns along the way. (East is always the direction of increasing right ascension.) M41 should be visible as a faint, sparkling haze. Chart reproduced from *Sky Atlas 2000.0* by W. Tirion.

every 1° in declination and 4 minutes in right ascension, then the polar alignment should be better than 1°. But it probably does not need to be better than about 1/2°. If, on the other hand, you wish to center small planetary nebulae at high power in a 10 arc-minute field of view, you need big, observatory-quality setting circles that can be read to about 2 arc-minutes, and the alignment should be better than about 5 arc-minutes. Such alignment will take time and better methods than described here, and is probably not feasible with a portable instrument.

Once the mount is aligned, the setting circles must be set to read correctly. This is easily done by centering the telescope on a

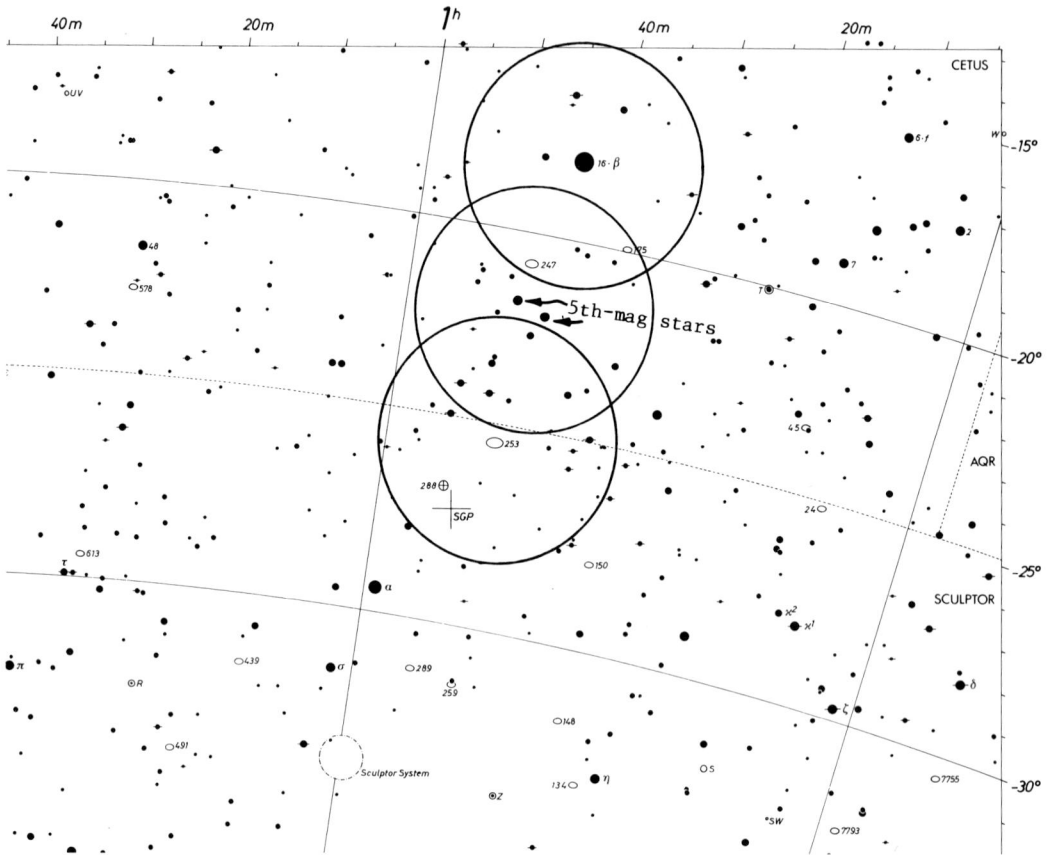

Figure 3.10b. Finding faint NGC 253 in the constellation Sculptor is more difficult than finding M41. The galaxy is in a star-poor region only about two degrees from the south galactic pole. The brightest star in the region is Beta Ceti, 7° to the north and slightly west of NGC 253.

Start by placing Beta Ceti in the center of the field of view. (You must know the constellations well enough to locate Beta Ceti!) Just outside the southern edge of the 6° field is a pair of 5th-magnitude stars. Find them and center them. Now NGC 253 is just south of the field of view and a little east.

Move south until the two 5th-magnitude stars are just outside the north edge. NGC 253 should be just east of center. Moving a fraction of a degree east should put NGC253 in the center of the field of the main telescope as well as the finder. The correct spot can be estimated by comparing the positions of the faint stars in the finder with those on the chart. Chart reproduced from *Sky Atlas 2000.0* by W. Tirion.

star whose coordinates are known and then moving the setting circles until they read the correct position.

If the polar alignment was not adequate, you might have difficulty finding objects. One way to circumvent this problem when observing in a localized region of the sky is to aim at a star in the region and turn the setting circles to read that star's coordinates. This way, any error in pointing nearby is minimized. When another area of the sky is to be observed, the setting circles would need to be reset using a star in that area. The time spent resetting the circles takes away from precious observing time.

Note: the right ascension circle must be carried along with the telescope by a clock drive, or it will give false readings just a minute or two after being adjusted to the coordinates of a known star.

In this author's opinion, setting circles should not be used except on permanent mountings by very experienced astronomers. The beginning amateur needs to learn the sky, and the learning will be impeded if setting circles are used. Many amateurs find that they remember object's positions among the stars (but not their coordinates) after only a year or two of active observing, and can position a telescope on one of those objects before most people would have time even to look up the coordinates. While examining a

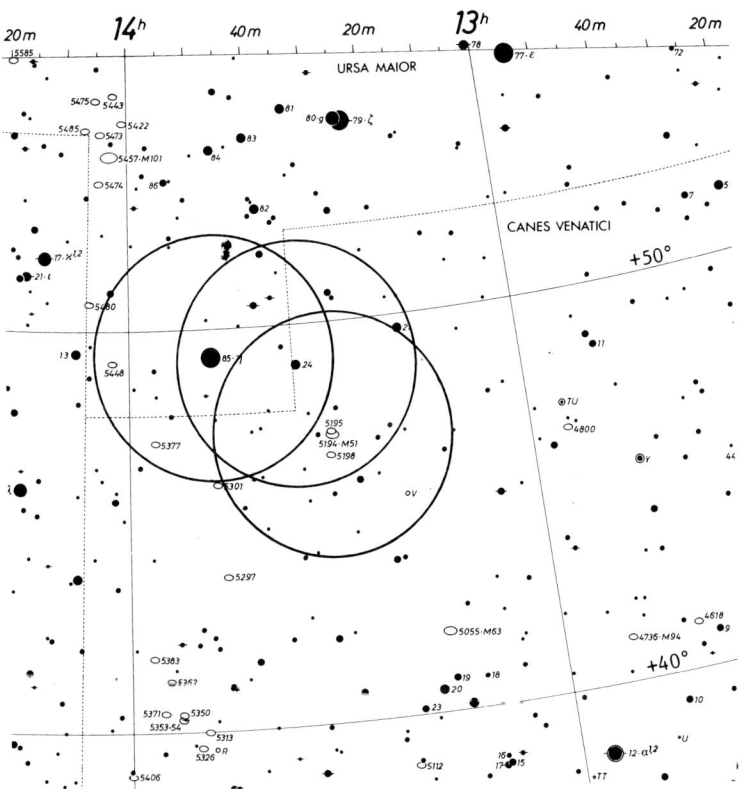

Figure 3.10c. M51, the Whirlpool Galaxy, is a favorite of amateur astronomers. Finding it is not too difficult as it lies near the end star of the Big Dipper (Eta Ursa Majoris). Start by centering the finder on that star. Move west nearly half a finder field to center the 5th-magnitude star 24 Canum Venaticorum. Now move south about a third of a field and west a little. M51 is just west of a 7th magnitude star. Note the many other galaxies in this region, and how one would star hop to them. The brightest galaxies are the Messier objects: M51, M63, M94, and M101. Chart reproduced from *Sky Atlas 2000.0* by W. Tirion.

star chart to star hop to an object, the amateur often sees other objects nearby worth searching out too. And who knows? By constantly searching for objects and comparing views to star charts, novae, supernovae, comets, or asteroids may be discovered.

FINDERS

The finder is a small low-power, wide-field telescope attached parallel to the main telescope. The main telescope has a very small true field of view, usually less than 2° and often less than 1/2°. A wider-field telescope is essential to point such a tiny field at something – otherwise you'd be searching for a needle in a haystack with a microscope.

Many telescopes come with finders too small and cheap for easily locating deep-sky objects. For star hopping, the finder must have a wide enough field of view to easily be pointed at objects (stars) seen with the unaided eye. It must also have enough magni-

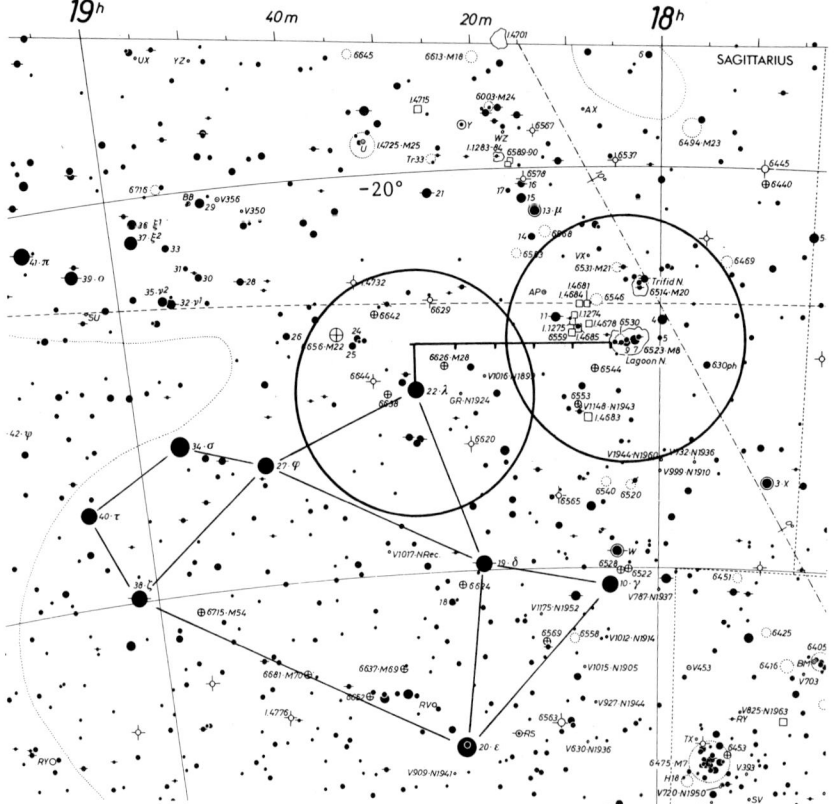

Figure 3.10d. Another way to locate objects is by "offsetting:" moving the telescope a known amount of right ascension and declination from a bright object to a faint one. Use the known size of your finder's view (in this case 6°) as your measuring device. M8, the Lagoon nebula in Sagittarius, is a favorite target of amateurs in the Northern-Hemisphere summertime. Offsetting to M8 and many other objects in the area can be done by first locating the "Teapot" of Sagittarius, outlined here. In dark skies, the Milky Way star clouds near the galactic center appear like steam rising from the spout of the Teapot. Start at the Teapot's top, Lambda (λ) Sagittarii, and move north 1°. (In offsetting it is usually best to do the shortest direction first because any distance errors will be minimized.) Next move west (the direction of decreasing right ascension) by 5.3°. M8 will now be centered. Chart reproduced from *Sky Atlas 2000.0* by W. Tirion.

fication to accurately aim at the object meant for viewing in the main telescope. A good finder is essential. If you can't find an object, what good is the main telescope?

A good finder has the characteristics of good binoculars. The smallest binoculars really useful for astronomy are 7×35 (7 power and 35 millimeters aperture), so a 7×35 finder is adequate only on the smallest telescopes, such as 2.4-inch refractors. Most amateur telescopes should be considered underequipped if they have anything less than an 8×50 finder.

A finder also needs to be of high optical quality. Since many of the brighter deep-sky objects appear as fuzzy stars at low power, real star images should be very sharp to help distinguish them.

The magnification of the finder is important too. If the main telescope is to be positioned very accurately, the magnification of the finder should be at least one tenth that of the main telescope. If we call the magnification of the main telescope divided by that of the finder the Finder Magnification Ratio (FMR), then the FMR should be less than 10. An FMR less than 5 is even better.

For example, suppose identical eyepieces were used on the main and finder telescopes. The apparent fields would be the same, but the true field of view of the higher power main telescope would be much less. If the finder has a power of 7× and a 7° field of view, and the main telescope has a power of 70×, the latter has a 0.7° field of view. An object must be placed within one twentieth of the finder's field of view from the center in order for it to appear in the field of the main telescope at all!

But the finder should be able to do better than place the object just anywhere within the main telescope field. It should really put it no farther than half the field of view (0.35° in this example) from the center. That is an

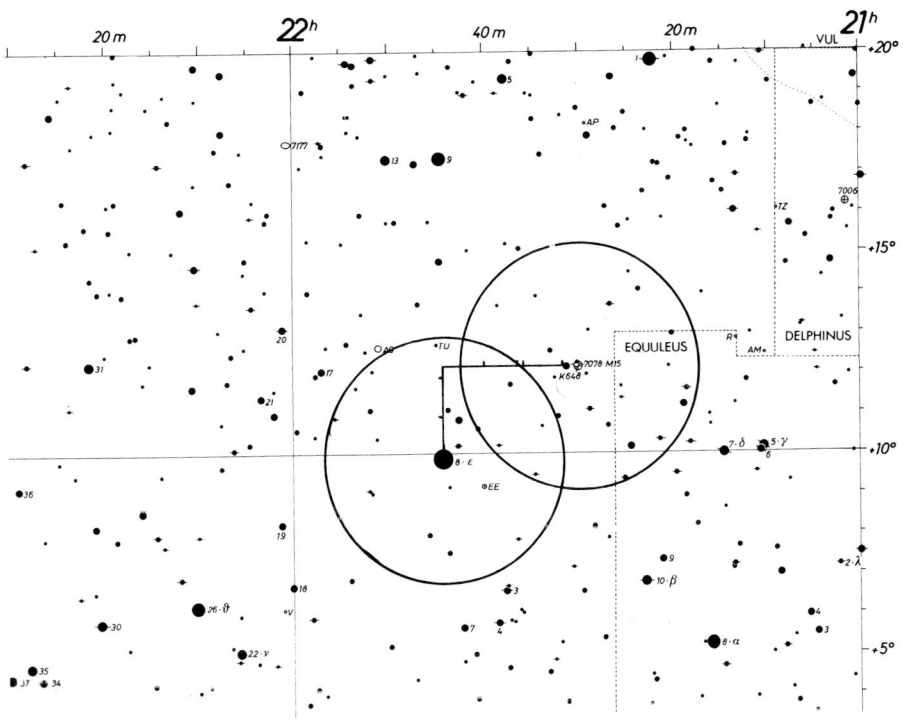

Figure 3.10e. In another example of offsetting, M15, a globular star cluster, can be easily found northwest of the 2nd-magnitude star Epsilon Pegasi. Move north 2.3°, then west 3.4°. Note that M15 is about 1/3° west of a 6th-magnitude star. Chart reproduced from *Sky Atlas 2000.0* by W. Tirion.

error of only one quarter the field diameter, or 10.5 arc-minutes for this case.

A 7-power finder can do this if it is accurately aligned with the main telescope and has thin cross hairs. (An 8-inch focal length finder objective and a 7× eyepiece would require cross hairs 0.01 inch or less in thickness – a reasonable value.) But now consider that after an object has been acquired, the magnification may be increased to 150× or 200×. If the object is lost from the field of the main telescope, where the true field is now only 20 or 15 arc-minutes, it would be very difficult to reacquire the object. A low-power eyepiece would have to be reinserted in the main telescope and you'd have to start all over again.

Because this book advocates observing at magnifications well above 100× on some deep-sky objects, a relatively high-power finder is also a good idea. Higher power, however, means a reduced field of view, and that can hinder comparisons with star charts. The field of view must not be less than about 5° or this becomes a real problem.

The solution to this dilemma is one of two: 1) a finder with exchangeable eyepieces, so that wide fields can be used for finding an object and higher magnification for precise aiming, or 2) two finders, one low power, one higher. Option 2 has the advantage that the higher-power finder could also be of greater aperture, and then very faint objects become easier to find. Disadvantages are the added expense and weight.

The ease of locating objects with two finders, a small one of low power and a larger one of higher power, cannot be overstressed. For many years I had a single finder on the 8-inch Cassegrain used for most of the drawings in this book. It is far better than most, because it has a 60 mm objective of 300 mm focal length (from an old telephoto lens) and a war-surplus Erfle eyepiece of 38 mm focal length and a 65° apparent field of view. The combination gives a breathtaking 8° field on the sky and 7.9×. The cross-hairs are graduated in 1° increments for easy star-hopping. However, after much of the research for this book was completed, and I was using magnifications consistently higher than in my early years of observing, I added a 3.1-inch, f/7.9 refractor. With a 20 mm Erfle, this telescope gives a 2° field of view at 31×.

The ease of getting the general region of the object in the low-power finder, then moving to the 3-inch and almost always seeing the object and centering it, and then moving to the 8-inch, makes the added cost well worth it. Except for a very few objects, the 8-inch is used at magnifications of 117× and higher. The FMR ratios are then only 3.9 from the 7.9× finder to the 31× finder, and only 3.8 from the 31× finder to the 117× of the main telescope. The 8-inch would have to be used at a magnification greater than 310× for the FMR ratio to be greater than 10, but higher-power eyepieces can also be used in the 3-inch refractor at such times. In addition, the 3-inch is a richest-field telescope, giving some marvelous views of the sky not possible with the 8-inch Cassegrain.

With these thoughts in mind, Table 3.5 was drawn up to recommend finder configurations.

MISCELLANEOUS TOPICS
Caring for optics

If the optics are to perform to their limit, they must be clean and free from scratches. This is true of eyepieces, finders and the main telescope. All optics get dirty with use, and they must be cleaned with great care.

The worst problem is invisible: fine abrasive dust that will scratch surfaces during cleaning. Before anything else, very gently brush off the surfaces with a camel's hair brush (which like other cleaning items is available at camera shops). Some companies sell cans of pressurized gas for blowing dust off surfaces such as lenses and mirrors, but before using one of these be sure it says it will not leave a residue and is safe for optics.

To clean small lenses such as eyepieces, the next step is to use lens-cleaning paper (obtainable at most camera stores) or sterile cotton swabs and a mild soap, alcohol, or lens cleaning liquid (also obtainable from some camera stores). Use only genuine sterile surgical cotton, not synthetic "cotton" balls. The glass should only be patted and not rubbed because any rubbing action will tend to scratch dust across the surface.

Mirrors can be cleaned the same way, but the aluminum coating is extremely easy to scratch, so the utmost care is required. If the mirror is removed from its cell, it can be

Table 3.5. *Recommended two-finder combinations*

Main telescope aperture		The lower-power finder			The high power finder	
		power	aperture		power	aperture
2- to 3-inch		7×	35 mm		none	
3- to 7-inch	1)	8×	50 mm	and	15×	50 mm
	or 2)	12×	50 mm		none	
8- to 11-inch		8×	50 mm	and	25×	60 mm
12- to 16-inch		8×	60 mm	and	40×	80 mm
16- to 24-inch		8×	60 mm	and	50×	100 mm

immersed in lukewarm water and mild, unscented pure soap for cleaning. First run lukewarm tap water over the mirror for several minutes to dislodge dirt on the surface. Then soak it in a plastic tub or sink full of lukewarm soapy water. (Place a towel in a sink to cushion the mirror if it is dropped.) Now swab the mirror very gently underwater with sterile cotton. Turn the wad of cotton in a backward-rolling motion so that as soon as part of it rubs the surface, it is carried away and won't touch it again. When the wad has been completely turned, throw it out and use a new one. Rinse by running tap water over the mirror again for several minutes. Finish with a rinse of distilled or deionized water (this does not leave stains), stand the mirror on its side on a towel, and let it dry. If your tap water contains many impurities, you should use distilled or deionized water for the whole process.

Eyepieces should be stored in a safe case. Some companies sell eyepiece cases, or camera cases can be adapted, or you can build your own. Most cases are foam lined, and many amateurs simply place the eyepieces in slots or holes in the foam. Such practice can damage the eyepiece if any lens surface touches the foam. The extreme case of damage would occur if the case was exposed to sunlight; the foam can melt onto the lens! The eyepieces should have some form of lens cover, like a plastic cap, before being placed in a foam box. Or keep them in plastic sandwich bags.

Dew

Dew is extremely hazardous to optics. Repeated dewings can turn mirror coatings brown because when dew settles, dust settles with it. The presence of water can also start a chemical etching process by impurities.

A telescope brought into a warm house from the cold will immediately dew up. The optics should be sealed before entering the house so that little water will condense on them, and after reaching room temperature, the seals should be loosened so any trapped moisture can escape.

Dew can also form on a telescope while it is being used outdoors, fogging the view. This can be prevented to some degree by having a long tube in front of the objective, about twice as long as the tube's diameter. Reflectors only dew in the severest conditions, because the mirror is at the end of a very long tube. Finder telescopes should have a long shield in front of the objective to prevent dew as well as to keep out stray light. Eyepieces are more difficult to keep from dewing up, especially since they are near the observer's breath. Eyepieces should not be left uncovered, but put in a protective box when not in use.

In really wet conditions, the box could be heated with two to three watts to drive off dew. The objective can also be kept dew free by placing one to three watts of heating elements around it. Such heaters can be made with small resistors or resistive wire and a low voltage source such as a car battery. (Do not use the car's own battery because after heating all night it might not have enough power to start the car.)

SUMMARY

The purpose of a telescope is to gather light and focus it to form an image. An eyepiece magnifies the image and directs the light to the eye, where it is refocused on the retina. Two basic parameters govern a telescope's ability to show an object: the amount of light gathered (determined by the size of the primary mirror or lens) and the total magnification at the eye.

The magnification and the eyepiece's apparent field of view determine the true field of view on the sky. A short-focus telescope (i.e. f/4.5) and a long focus instrument (such as f/15) of the same aperture will give essentially identical views of deep-sky objects when used at the same magnification with eyepieces having the same field of view. The short-focus telescope has an advantage in that it can achieve low powers and wide fields with standard eyepieces. The long-focus telescope has the advantage when working at high powers, as they can be achieved with eyepieces having reasonable focal lengths and comfortable viewing positions for the eye. Optical aberrations are also reduced.

The differences among telescope types are not as important for viewing deep-sky objects as for planetary or double-star observing, since most of the limitations for deep-sky work are in the human eye.

Even in perfect seeing conditions, an upper limit to the useful magnification is set by the resolution of both the telescope and the eye. The eye can resolve about one arc-minute when an object is bright, but resolution decreases as the object becomes fainter, and at the limit of detection, the eye's resolution is only about 0.5°. In this situation very high powers are necessary to magnify the finest details a telescope itself resolves so that they become visible to the eye.

For bright subjects, the magnification limit is often said to be about 60x per inch of the telescope's objective diameter. (A 10-inch telescope is said to have a limit of 600 power). However, for faint objects the limit is as much as 330× per inch. The lesson to remember is to use however high a magnification seems to work best.

Special filters can partially reject both natural and manmade light pollution. These nebula filters work by increasing contrast between certain objects and the sky background. Even though the filters decrease the light of all objects, if the background is reduced even further, the improved contrast may more than compensate, and the object will actually appear brighter to the eye. Nebula filters work best on emission and planetary nebulae, rather than on stars, galaxies, and reflection nebulae. The filters generally should be used between the eyepiece and the telescope objective, because many such filters do not work well if the light is more than a few degrees off-axis. The effectiveness of filters from different manufacturers varies considerably, so try to use several in actual observing sessions, and choose the best one to purchase.

For finding objects in the first place, the best method for the amateur to learn is star hopping. First the brightest stars must be learned with the aid of an all-sky map or planisphere. Then, using a detailed star atlas, the telescope's finder is pointed to a known naked-eye star. By using the patterns on the star chart, the finder telescope, along with the main telescope, is moved step by step to the object of interest.

The finder should have an aperture of at least 35 mm for main telescopes of 3 inches aperture or less, and at least 50 mm (2 inches) for telescopes with apertures over 3 inches. The finder magnification should be no less than one-tenth the magnification used on the main telescope. Otherwise the main telescope cannot be pointed accurately. Amateur telescopes larger than about 8 inches should have two finders, the second with an aperture and magnification midway between the small finder and the main telescope.

4
The faintest star visible in a telescope

INTRODUCTION

The faintest point of light detectable by the unaided eye was derived in Chapter 2. The eye's fundamental limit is around 50 to 150 photons of green light arriving over a several-second period, corresponding to a star as faint as magnitude 8.5. Seeing such a faint star requires perfect conditions and dark adaptation as well as exclusion of all extraneous light, including all other stars in the sky.

Thus it's not surprising that no one sees 8.5-magnitude stars with the naked eye in real life. A more typical limit is 7 or 7.5 for a skilled observer in excellent country skies.

Stars and other small objects form a special case for detection by the eye in the telescope. As we have seen, any object less than about 0.5° across as presented to the eye can be considered a point source if it is so dim it's at the threshold of detection. A star image is actually a diffraction disk, but it is so small that, if faint, the disk is a point to the eye at any reasonable magnification at all.

Magnification does not change the brightness of a point source in the telescope, but it does decrease the surface brightness of the background and reduces the field of view so other stars do not interfere. Therefore, the fundamental limits of the eye can be reached when a telescope is used. Here it really is possible to see the equivalent of 8.5-magnitude stars naked-eye.

MAGNIFICATION

The fundamental magnitude limit M_t of a telescope is given by

$$M_t = M_e + 2.5 \log_{10}(D^2 t / D_e^2) \quad \text{(equation 4.1)}$$

where M_e is the eye limiting magnitude (8.5 for the ideal case), D is the telescope diameter, D_e is is the eye diameter, taken to be 7.5 millimeters, and t is the telescope transmission factor (which is usually about 0.7). This equation reduces to

$$M_t = 3.7 + 2.5 \log_{10}(D^2) \quad \text{(equation 4.2)}$$

where D is expressed in millimeters. This formula was used to list the ideal limiting magnitudes for telescopes of various apertures in Table 4.1.

The surface brightness M_b of the sky or an extended object (in magnitudes) is darkened by the telescope magnification and transmission factor as follows:

$$M_b = -2.5 \log_{10}(D^2 t / m^2 D_e^2) \quad \text{(equation 4.3)}$$

where m is the magnification. This is why magnification helps to detect faint stars when the sky is bright, or even under dark country skies compared to when low power is used. High magnification also increases the apparent angle between field stars.

Consider a dark country sky with a surface brightness of 24 magnitudes per square arc-second. At the minimum usable magnification, m_m, computed from the equation

$$m_m = D / D_e, \quad \text{(equation 4.4)}$$

which is 27× for an 8-inch telescope, the sky brightness is reduced from its naked-eye level by

$$M_m = -2.5 \log_{10}(t), \quad \text{(equation 4.5)}$$

or if t is 0.7, 0.39 magnitudes / sq. arc-second.

Examining Figure 4.1, which shows the faintest star detectable by the eye, we find that at a sky background of 24.4 mag./sq. arc-sec., the naked-eye limit is 7.6. Using this value for M_e in equation 4.1, we find a limiting magnitude of 14.4 for an 8-inch telescope at its lowest usable power of 27×.

Table 4.1. *Ideal limiting magnitudes*

Aperture (inches)	Visual magnitude	Aperture (inches)	Visual magnitude
1	10.7	12	16.1
2	12.2	14	16.5
3	13.1	16	16.7
4	13.7	18	17.0
5	14.2	20	17.2
6	14.6	22	17.4
7	14.9	24	17.6
8	15.2	30	18.1
10	15.7	36	18.5

Note: Large refractors have brighter limiting magnitudes due to absorption in the thick objective glass.
Computed from equation 4.2

For higher magnifications, equation 4.3 gives the reduction in sky brightness. For example, choosing a power of 90× and $t=0.7$ in equation 4.3, we see the sky background is 3 magnitudes fainter than that at 27×! Thus, the background is reduced to 27 magnitudes/sq. arc-second, improving the eye limit, M_e, to its best possible value, 8.5. The magnitude limit is then 15.2 in the 8-inch telescope, an increase of 0.8 magnitudes over that seen at 27 power.

A more dramatic example can be seen by observing stars in a suburban sky with a surface brightness of 20 magnitudes/sq. arc-sec., where the naked eye limit is 5.5. At a magnification of 27× on the 8-inch, stars of magnitude 12.5 would be just detectable. If the magnification were increased to 200, the sky surface brightness in the telescope would be reduced to 24.7 magnitudes / sq. arc-sec. and stars as faint as 14.6 would be within view. At 570 power, the sky surface brightness in the telescope would be reduced to 27.0 magnitudes / sq. arc-sec. and the 8-inch would reach its limit of magnitude 15.2.

This would be true even in bad seeing, because a very faint star image blurred to a diameter of three arc-seconds is still a point as far as the eye is concerned, even at high magnification.

At the very limit of detection, there must be no other stars within at least a couple of degrees (as seen by the eye) of the star being scrutinized. Greater distances are required if the neighboring stars are bright.

FINDING TONIGHT'S MAGNITUDE LIMIT

Because atmospheric seeing does not affect visibility of faint objects at the eye's threshold, the deep-sky observer's main measure of sky quality is merely the sky surface brightness. However, this is difficult to determine. Instead, observers traditionally take note of the faintest star visible to the naked eye.

The atmosphere itself absorbs light and thus affects the faintest star observable. This *atmospheric extinction* increases farther from the zenith, as described in Appendix C. Thus, when the faintest star visible to the naked eye or in a telescope is determined, its altitude above the horizon should be factored in – at least if the altitude is less than about 60°.

Making such a sky-quality estimate requires adequate star charts with visual magnitudes recorded. For naked-eye determinations many charts are available, some of which are listed in Appendix A.

For telescope work, open star clusters provide many stars in one view for easy limiting-magnitude determinations. Appendix B presents a number of clusters around the sky for this purpose. They were plotted from data gathered by Hoag et al. (1961) and are shown in two charts per cluster: one with stars only, and on the facing page, the cluster with visual magnitudes given to the tenth with the decimal point omitted (129 is magnitude 12.9). Also given are the plate scale and the viewing distance from the book to show how the cluster will appear at various magnifications. Sparsely populated clusters were chosen so brighter field stars do not interfere. Clusters that have very bright members, such as the Pleiades, are not included since these interfere severely with viewing the fainter members. Table 4.2 lists the star clusters in Appendix B and the magnitude range of the stars plotted on the charts.

Unfortunately, open clusters are not evenly distributed around the sky, but are concentrated near the Milky Way. This caused gaps in the Hoag study, such as a lack of clusters between 7.5 hours and nearly 18 hours right ascension. Clusters do exist in this right ascension range, but they are far south and were not measured in the Hoag study. But at least one of the clusters should be visible at some time of night each night of the year.

An example of using the star clusters to monitor site quality is shown in Figure 4.2. I wanted to find the site closest to Denver, Colorado that could be considered free of light pollution. The star cluster NGC 6910 was chosen for the test. Sites were selected at different distances from the metropolitan area that were uncontaminated by smaller, closer towns. The best skies were found only at a distance of 70 to 80 miles away in the Rocky Mountains. A site over 200 miles from Denver produced the same result as one at 80 miles, though there were always small towns at distances of about 50 miles. The theoretical limit was never reached, possibly because of these small cities, or because of atmospheric dust or airglow during the season of observations (summer), or more likely, the telescope transmission factor is closer to 0.5.

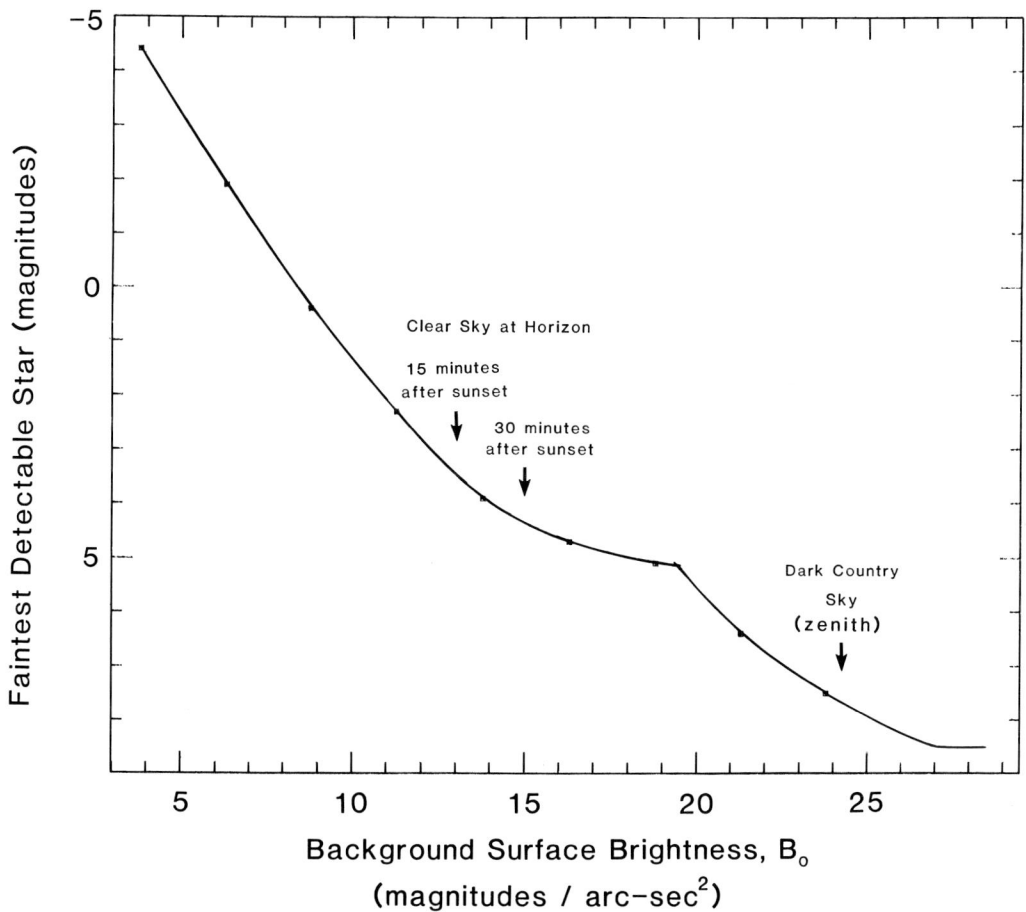

Figure 4.1. The faintest star that can be seen by a trained observer depends on the background surface brightness of the sky. A good dark country sky has a surface brightness of about 24.4 magnitudes per square arc-second, giving a visual limit of magnitude 7.8. High magnification on a telescope can reduce the background below 27 magnitudes per square arc-second, allowing the eye to reach its fundamental detection limit given in Table 4.1. Detection of faint stars in a nebula is more difficult since the background (surrounding the star) is greater than that of the night sky.

Table 4.2. *Star clusters illustrated in Appendix B for faintest-star determination*

Object	Position (epoch 2000.0)		Magnitude range
	R.A.	Dec.	
NGC 225	00^h 43.8	+61° 47′	9.3 – 15.9
NGC 1647	04 46.9	+19 05	8.6 – 16.3
NGC 2129	06 01.1	+23 18	7.4 – 16.6
NGC 2422 (M47)	07 36.3	−14 30	5.7 – 14.2
NGC 6494 (M23)	17 56.9	−19 01	6.5 – 14.2
NGC 6823	19 43.2	+23 19	8.8 – 15.9
NGC 6910	20 23.0	+40 48	7.4 – 14.8
NGC 7031	21 07.3	+50 50	11.3 – 17.2
NGC 7235	22 12.5	+57 15	8.8 – 15.8

All data from Hoag *et al.* (1961).

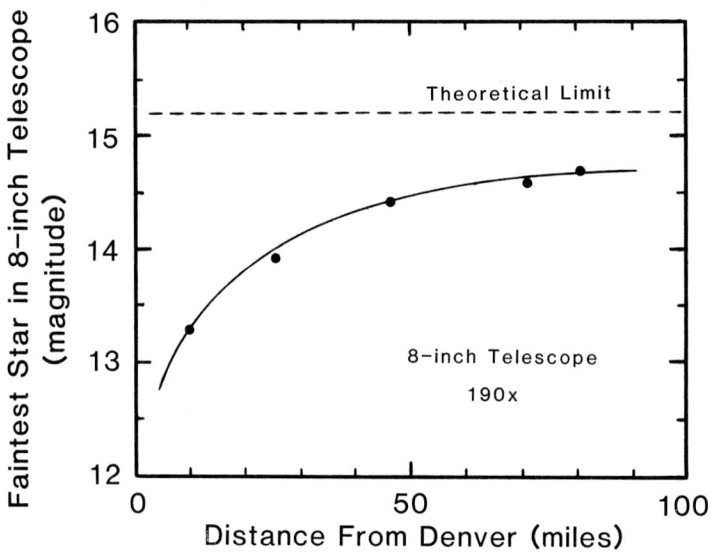

Figure 4.2. The faintest star visible at different distances from the Denver metropolitan area, population 1.8 million.

THE FAINTEST STAR VISIBLE IN A TELESCOPE

SUMMARY

The faintest star visible in a telescope depends on the aperture, power and sky brightness. High power reduces the surface brightness of both the sky and the star image (since the star is a diffraction disk in the telescope). However, since the eye sees faint star disks as points, the surface brightness of a faint star does not *appear* to change with increasing magnification. The background surface brightness does visibly darken, and so faint stars become easier to see.

The telescope, with its magnified and restricted field of view, can be used to reach the eye's fundamental detection limit, something a naked-eye observer can never achieve. This ideal condition is reached when the background is darkened sufficiently by magnification so that the faintest star possible can be seen. This can be done whether observing in superbly dark country skies or in city skies with light pollution. Similarly, faint stars in a nebula can be seen more easily at higher magnifications because the surface brightness of the nebula is reduced.

The faintest star observable near the zenith with a specific telescope and magnification is a fundamental parameter documenting the conditions for observing deep-sky objects.

5
Making drawings and keeping records

INTRODUCTION

Anyone who observes the sky should keep a logbook. Even the most casual celestial sightseeing becomes more meaningful if a few notes are jotted down in a permanent record. For more serious observing, complete notes are essential. They document the viewing conditions, provide a record of what was actually seen, and create a baseline for comparing future observations and gaining experience. And, of course, they will form a rich personal diary of starry nights.

Good records sometimes serve purposes unanticipated at the time. Finding a supernova in a galaxy, or an unexpected asteroid or comet, are examples that come to mind.

Amateur observers often just list the objects they see during a night. This has little value unless you just like to compile lists and figure out such things as how many galaxies you have seen. The next level of sophistication is to note the date, time, magnification and telescope used. This still has little value except to list-compilers.

At the next level, the observer might write a few comments such as:

Time	Object	Power	Comments
11:05	M63	36×	easy
11:10	M51	75×	maybe some spiral structure
11:15	M14	50×	partially resolved
11:25	M13	75×	mottled, spectacular
11:40	M57	100×	ring easily seen
12:10	Veil	36×	barely seen

Such a list has only a little more value. What is "easy?" Spiral structure means different things to different people. "Barely seen" does not tell what it actually looked like. Was it elongated? Were stars visible through it? Which parts of it were seen?

The list above is typical of some of my early records. In preparing this book, I was very disappointed to discover their lack of usefulness. Back then I also made many drawings a half inch to two inches wide, but rarely included field stars. Thus there is no way to tell the scale of a drawing or the size of what was seen. Is that fuzzy patch the full extent of a galaxy including its spiral arms, or only the bright central region? A couple of field stars would have allowed comparison with photographs to find out.

Even longer descriptions can seem inadequate on future reading. In some of my more exciting moments under superbly dark country skies, I wrote "just like a photograph", and "so spectacular and complex that a true drawing cannot be made." These comments recall fond memories but serve little other purpose.

To document the view fully, records must include the sizes of objects, the shapes and relative brightnesses of all their parts, and their placement with respect to stars. This chapter outlines ways to do so – either in writing or by one of three methods for making drawings.

WRITTEN DESCRIPTIONS

Any record of a sighting should include certain basic data: the observer, date, start and ending times of the observation (be sure to state the kind of time and date, such as Universal Time or Mountain Standard Time), the observing site, the eyepieces and their magnifications, a sky quality note such as the faintest star visible, whether a clock drive was used, and the approximate altitude or air mass of the object. (See Appendix C for a discussion of air mass.)

A written description should include estimates of the object's size, the directions and distances of its details from reference points, and the details' relative brightnesses. An example:

NGC 4565 *Edge-on galaxy in Coma Berenices 5/11/83 Start: 10:00 UT*

8-inch, f/11.5 Cassegrain   *Location: Waianae Ranch, Hawaii*

Eyepieces: 20mm Erfle (117×) 12.4mm Erfle (188×)

Faintest star at zenith: about magnitude 14.3 at 188×. Observer: R. Clark.

This object was hard to find, being in a star-poor area. Once found, the view was amazing! I had seen the galaxy in poorer skies in the early 1970s but did not think the dark lane was visible with an 8-inch. Now the dark lane could be seen at 117× and 188× by several observers easily (no averted vision).

The galaxy is about 11 arc-minutes long, extending northwest to southeast. The boundaries of the arms are fairly sharp, except at the tips where they fade slowly into the sky background. The central hub is oval (about 2 by 1.5 arc-minutes); the long axis is aligned with the arms. The edges of the hub fade first slowly, then quickly into the sky background. Best view is with the 12.4 mm Erfle at 188×, in which the galaxy extends over the whole field.

The dark lane is about 4 to 5 arc-minutes long and about 0.3 arc-minute wide. The inner nuclear region (the central 0.25 arc-minute) is much brighter and lies southwest of the dark lane.

On each side of the nucleus in the long direction is a bright spot about 0.25 arc-minute in diameter. Another bright spot was glimpsed occasionally about 4 arc-minutes from the nucleus along the galaxy's long axis.

A star lies 1.5 or 2 arc-minutes northeast of the nucleus. It is fainter than the nuclear region, probably 12th or 13th magnitude. A second star, a little fainter, is about 4 arc-minutes south of the nucleus and slightly west. A brighter one (mag. 10 or 11?) is 8 arc-minutes west and 3 arc-minutes south of the nucleus.

End of observation: 10:42 UT.

DRAWINGS

It's often said that a picture is worth a thousand words. Composing a thousand words takes much time and thought and still may leave the reader with the wrong mental image. So when it comes to accurate recording of scientific data, there's often no substitute for a picture.

In past centuries a scientist was necessarily a draftsman. Nowadays scientists in almost all fields rely on photography to record images, and the pencil and sketchpad no longer rank as essential scientific tools. Visual astronomy, however, remains an exception.

You'll need white paper, either plain or with light graph lines for judging distances, a soft pencil such as number 1 or 2, and a clean eraser. Some observers draw with chalk on black paper, but others find chalk hard to control. Drawings of deep-sky objects made at the telescope are generally done with scribbled pencil shading. Better-looking finished versions, made immediately afterward indoors, use what is called the finger-smudging technique. Using a dull pencil, lightly scribble a somewhat smaller area than the desired shading. Then use a clean, dry finger to smear the marks until they blend together into a uniform texture. If the patch is too faint, scribble and smear some more until the desired effect is produced. If the smear becomes too dark, the eraser is used in the same way as the pencil, followed by more smearing. But lightening a smudge is harder than darkening it, so sometimes it's necessary to start again.

Pencil drawings on white paper are negatives: the brighter a star or nebulous patch, the darker the pencil mark or smudge. Some people find it hard to get used to a negative image, but most quickly learn to mentally reverse dark areas on the paper to correspond to bright areas in the sky.

Doing anything at the telescope other than looking through it requires light. Any extra light reduces the eye's sensitivity to faint objects. But if the working light is too dim, drawings become less accurate. I recommend that a very dim flashlight be used. Red light is traditionally preferred, but the rod cells are more sensitive to *all* colors than cone cells so there is no advantage to using red light. Figure 2.3c (page 9) shows this. In fact many amateurs tend to use too bright a red light. They have the false impression that if it is red, night vision will not be impaired. In practice, however, I find that very dim green or yellowish light works fine. Strong colors should be avoided because they distort the eye's color perception at the telescope, if an object is bright enough to show color at all.

A good light level can be obtained from a two-battery flashlight by replacing the bulb with one meant for three, four, or five batteries. This both dims and reddens the light, while extending battery life. The flashlight should cast a smooth glow with no bright or dark spots. This can be ensured by placing one or two pieces of paper in front of the bulb to act as diffusers. The brightness can be fine-tuned by using two pieces of polarizing material between the bulb and paper to act as a variable filter. Buy a high-quality flashlight – it is very annoying to have to shake or bang it to make it work.

A drawing can cover any amount of sky and be any size on the paper you wish. Some authors recommend using just one eyepiece, drawing a circle to represent its field of view, and sketching what is seen in this field. This is a bad idea for several reasons. The most important is the need to use many magnifications to explore an object properly, one of the main thrusts of this book. Another is that you may want to draw a larger area than fits in the field, or just a small detail.

One rationale for the circle method is to give others a feel for what they might see in their eyepiece at the same magnification. But even at the same power, one eyepiece can show a field twice as big as another. The right way to indicate an object's apparent size is to give the distance from which to view the paper so the object looks as big as it did in the telescope. Angular distance on the sky corresponds to linear distance on a drawing or photograph. The relationship between them is called the scale (or frequently the "plate scale", a carryover from photographic plates). Often it is given in arc-seconds per millimeter. If an angle a_t in the sky corresponds to the distance l_s on the drawing or photograph, then the scale p_s is

$$p_s = a_t / l_s. \qquad \text{(equation 5.1)}$$

Once the scale is set, the appropriate viewing distance v to the paper depends only on the magnification m of the telescope:

$$v = l_s / (2m \tan(a_t/2)). \qquad \text{(equation 5.2)}$$

The drawings in Chapter 7 give both the scale and viewing distances for various magnifications. By holding your eye at these distances from the page, you see just how large an object appears at a given magnification.

If an eyepiece's true field in your telescope is known (see Chapter 3), you can find the linear size of the eyepiece field of view l_e on the drawing (or a photograph) from the equation

$$l_e = 2mv \tan(a_t/2), \qquad \text{(equation 5.3)}$$

where a_t is now the true field of view of the telescope with that eyepiece.

Even though many magnifications should be used to view an object, one will often give the best overall impression of the object. This magnification, along with a viewing distance of about 25cm (10 inches), will define a good scale for the drawing.

Any drawing takes considerable time, first at the telescope, then indoors doing a geometric correction and a finished version (discussed below). A very simple subject with only a couple of field stars, such as a faint, featureless galaxy, may take only 10 minutes at the telescope. Most subjects take over 30 minutes, and complicated ones like the Orion nebula (M42), several hours. The more complex a subject, the more you must go back and forth from eyepiece to paper, and each time you return to the telescope it takes five minutes or so to regain full night vision even when using a very dim flashlight.

After the drawing is completed at the telescope, a finished version is made indoors by the finger smudging method. For this finished drawing, a geometric correction should be made using known positions of stars to "straighten out" positional errors. This drawing usually takes longer than the one made at the telescope. In general, the more time spent, the more accurate the portrayal is likely to be.

Good drawings do not require special artistic talent or experience, but they do demand close attention, much time at the telescope, much time redrawing with finger smudging, and honesty in not recording details remembered from photographs but not positively seen. All this can require more time than taking a long-exposure photograph. But the result is often as satisfying or more so.

Drawing even has certain advantages over photography. One is low cost. The simplest, bare-bones telescope is all you need – no clock drive, alignment devices, or electronic guider, not to mention camera, film, and a photographic darkroom. Secondly, the eye has a far greater latitude, or dynamic range, than film. This means bright and faint details can be seen at the same time, which is something film cannot do (without very special techniques). For example, have you ever seen a photograph of M42 that shows the large, faint outer wisps *and* the bright Trapezium? See Chapter 7 for many examples.

Drawing method 1: the initial-blind method

For this method you start "blind" with no information about the subject that might bias you; a photograph should not be at the telescope.

First the field stars are drawn, then the object itself is sketched in, using the stars to position its parts correctly. Finger smudging is not used, just gentle shading with the pencil. The working light should be so dim that you can't tell pencil shading from finger smudges anyway; the smallest detail visible in very dim light at a distance of 10 inches (25.4 cm) is about a sixteenth of an inch (1.6 mm).

Because of this, a drawing made at the telescope will look terrible in normal room lighting. More to the point, a subject cannot be rendered very faithfully under dim illumination. But detail will be remembered for a short while after leaving the telescope. This is the time to redraw the rough sketch indoors.

While doing so, a geometric correction can be performed. This is done by finding a suitable photograph of the object and tracing only the field stars (and, if the object has very bright portions with sharp boundaries, these as well.) But remember that brightnesses may appear different on the photo than perceived by the eye. Draw the star brightnesses as remembered and recorded, not as they appear on the photograph.

Lastly, the photograph should be put away and the hazy object finger-smudged in. Here one needs to be very honest and not biased by the photo just seen. Only when the drawing is complete may it be compared with the photo.

The final drawing should be checked for accuracy during another observing session, preferably under similar skies at the same time the next night. The object should first be studied in the telescope, then a bright enough light should be used to confirm the details in the drawing.

Drawing method 2: filling in details

In this method you start with a pre-drawn, geometrically correct sketch and fill whatever details you can see in the telescope. Such a sketch could be traced from a photograph, but trouble may result because photos usually record stars that are too faint for visual use and the proper magnitude cutoff can't be told. A better source is a drawing already made by method 1. Such a drawing can be traced or copied, leaving out the detail at the limit of detection.

Using this preliminary sketch, you add any newly observed detail. Then a new final drawing is made indoors using finger smudging as in method 1.

Drawing method 3: the double-blind method

In this procedure no photographs are used before, during, or after the drawing at the telescope is made. This method, the freest from bias, requires an accurately driven telescope and a grid reticle in the eyepiece.

The stars and subject are drawn on graph paper, where the squares match the reticle grid seen in the eyepiece. Everything can be plotted accurately this way; there is no need for a geometrical correction in room light. However, nebulosity will still need to be redrawn by finger smudging. The grid may be retained in the final version, or removed.

This is the most difficult drawing method, and reticle eyepieces are not available in today's telescope market. They have to be custom made. A grid is not needed in all eyepieces, but in at least one so that some star positions can be accurately plotted.

SUMMARY

Good astronomical records provide a useful baseline for seeing the same and possibly more detail in the future. They add meaning to a night's work and improve the observer's perceptiveness. Detailed records provide reminders of enjoyable and sometimes awe-inspiring observing experiences. Careful drawings can also be displayed in the same manner as photographs, showing others, including non-astronomy enthusiasts, the beauty that can be seen in the night sky.

6
A case study: the Whirlpool Galaxy Messier 51

INTRODUCTION

In this chapter we will see how the principles discussed up to now – including telescope size, sky quality, and magnification – affect the features that can be seen in a particular galaxy.

Our case study is an object that nearly every amateur tries to observe not long after getting a telescope: the Whirlpool Galaxy M51, with its spiral arms. This was the first galaxy in which spiral structure was discovered, in 1845 by the Earl of Rosse using a 72-inch speculum-metal reflector. It still has the reputation as the galaxy with the most easily visible spiral arms. What telescope size, magnification, and sky conditions are required to see them?

BRIGHTNESS PROFILE OF M51

Figure 6.1 shows a photograph of M51 with a line from the nucleus to the edge. At 15 places along this line, the galaxy's light was measured photometrically by François Schweizer in 1976 as part of a study of the spiral structure of galaxies, providing us with an accurate brightness profile to analyze. The light levels along the line are plotted in Figure 6.2 – first the actual levels that were measured, then with various amounts of light pollution added to illustrate the situation under increasingly poor skies. The same data are listed in Table 6.1.

Note the dip in luminance (surface brightness) 65 arc-seconds from the nucleus. This represents the background level between two spiral arms. What conditions are needed to see the increase in brightness marking the next spiral arm out, at 77 arc-seconds?

Contrast is defined as

$$C = (B - B_o) / B_o \quad \text{(equation 6.1)}$$

where B and B_o are the surface brightnesses (in linear units) of the object and background, respectively. In stellar magnitudes, this becomes

$$C = 10^{[-0.4(M - M_o)]} - 1 \quad \text{(equation 6.2)}$$

where M and M_o are the surface brightnesses in magnitudes per unit area (such as per square arc-second). In the case we're considering in M51, $M = 20.41$ and $M_o = 21.24$ when the sky is very dark (24.25 magnitudes / sq. arc-sec; see Table 6.1). This corresponds to a contrast C of 1.33. Note that the "background" in this case is not the sky background, but the galaxy's glow between spiral arms.

If we switch to higher powers, the surface brightness dims because the light is spread out over a larger area. But the *contrast* remains the same. This fact is represented in Table 6.2 – which was drawn up for an 8-inch telescope at various magnifications for the six sky conditions in Table 6.1.

VISUAL DETECTION

The visibility of the spiral arm at a given contrast and surface brightness depends on its apparent size – that is, the telescope's power – and the eye's sensitivity. Take a minute to examine Figure 6.3. At 27×, the spiral arm appears 13 arc-minutes wide to the eye. Note where the vertical line for 27× meets the horizontal line for sky condition A (contrast = 1.33). Here, the sloping lines showing background surface brightness show that the eye can only detect such a small object if it is as bright as 21.3 magnitudes per square arc-second. This point falls below the heavy, roughly horizontal arc showing the detection limit for an 8-inch telescope. So the spiral arm will not be visible.

Table 6.1. *Radial brightness profile of M51 for different sky conditions*

Distance from nucleus (arc-sec.)	Brightness in magnitudes/sq. arc-second					
	Darkest sky					Brightest sky
	A	B	C	D	E	F
0	18.00	17.98	17.98	17.94	17.84	17.64
25	20.56	20.48	20.33	20.03	19.51	18.78
31	20.73	20.64	20.47	20.13	19.57	18.81
38	21.07	20.95	20.73	20.31	19.67	18.86
46	21.39	21.23	20.95	20.46	19.75	18.89
55	21.32	21.17	20.90	20.43	19.73	18.89
65 M_o	21.40	21.24	20.96	20.46	19.75	18.89
77 M	20.48	20.41	20.27	19.98	19.47	18.76
92	21.37	21.22	20.94	20.45	19.75	18.89
110	22.39	22.03	21.51	20.78	19.91	18.96
130	22.60	22.18	21.60	20.82	19.93	18.97
150	21.98	21.72	21.31	20.67	19.86	18.94
170	22.11	21.82	21.38	20.71	19.87	18.95
190	23.50	22.70	21.87	20.95	19.98	18.99
210	24.25	23.00	22.00	21.00	20.00	19.00
Contrast M, M_o	1.33	1.15	0.89	0.56	0.29	0.13
Log contrast	0.124	0.061	−0.051	−0.252	−0.538	−0.886

Data in column A are from Schweizer (1976), *Astrophysical Journal* Supplement **31**, pp. 313–32, from his Figure 5b at position angle 135° from the galaxy's nucleus. Data are B_3 magnitudes, which are reasonably close to visual magnitudes.

If we boost the power to 120×, the apparent width of the spiral arm swells to 60 arc-minutes. Its surface brightness is now fainter, 25.0 magnitudes per sq. arc-second. But the eye is much more sensitive both to faint objects and low-contrast objects *if they are large*. Under sky condition A, the eye could see something 60 arc minutes wide even as faint as 25.7 magnitudes per sq. arc-second. So the arm is now quite visible in an 8-inch scope. It could even be detected in a sky as poor as condition D.

Interestingly, the graph shows that boosting the power from 120× to 200× would bring the spiral arm just into detectability in a 6-inch scope under sky A. Even though the arm would be more apparent in the 8-inch than the 6-inch at all powers, it is less apparent at 200× than it was at 120×.

Note that a faint object becomes harder to detect at both very high *and* very low powers. This is why the thick arcs for various telescopes curve up at both ends. Detection is easiest when the object is at the bottom of each arc – at the *optimum magnified visual angle* discussed in Chapter 2. The optimum magnified visual angle is shown by the somewhat wobbly line climbing up the graph from lower left to upper right.

DISCUSSION AND SUMMARY

The sizes and contrasts of deep-sky objects vary greatly. So does the size and contrast of detail within a single object. So a large range of magnifications is needed to see all possible detail. For any feature, there is an optimum magnified visual angle at which it can best be detected. Fortunately, the decrease of detectability on each side of the optimum is small, so fine gradations of magnification are not needed.

The eye's response to light, like the responses of our other senses, is logarithmic. So the sequence of magnifications used should follow a logarithmic trend. For example, you

A CASE STUDY: THE WHIRLPOOL GALAXY, MESSIER 51

Figure 6.1. Messier 51. The line shows where the photometric brightness measurements analyzed in this chapter were made. (Courtesy Palomar Observatory.)

VISUAL ASTRONOMY OF THE DEEP SKY

Table 6.2. *M51 background brightness (M_0) in spiral arm for an 8-inch telescope*

Contrast (C)		M_o					
C	Log(C)	27×	60×	120×	200×	400×	600×
1.33	0.125	21.78	23.52	25.02	26.13	27.64	28.52
1.15	0.061	21.62	23.36	24.86	25.97	27.48	28.36
0.89	−0.051	21.34	23.08	24.58	25.69	27.20	28.08
0.56	−0.252	20.84	22.58	24.08	25.19	26.70	27.58
0.29	−0.538	20.13	21.87	23.37	24.48	25.99	26.87
0.13	−0.886	19.27	21.01	22.51	23.62	25.13	26.01
Apparent size of spiral arm:		13.5′	30′	60′	100′	200′	300′
Log size:		1.130	1.477	1.778	2.000	2.301	2.477
Reduction in surface brightness (mags)		0.38	2.12	3.62	4.73	6.24	7.12

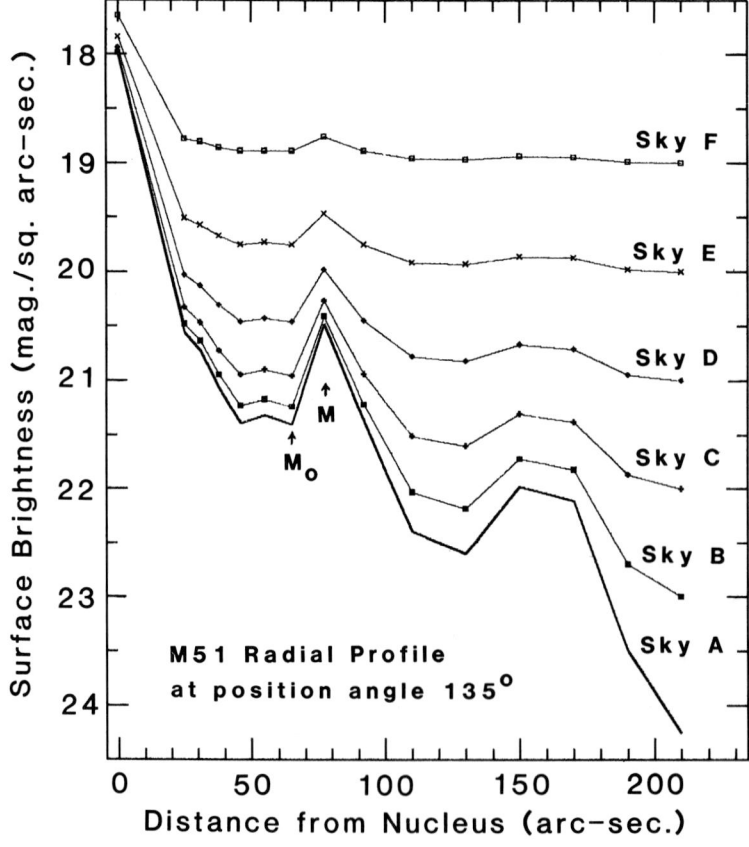

Figure 6.2. The data from Table 6.1 are plotted to show the contrast of M51's spiral arms in different amounts of light pollution. Sky A is for a dark country site, while F is typical of a large city. The spiral arm analyzed in the next figure is the one 77 arc-seconds from the nucleus.

might want a series of eyepieces that each give about 1.6 times higher power than the last. If your telescope's minimum useful power is 30×, then a reasonable series of magnifications might be 30×, 50×, 75×, 125×, 200×, 315×, 500×, and so on.

An astonishing fact emerges from analyzing the eye's sensitivity under astronomical conditions. A small telescope requires *higher* magnification to detect a faint object than a large telescope! This effect is discussed further in Appendix F.

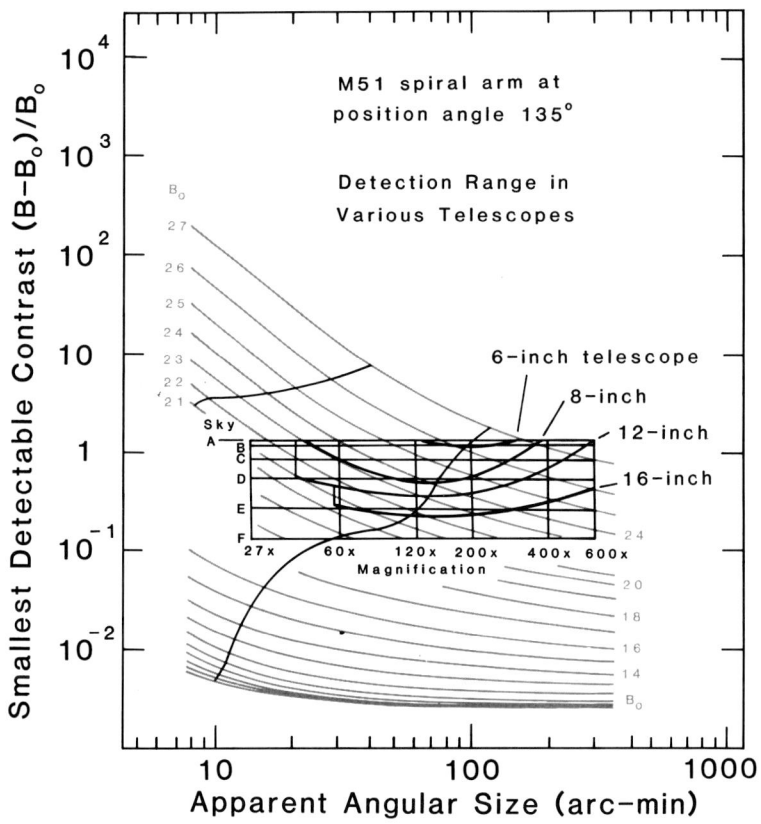

Figure 6.3. The visibility of the main spiral arm in M51 is analyzed for various telescopes, powers, and sky conditions, using the data in Tables 6.1 and 6.2. For different amounts of light pollution (horizontal lines labeled "A" through "F"), the contrast between the arm and its background was computed, using the values of M and M_o in Tables 6.1 and 6.2. Next, for each magnification on a telescope of a given aperture, the apparent background surface brightness (M_o) was computed. Each contrast (converted to a logarithmic scale) and each magnification (which determines the spiral arm width) was used to plot a point. If the background surface brightness M_o is brighter than the detection limit B_o (the lines sloping downward to the right), the spiral arm can be seen. Heavy black arcs show the detection limits for different sized telescopes. The spiral arm can be seen at brightnesses and contrasts greater than each telescope's limit; that is, above the arc.

The arcs illustrate several points. First, large telescopes do better than small ones for detecting low-contrast features – or the same feature in a more light-polluted sky. The minimum useful magnification on a telescope will *not* show all the detail. Instead, the optimum magnified visual angle should be used. If the magnification is raised too high, the detail is again lost from view.

7
A visual atlas of deep-sky objects

This chapter illustrates many of the best galaxies, star clusters and nebulae as they appear in a modest amateur telescope. Most of the observations were made by the author with a homemade 8-inch f/11.5 Cassegrain reflector from observing sites no better than those near any large American city. The drawings have good, uniform quality control, so an observer should be able to tell, after a few trials, whether he or she can expect to see less or more in any object illustrated, given his or her particular telescope and sky quality.

Facing each drawing is a photograph at the same scale and orientation. Thus the viewer can readily determine which features in photographs can and cannot be seen. Also given are the distances from which to view each drawing so the object looks the same size as at various magnifications in a telescope.

Full data are given for each observation, including how long it took to make the preliminary drawing at the telescope. Much additional time was spent preparing each final drawing, as discussed in Chapter 5. All drawings were made by either method 1 or 2 described in that chapter.

THE PERSONAL EQUATION

Astronomers have long spoken of the "personal equation" to account for differing results by visual observers. The personal equation is a correction to be applied to an individual's data to bring it to some impersonal standard. The differences in what people see probably depend more on their experience than on actual differences in their eyes. The fundamental capacities of the eye are about the same for most people. For example, in controlled tests of the faintest visible star, the difference from one person to the next is probably less than one magnitude, and even this may be due largely to how well someone has learned to use averted vision.

Focusing a telescope corrects for nearsightedness and farsightedness. Other eye problems aside, visual acuity depends on the density of rods and cones in the retina; sensitivity depends on their photochemical action and links to the brain. This neural architecture is probably much the same in most people. On the other hand, years of practice can make a great difference in fine tuning the techniques of visual observing.

Hopefully the information in this book will greatly shorten that time. My own growth in observing ability is interesting in this regard. Starting as a very active amateur in 1968, I had observed all the Messier objects and many NGC objects by 1971. I located many supposedly difficult ones that turned out to be not very hard. Even the notoriously elusive Horsehead Nebula was easy in a dark country sky. By early 1982 I was making detailed drawings of everything observed, and that summer I decided to write this book. That autumn I did most of the research and analysis for the previous chapters.

While doing so, I realized I had not been reaching the fundamental limits of the eye. I had not known the concept of the optimum magnified visual angle: how to match the telescope power to the eye's detection characteristics. The result of this increased understanding can be seen in the drawings of the Orion Nebula (M42) made in January 1982 and January 1983, on pages 101 and 103. The second drawing shows much more detail. Although sky conditions were slightly better, most of the improvement resulted from using the eye and telescope together more effectively. A greater range of magnifications, and spending more time studying the object, resulted in such features as faint arcs of nebulosity coming into view. Now, when examining M42, these details are quickly seen.

Previously each observer had to discover such techniques by hit or miss, which often took many years or a lifetime. With an understanding of the material presented in this book, the time should be shortened to perhaps a year or so. It does take considerable practice to develop good techniques. After all, one can read a book on how to drive a car, but learning to drive happens behind the wheel. With these thoughts in mind we will now explore the variety and beauty that can be seen through small amateur telescopes.

AVERTED VS. DIRECT VISION

The appearance of deep-sky objects depends strongly on whether direct or averted vision is used. Direct vision has sharper resolution but lacks sensitivity. Thus, looking straight at an object will show its brighter parts in detail, while the fainter parts may be totally lost. Changing from direct to averted vision and back can produce some interesting blinking effects. For example, when looking directly at a globular star cluster, the bright central mass of stars may be partially resolved into individual pinpoints, but the fainter outer regions are invisible. Averted vision will show a fuzzy, unresolved (or less resolved) central region, but the outer parts come into view.

Some open clusters show similar effects. While using averted vision, fainter stars may be seen in the center of the cluster but they will not be clearly resolved. Thus they give the impression of a faint nebulosity in the cluster. This effect is shown in some of the drawings, such as of M11 and M67, and it crops up in many old descriptions of clusters by early visual observers.

In the case of diffuse nebulae and galaxies, if you can see the object with direct vision at all, you can probably increase the power to see more detail. Low powers should be used if the nebula is very large and already at the limit of averted vision. If all you want is to detect the object's existence, then a power should be used that magnifies its apparent size to about 3° or 4°.

But usually the nebula is brighter than the detection limit and powers many times more than that required for simple detection can be used in an effort to see detail – perhaps swirls and dark spots in a nebula, or faint mottlings, spiral arms, or dark lanes in a galaxy. The range of visual experiences is actually quite large with modest amateur telescopes.

In some cases, stars coincide with the nebula or galaxy in view. A good example is the planetary nebula M27, which is in a rich region of the Milky Way with many foreground and background stars plus a 13th-magnitude central star. These stars can be hard to detect against the nebula's light. The central star, for instance, appears fainter than 13th-magnitude field stars; it is difficult in an 8-inch telescope at low powers or with mediocre sky conditions. But high powers spread out the nebula's light while stars remain point sources so far as the eye is concerned. So the central star becomes much easier.

One of the hardest aspects of averted vision to master is holding the eye motionless on one point for six seconds or more while trying to grasp detail in the periphery of your vision. The eye tends to jerk, especially if fatigued. On the other hand, in some conditions moving the eye (or gently jiggling the telescope) helps bring an object into view, because peripheral vision is highly sensitive to anything moving.

THE OBSERVATIONS AND DRAWINGS

The drawings in this chapter are the product of great effort to detect all the detail that could possibly be seen. This detail is necessarily portrayed more prominently than it actually appeared. If a true representation were drawn, it would take the reader similar time and effort to discern it on the printed page. This would amount to several minutes for simple objects, and hours for something complex like the Orion Nebula.

It's worth remembering, however, that the true detail and contrast in deep-sky objects is not subtle at all – as the camera proves. Only the detection limits of the eye make them seem so. After all, our eyes were not designed for astronomy but for the very different job of day-to-day survival on Earth.

Since you are probably reading this book in adequate room light, the detail in each drawing will be obvious and quickly seen. Experiment by viewing the book in very dim light – so dim that some details need averted vision. One thing you'll notice right away is that the photographs become much more like the drawings.

I find that the drawings of M51 are particularly interesting for practising averted vision techniques this way. For example, try to see how far you can dim the lights and still see the spiral arms from various distances, using direct and averted vision. The viewing distances represent magnifications – though the surface brightness of a picture does *not* diminish as you move closer, unlike the case with magnification in a telescope. Such practice is good training before going out into the field.

Included with each illustration is the object's Messier, NGC, IC, or other catalogue number, its right ascension and declination for equinox 2000.0, its constellation, and its type. Also given are the Universal Times and dates when the observation for each drawing began and ended, along with the observing location, telescope, eyepieces, magnifications, air mass (see Appendix C), and the faintest visible star (usually at the zenith at a given magnification). The scale of the drawing is indicated, as is the orientation (direction of north) and in some cases the drawing method used.

There is a brief description of each object's physical nature as currently known, but that is not the primary purpose of this book. Excellent companion guides that cover such material are *Burnham's Celestial Handbook* and *The Messier Album*. A more up-to-date listing of data for thousands of deep-sky objects is *Sky Catalogue 2000.0, Volume 2*. See Appendix A.

The objects are presented in order of increasing right ascension. Distances from which to view the page are given to represent a range of telescopic magnifications. At any other magnification m, the viewing distance v in centimeters can easily be computed from this simplified version of equation 5.1:

$$v = 1/(2m \tan(p_s/120)) \qquad \text{(equation 7.1)}$$

where p_s is the scale in arc-minutes per centimeter and the argument to the tangent is in degrees.

My observing sites and conditions varied from essentially perfect to quite poor. The best were at Manastash Ridge in the Cascade Mountains of Washington, site of the University of Washington's 30-inch telescope, and in the Colorado Rockies. Observations at the Washington site were made before 1972. The dark sites there were truly spectacular. The naked eye could see to magnitude 7.5, and the summer Milky Way cast a diffuse shadow! The faintest star I documented was magnitude 14.8, in the star cluster NGC 6910. This star was seen very easily by four observers in three 8-inch telescopes. Moreover it was close to other, brighter stars near the center of the cluster (see the NGC 6910 magnitude chart in Appendix B), so the true 8-inch magnitude limit must have been 15.0 and maybe slightly fainter.

The rest of the observing sites were near sea level and close to major cities. One is Patten's Observatory, the old location of the Seattle Astronomical Society's 12.5-inch Newtonian telescope, where observations were carried out in the late 1960s and early 1970s. The rest of the sites are on the island of Oahu in the state of Hawaii. The best location there to which amateurs had intermittent access was a private ranch in the Waianae area, on the west coast of the island. This spot was slightly better than that at Patten's, having an 8-inch limiting magnitude of about 14.5 on a good night. (Haze from ocean spray is a problem at all low-elevation Hawaiian sites.) The next best location was on an abandoned runway at Barbers Point Naval Air Station. Here the 8-inch limiting magnitude was around 14.0. Somewhat poorer was my home in Hawaii Kai, at the east end of the island, with 8-inch limits of 13.6 to 13.8. Finally, some observations were made at Ewa Beach, where the 8-inch limit was about 13.4. These limits are for good nights; they could be considerably worse, owing to haze.

Haze and volcanic smoke are two seemingly constant problems in Hawaii. The ocean kicks up a lot of spray, which is carried all over the islands by strong trade winds. In the spring of 1982, the Mexican volcano El Chichon injected much dust high into the atmosphere, which had a devastating effect on observations by both professionals and amateurs. In April 1982, while at Mauna Kea Observatory, I found the naked-eye limit at new moon to be around 4.0; normally it is 7 or better. By late June, on Oahu the faintest star observable was about 0.6 magnitude worse than normal. In January and February 1983, Kilauea volcano was erupting, covering the Hawaiian islands with what is called "vog" (volcanic smoke plus fog), which worsened the limiting magnitude somewhat.

Most observers don't have to contend with such problems, but there are always variations from night to night. So the sky quality should be documented at the time of each observation.

Definitions

Several words and phrases used in the descriptions of the objects need to be defined:

Small telescope or **small amateur telescope:** one less than 6 inches (15.2 centimeters) in aperture.

Medium size telescope or **medium amateur telescope:** one 6 inches (15.2 centimeters) to 12 inches (30.5 centimeters) in aperture.

Large telescope or **large amateur telescope:** one more than 12 inches (30.5 centimeters) in aperture.

Very poor skies: those typical of the middle of a large city with a population of a million or more. Such a sky is brighter than a country sky at full Moon, with a surface brightness overhead of about 19 magnitudes per square arc-second or brighter.

Poor skies: those typical of the edge of a large city with a population of a million or more. The sky surface brightness overhead is about 20 magnitudes per square arc-second or slightly brighter. The naked-eye limit is near magnitude 5.

Moderate skies: those typically found a few miles from large cities. The sky surface brightness is 20 to 21 magnitudes per square arc-second, and the naked-eye limit is magnitude 5.5 to 6.0.

Good skies: those typical of the country a few tens of miles from a large city, with few lights in the vicinity. The sky surface brightness is 21 to 23 magnitudes per square arc-second, and the naked-eye limit is magnitude 6.0 to 7.0.

Excellent skies: those typically found only 90 miles (150 km) or more from a large city, with no bright lights within a few miles. There is no haze in the sky. The sky surface brightness overhead is fainter than 23 magnitudes per square arc-second. These skies are common in the mountains of the western United States, but are rare along the east coast of the U.S. in the summer because of haze.

North, **south**, **east** and **west** are used in the astronomical sense: north means the direction to the north celestial pole, east means toward increasing right ascension. If an observer is in the Northern Hemisphere and facing south, astronomical east is to his or her left (which is also east on land).

Very low power: magnifications less than about 30×.

Low power: magnifications from about 30× to 70×.

Medium power: from about 70× to 200×.

High power: from about 200× to 400×.

Very high power: over 400×.

M31, THE ANDROMEDA GALAXY (NGC 224), M32 (NGC 221), M110 (NGC 205), GALAXIES IN ANDROMEDA

M31: R.A. $00^h\ 42.7^m$ Dec. $41°\ 16'$ (2000.0)
M32: R.A. $00^h\ 42.7^m$ Dec. $40°\ 52'$
M110: R.A. $00^h\ 40.3^m$ Dec. $41°\ 41'$

Technical. Messier 31, commonly called the Great Galaxy in Andromeda, is a large spiral tilted 13° from edge-on. It is the brightest galaxy we can view in the sky apart from our own Milky Way. Arab astronomers recorded it in the 10th century as a small, diffuse patch, but under the murky skies of Europe it eluded notice with the unaided eye. No European recorded it until 1612, when Simon Marius saw it with the help of a telescope.

The Andromeda Galaxy is part of the Local Group of galaxies, which is about 5 million light-years in size; the four main members are M31, M33, Maffei 1, and our own galaxy. About 30 galaxies are currently recognized as members of the Local Group, most of them dwarf ellipticals or irregulars associated with our galaxy or M31.

M31 has approximately twice the mass of our galaxy, or about 300 (Amer.) billion times the mass of the Sun, and puts out about 11 billion times the Sun's light. Its distance is usually given as 2.3 million light-years, and it is approaching us at about 80 kilometers per second.

On deep-sky photographs M31 covers fully 1.25° by 4.1°, which corresponds to a diameter of about 100,000 light-years. In his book *Galaxies*, Timothy Ferris gives an interesting observation on space and time. Since M31 is 2.3 million light-years away, we are seeing 2.3 million years into the past. However, since one side of the galaxy is about 100,000 light-years more distant, a thousand centuries are spanned from one side to the other. Thus we are not only looking back in time, we are seeing many times at once!

Visual. M31 appears somewhat less than 4° long through most instruments, though some observers have reported seeing a 5° extent with fine binoculars, under excellent skies. These outer parts are extremely faint.

Unlike many bright objects, M31 often leaves a disappointing impression at first sight. The central region is very bright but fades quickly and smoothly toward the edges. The beautiful details seen in photographs are difficult visually in any telescope. In fact, to see the outer edge corresponding to about 2° length requires a very dark sky. Because of its size, very low power is needed to see all of the galaxy in one view. But at such magnifications, almost no detail appears.

With good skies and a moderate aperture, the dark lane to the northeast is easily seen at 30x and up. An 8-inch telescope will show this lane under moderate skies, as well as the bright star cloud to the southwest (upper right in the photograph) at higher powers. This star cloud is known as NGC 206. There are many dark patches in the galaxy, especially near the bright central portion, but none have been reported by amateur observers.

The brightness of M31 varies tremendously from the nucleus to the outer edges. Assuming a size of 150 by 50 arc-minutes and a total magnitude of 4.0, the mean surface brightness is 22.3 magnitudes per square arc-second. However, the nucleus is many magnitudes brighter than this, while the edges are fainter.

Near M31 are two of its companion galaxies: M32 and M110. These are dwarf elliptical systems that have a relationship with M31 similar to that of the Magellanic Clouds with our own galaxy.

M32 is south of M31's nucleus. It appears on the edge of a spiral arm whose boundary is evident in good skies with a 6- or 8-inch telescope. M32 was first seen by Le Gentil in 1749; it can be found by careful observers with a 2-inch telescope. Its angular size is 3.6 by 3.1 arc-minutes, and its total magnitude of 9.5 yields an average surface brightness of 20.7 magnitudes per square arc-second.

M110 (NGC 205), northwest of the nucleus of M31, is considerably more difficult than M32 because it is spread over a larger area and has a fainter total magnitude, 10.8. It is 8 by 3 arc-minutes in size, and has an average surface brightness of 22.9 magnitudes per square arc-second. This galaxy has been seen with telescopes as small as 2.4 inches. In fact, a good friend of mine, Jon Seamans, has observed all the Messier objects with a 2.4-inch refractor.

M31 has two more companion galaxies about 7° to the north: NGC 147 and NGC 185. Both are about the same difficulty visually as NGC 205. They are described in Appendix E and in *Burnham's Celestial Handbook*.

Photograph of M31, M32 and M110. The Great Galaxy in Andromeda, M31, is the main object at center. M32 is at upper center, and M110 (NGC 205) is the small object about 6 cm to the lower left of the nucleus of M31. South is up. (Courtesy Palomar Observatory.)

A VISUAL ATLAS OF DEEP-SKY OBJECTS

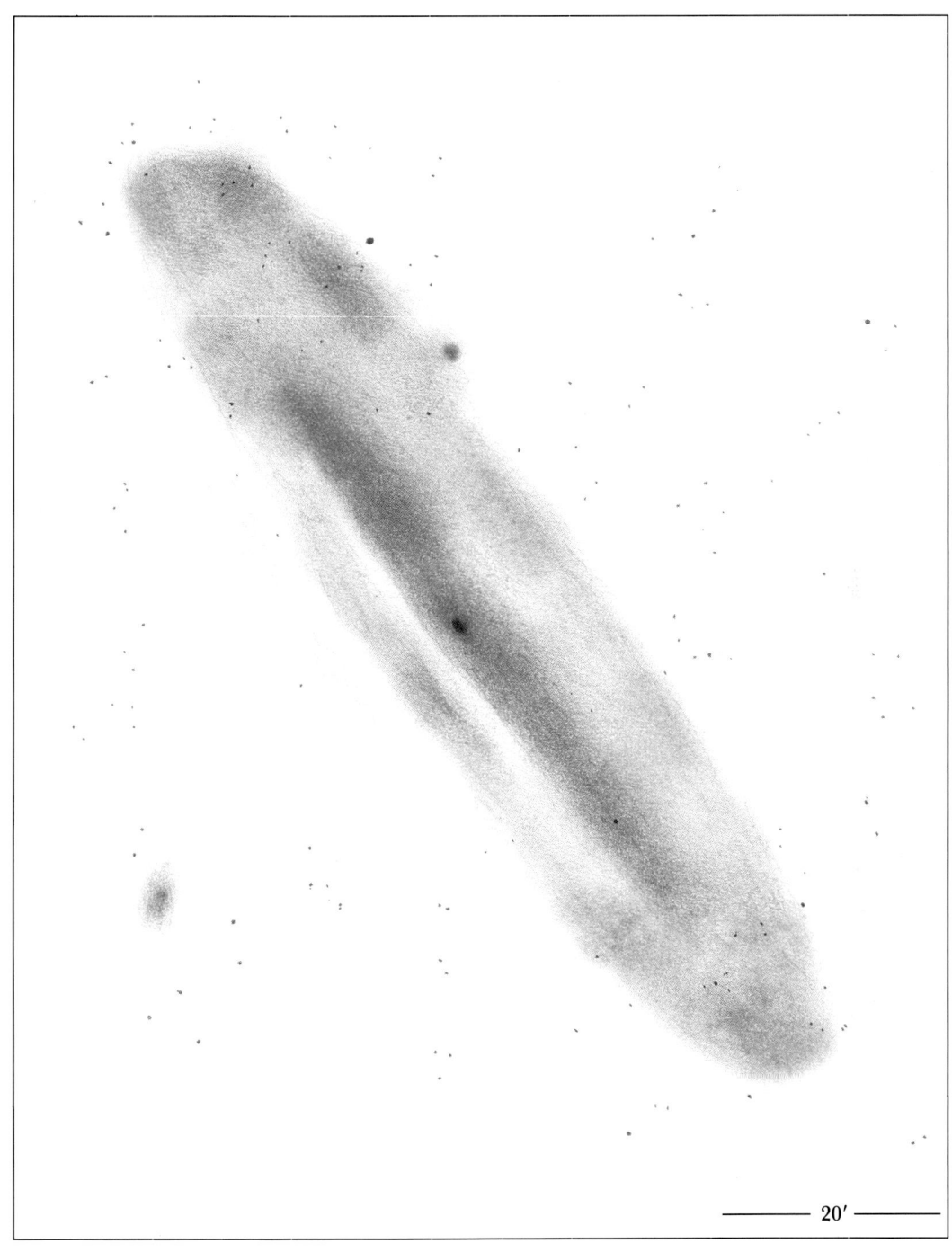

Drawing of M31.

Scale: 6.0 arc-min/cm	Viewing Distance (cm)		air mass: 1, faintest star: 13.5 at zenith, 52×; tracking
8-inch f/7.25 Newtonian			
28mm Kellner (52×)	25×:23	200×:3	8/17/69 UT (time not recorded) at Pattens
	50×:11	300×:2	Observatory, Wash.; R. Clark
	100×: 6	400×:1	

NGC 246, PLANETARY NEBULA IN CETUS

R.A. $00^h 47.1^m$, Dec. $-11° 53'$ (2000.0)

Technical. NGC 246 is a planetary nebula about 1,500 light-years distant and 1.7 light-years in diameter. It is elliptical in shape, but the rim is not quite complete on the west side.

Visual. With a total magnitude of 8.5 and an angular size of 2.5 by 4 arc-minutes, the nebula's mean surface brightness is 19.6 magnitudes per square arc-second. It has a 12th-magnitude central star. In 7 × 50 binoculars the nebula appears very small but not quite stellar. In small telescopes it is a featureless disk.

The 8-inch under good skies showed the central star as well as two field stars superimposed on the nebula, but no other features were made out. The west side faded gradually into the sky background, while the other edges appeared sharp. Large amateur telescopes under excellent skies may be able to bring out some details.

A VISUAL ATLAS OF DEEP-SKY OBJECTS

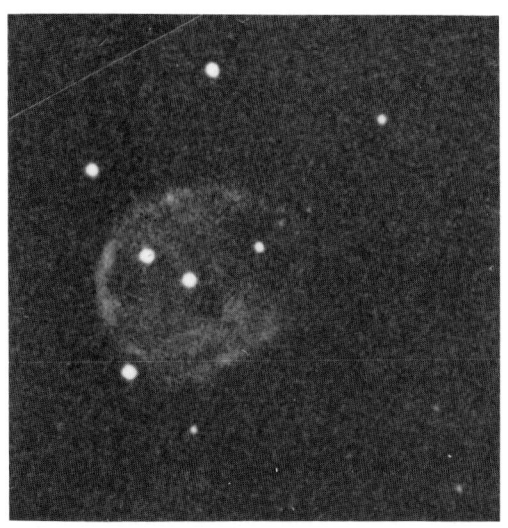

Photograph of NGC 246. South is up. (Courtesy Jack B. Marling.)

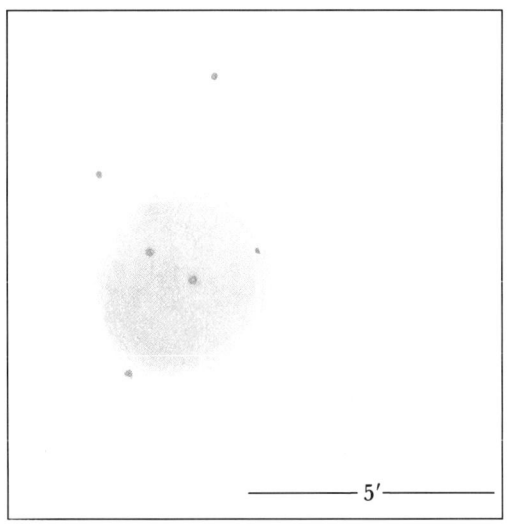

Drawing of NGC 246.

Scale: 1.5 arc-min/cm	Viewing Distance (cm)	
8-inch f/11.5 Cassegrain		
28mm Kellner (82×)	25×:92	200×:11
20mm Erfle (117×)	50×:46	300×: 8
12.4 Erfle (188×)	100×:79	400×: 6

air mass: 1.20, faintest star: 14.2 at zenith, 188×; no tracking.
10/8/83 9:50–10:10 UT at Waianae ranch, Hawaii; R. Clark

NGC 253, GALAXY IN SCULPTOR

R.A. $00^h\ 47.6^m$, Dec. $-25°\ 18'$ (2000.0)

Technical. NGC 253 is an unusual galaxy in that its nucleus appears to be ejecting material, much in the manner of the nucleus of M82 but less violently and in a way not obvious on most photographs. The core of the galaxy radiates about half the system's total energy. Most of this emission is at infrared wavelengths, amounting to about 100 billion times the energy of our Sun. Gas is being spewed from the nucleus at a velocity of 120 kilometers per second.

NGC 253 is part of the Sculptor cluster of Galaxies: a group over 20° in diameter and centered near the south galactic pole. The distance to NGC 253 is somewhat uncertain, but it's probably only a little beyond the farthest members of the Local Group. The galaxy is thought to be around 40 000 light-years in diameter.

Photograph of NGC 253. South is up. (Courtesy Palomar Observatory.)

A VISUAL ATLAS OF DEEP-SKY OBJECTS

Visual. NGC 253 is large and bright but often overlooked by Northern-Hemisphere amateurs because it is rather far south in a nondescript constellation. It contains dust lanes of great complexity in all areas. The galaxy is fully 22 by 6 arc-minutes in size and is inclined 17° from edge-on. At a total visual magnitude of 7.0, its average surface brightness is 20.9 magnitudes per square arcsecond.

When high in a good sky, NGC 253 shows considerable detail in medium size telescopes. The 8-inch under moderate skies shows distinct brighter portions that correspond to spiral arms. The galaxy is bright enough to be seen in a 2-inch. In the 3-inch finder at 31×, it is a smooth oval. But even large telescopes show no nucleus. This galaxy must rival M31 in beauty for Southern Hemisphere observers, although it is somewhat smaller.

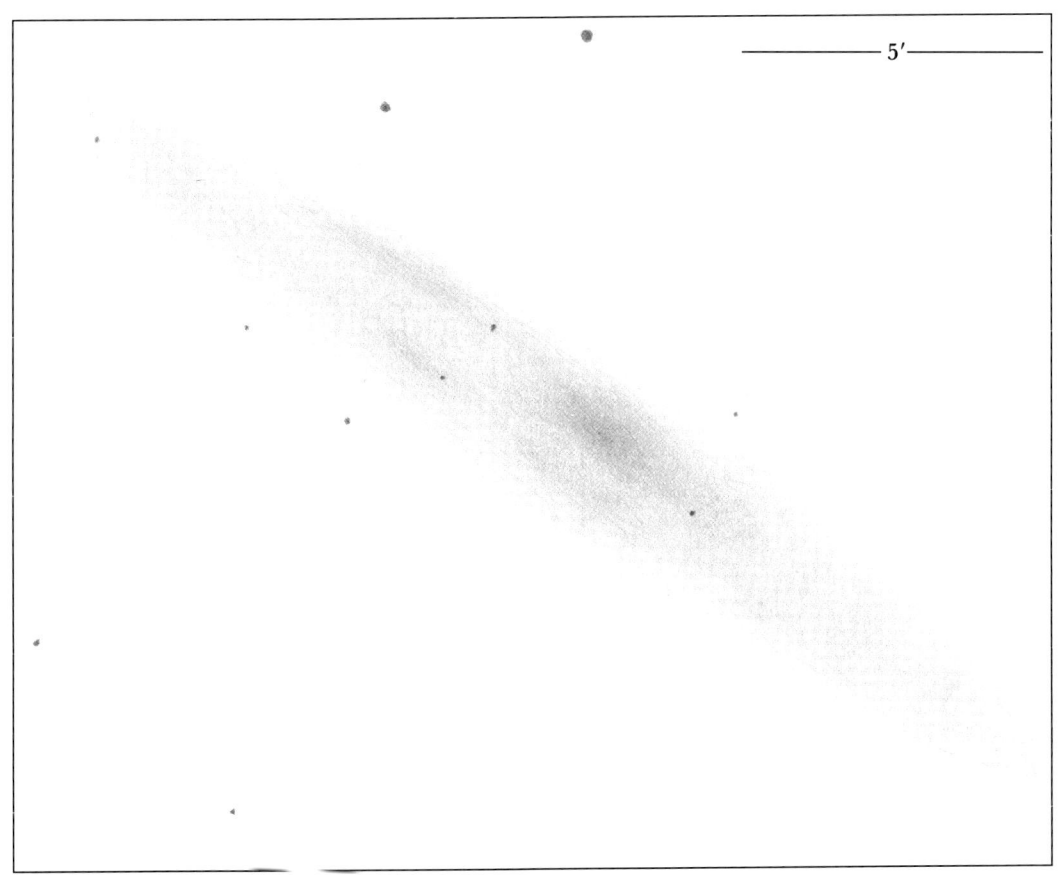

Drawing of NGC 253.

Scale: 1.2 arc-min/cm
8-inch f/11.5 Cassegrain
20mm Erfle (117×)
12.4mm Erfle (188×, best view)
7mm Erfle (334×)

Viewing Distance (cm)

25×:115   200×:14
50×: 57   300×:10
100×: 29   400×: 7

air mass: 1.51, faintest star: 13.8 at zenith, 188×; no tracking
9/5/83 13:00–13:40 UT at Hawaii Kai, Hawaii; R. Clark

M33 (NGC 598), GALAXY IN TRIANGULUM

R.A. $01^h 33.9^m$, Dec. $30° 39'$ (2000.0)

Technical. M33 is one of the four spiral galaxies in the Local Group, but it is a small one, estimated to hold fewer than 20 billion stars compared to the 300 billion or so in the Andromeda Galaxy (M31). M33 is about 40 000 light-years across and tilted 55° from edge-on. It is probably only about 700 000 light-years from M31, which would appear three times bigger from a planet in M33 than from Earth. M33 is about the same distance from our galaxy as M31 or possibly a little closer. It was discovered by Messier on August 25, 1764, and today is often called the Triangulum Galaxy or the Pinwheel.

Visual. M33 is about 1° across, including the faint extensions, but it has a low surface brightness. Amateurs often miss it because they are looking for a smaller object. With a total magnitude of 5.3, its average surface brightness is a dim 22.8 magnitudes per square arc-second.

M33 is a challenge to the novice. Experienced observers seem to have little trouble finding it, and it can even be seen with the naked eye under excellent skies. The brighter portions cover an area 20 by 30 arc-minutes. Within 1° of the nucleus are many NGC and IC objects, the brightest of which is NGC 604. This is an H II region (a cloud of ionized hydrogen), similar to the Great Nebula in Orion but much larger, with an embedded star cluster. NGC 604 can be seen as a round spot at the end of the spiral arm to the northeast of the nucleus; it is at lower right in the drawing. I have seen it in the 3-inch finder at 89×.

Most observing guides recommend very low powers for M33. Although low powers are fine for locating it, they will not show any detail. This probably accounts for the galaxy's often-reported lack of features. Medium powers (100× to 200× on an 8-inch telescope) will show the two main spiral arms and NGC 604 even under only moderately good skies. Excellent dark country skies should show a wealth of detail with careful observation using only a 6-inch telescope.

Photograph of M33. South is up. (Courtesy National Optical Astronomy Observatories.)

Drawing of M33.

Scale: 1.9 arc-min/cm	Viewing Distance (cm)		air mass: 1.03, faintest star: 13.8 at zenith, 188×;
8-inch f/11.5 Cassegrain			no tracking
20mm Erfle (117×)	25×:72	200×:9	9/5/83 12:00–12:50 UT at Hawaii Kai, Hawaii; R.
12.4mm Erfle (188×)	50×:36	300×:6	Clark
	100×:18	400×:5	

M74 (NGC 628), GALAXY IN PISCES

R.A. $01^h\ 36.7^m$, Dec. $15°\ 47'$ (2000.0)

Technical. M74 was first seen in 1780 by Pierre F. A. Mechain. It is a face-on spiral about 30 million light-years distant and 80 000 light-years wide. Its mass is estimated at about 40 billion suns. On deep photographs M74 shows beautiful spiral arms about 3000 light-years thick, with dust lanes tending to outline their inner edges.

Visual. M74 is often regarded as the most difficult Messier object to locate. With a diameter of 9 arc-minutes and a total magnitude of 9.0, M74 ought not to be hard. The difficulty is probably due to improper use of magnification. Again, guides often suggest using low powers, since the average surface brightness is low at 22.4 magnitudes per square arc-second. But most of the light is concentrated in the small nuclear region about 40 arc-seconds in diameter. So at low power the galaxy can look like a field star. Higher powers show the nuclear region as a disk.

In the 8-inch under only moderate skies, 188× gave the best view. The bright central region was surrounded by a faint, uniform glow. M74 can be seen in telescopes as small as a 2.4-inch refractor. Under excellent skies, I have seen it in the 3-inch finder at 31× and even the 2.4-inch finder at 7.9×.

A VISUAL ATLAS OF DEEP-SKY OBJECTS

Photograph of M74. South is up. (Courtesy National Optical Astronomy Observatories.)

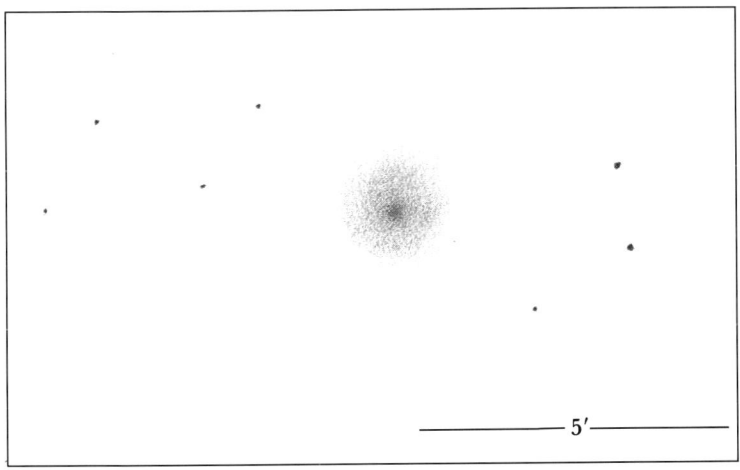

Drawing of M74.

Scale 1.2 arc-min/cm Viewing Distance (cm)
8-inch f/11.5 Cassegrain
12.4mm Erfle (188×,
  best view)   25×:115   200×:14
20mm Erfle (117×)   50×: 57   300×:10
                           100×: 29   400×: 7

air mass: 1.21, faintest star: 14.0 at zenith, 188×;
no tracking
1/16/83 6:45–6:55 UT at Barbers Point, Hawaii;
R. Clark

M76 (NGC 650–651), PLANETARY NEBULA IN PERSEUS

R.A. $01^h\ 41.9^m$, Dec. $15°\ 34'$ (2000.0)

Technical. M76 is a small planetary nebula similar in appearance to the Dumbbell Nebula, M27. Like all planetaries, its distance is poorly known; published values range from 1700 to 8000 light-years. The central star, magnitude 16.5, is extremely hot, about 60 000 Kelvin. Its light output is less than the Sun's, while the nebula itself emits two or three times the Sun's light. The nebula is 1 to 4 light-years across, depending on the accepted distance.

Visual. M76 is called the "Little Dumbbell" and is probably the faintest Messier object. Visually, it looks like the bright portion of M27 but much smaller. With a total magnitude of 10, the average surface brightness is 18.7 magnitudes per square arc-second, high compared with most deep-sky objects. This surface brightness, due to the small size of 1.5 by 0.7 arc-minutes, allows high powers to be used to search for detail.

M76 is brighter at its ends than in the middle, which caused 19th-century observers to catalogue it as two objects, NGC 650 and NGC 651. The 8-inch showed considerable detail at powers around 200× and up under only moderate skies. The small size and high surface brightness mean that good, steady atmospheric seeing is needed for viewing fine detail within the nebula – an unusual situation for a deep-sky object. The 8-inch showed a distinct difference between the two components; the southern portion had a marked point, aimed south. There are many field stars near M76 even in a high-power view.

Photograph of M76. South is up. (Courtesy Laird A. Thompson. Canada–France–Hawaii Telescope Corporation.)

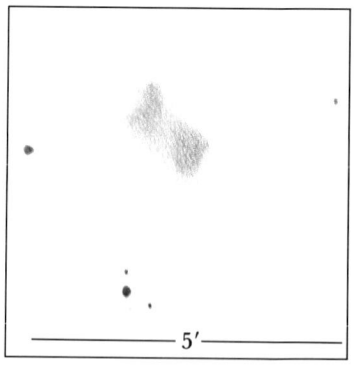

Drawing of M76.

Scale: 1.2 arc-min/cm
8-inch f/11.5 Cassegrain
12.4mm Erfle (188×, best view)
9mm Kellner (260×)

Viewing Distance (cm)	
25×:115	200×:14
50×: 57	300×:10
100×: 29	400×: 7

air mass: 1.44, faintest star: 14.0 at zenith, 188×; no tracking
1/16/83 7:20–7:42 UT at Barbers Point, Hawaii; R. Clark

NGC 891, GALAXY IN ANDROMEDA

R.A. $02^h 22.4^m$, Dec. $42°21'$ (2000.0)

Technical. NGC 891 is a beautiful edge-on galaxy with a dark lane extending from one end to the other. This is one of a small group of galaxies that includes NGC 1023 and NGC 925. The distance to NGC 891 is estimated to be between 20 and 40 million light-years. The galaxy's total light is roughly 1 or 2 billion times the light of our Sun. The dark lane consists of dust clouds mainly along the galactic equator, much like those in our Galaxy. In fact, all-sky photographs of the Milky Way appear remarkably similar to photographs of NGC 891.

Visual. NGC 891 has a total magnitude of 12.2 and an angular size of 12 by 1 arcminutes. Its mean surface brightness is somewhat low at 23.5 magnitudes per square arcsecond. Although NGC 891 may be glimpsed through a 3-inch telescope under excellent skies, a medium size telescope is needed to begin to show detail.

Through the 8-inch under good skies, the overall size and shape were seen well at 117×, but the dark lane could be detected only at 188×. The dark lane was visible only near the nucleus. Larger telescopes under similar skies show the lane extending farther into the edge-on spiral arms. It has a surface brightness about 0.6 to 0.9 magnitudes fainter than the bordering bright zones. This magnitude difference implies a contrast of 0.57 to 0.44 (log contrast of –0.24 to –0.36). Such a contrast and surface brightness indicate that at least an 8-inch telescope and good skies are required to detect the dark lane.

Photograph of NGC 891. South is up. (Courtesy Mount Wilson and Las Campanas Observatories, Carnegie Institution of Washington.)

A VISUAL ATLAS OF DEEP-SKY OBJECTS

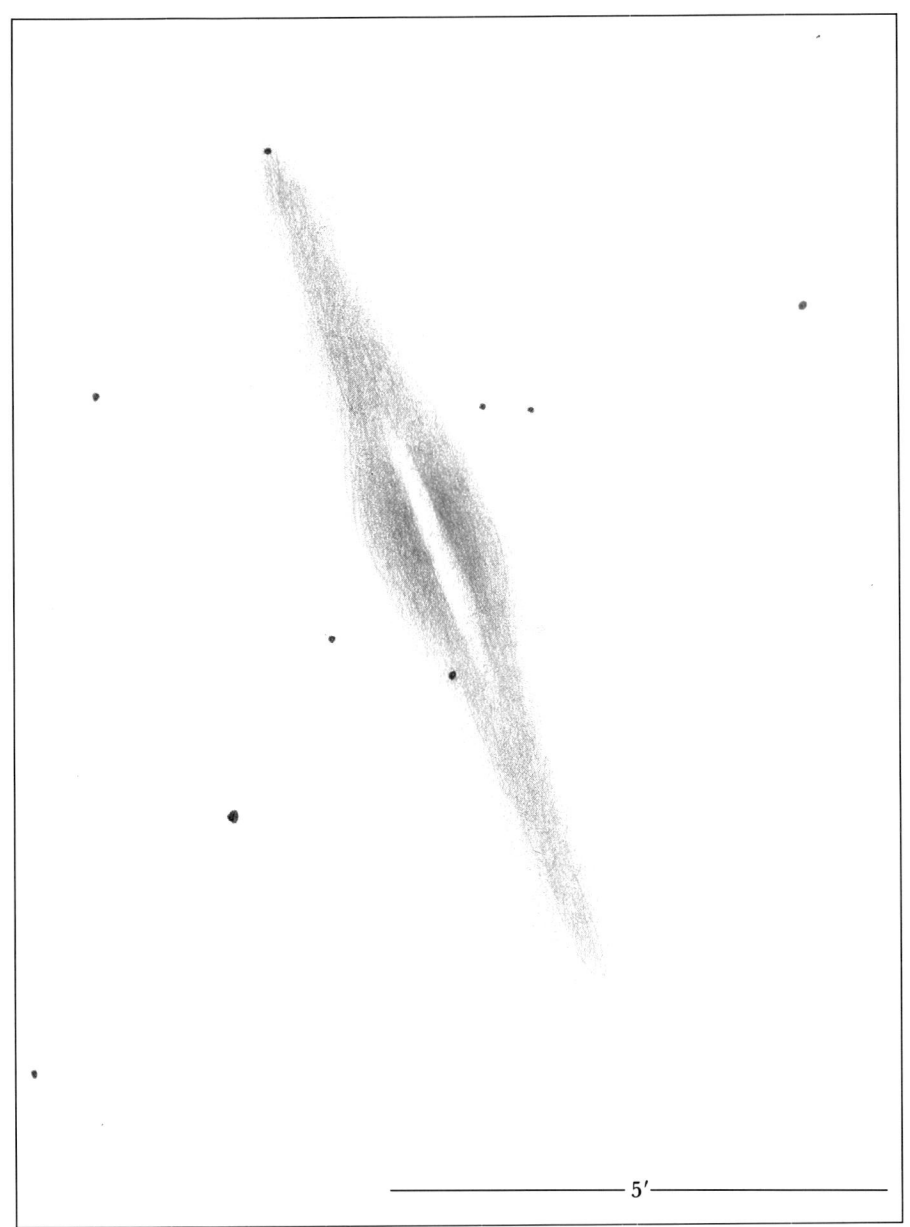

Drawing of NGC 891.

Scale: 0.75 arc-min/cm	Viewing Distance (cm)	
8-inch f/11.5 Cassegrain		
20mm Erfle (117×)	25×:183	200×:23
12.4mm Erfle (188×)	50×: 92	300×:15
	100×: 46	400×:11

air mass: 1.17, faintest star: 14.2 at zenith, 188×;
no tracking
10/8/83 13:33–13:50 UT at Waianae ranch,
Hawaii; R. Clark

M77 (NGC 1068), SEYFERT GALAXY IN CETUS

R.A. 02^h 42.7^m, Dec. $-00°$ 01' (2000.0)

Technical. M77 is one of a class of galaxies with bright, very active nuclei. These are the Seyfert galaxies, named after Carl Seyfert who studied them in the early 1940s. The nucleus of M77 is ejecting clouds of gas at a velocity of 600 kilometers per second, and each of these clouds is estimated to contain as much mass as 10 million suns. The nuclei of some Seyfert galaxies vary in brightness, but that of M77 does not.

The distance to M77 is about 75 million light-years, and the diameter of its outermost region is around 100 000 light-years. The total mass is estimated to be about 100 billion suns; the light output, 40 billion suns. Like most Seyfert galaxies, M77 is also a strong radio source. Radio astronomers know it as 3C 71.

Visual. M77 is magnitude 10.0 and about 2.5 by 1.7 arc-minutes in size. Its mean surface brightness is 20.2 magnitudes per square arc-second. Faint spiral arms extend to a diameter of 6 arc-minutes but have not been reported visually. The galaxy has a bright inner spiral pattern 40 by 20 arc-seconds in size, which is not seen in small amateur telescopes. Some observers have reported a mottled effect in large telescopes. A second, larger spiral pattern extends to 2.5 by 1.7 arc-minutes.

The visual impression is that of a bright central region surrounded by a diffuse oval, which in turn is surrounded by a larger and fainter oval. In the 8-inch the bright inner region did show a brighter spot southwest of the nucleus. This spot corresponds to a bright part of a spiral arm, which might be recognizable as such in a large telescope under very dark skies at 200× or more.

Photograph of M77. South is up. (Courtesy Lick Observatory.)

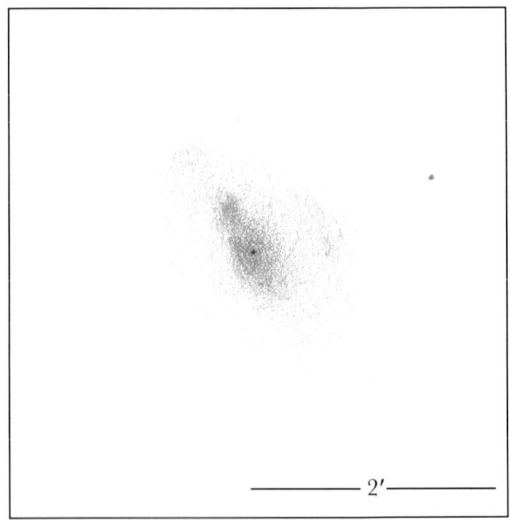

Drawing of M77.

Scale: 0.6 arc-min/cm
8-inch f/11.5 Cassegrain
12.4mm Erfle (188×)
9mm Kellner (260×)
6mm Orthoscopic (389×)

Viewing Distance (cm)

25×:229 200×:29
50×:115 300×:19
100×: 57 400×:14

air mass: 1.64, faintest star: 13.5 at zenith, 188×; no tracking
1/13/83 8:45–9:18 UT at Hawaii Kai, Hawaii; R. Clark

NGC 1365, BARRED SPIRAL GALAXY IN FORNAX

R. A. 03^h 33.7^m, Dec. –36° 08' (2000.0)

Technical. NGC 1365 is a beautiful barred spiral galaxy, probably the finest example of its class in the southern sky. It is the third brightest member of the Fornax Cluster of Galaxies, which is some 50 million light-years distant. The bar is about 45 000 light-years long; the arms stretch from the bar's ends for at least twice that distance.

NGC 1365 is also one of the most luminous of all known barred spirals, and in the 1970s was found to emit X-rays. The X-ray emission comes mainly from the region of the nucleus, as is the case with Seyfert galaxies. But NGC 1365 seems to be a new type. The X-rays from Seyferts are associated with fast-moving gas, while none is observed in this galaxy.

Visual. NGC 1365 has a total magnitude of 11.2 and an angular size of 8 by 3.5 arcminutes. The mean surface brightness is 23.4 magnitudes per square arc-second, which is, unfortunately, somewhat low. The galaxy should be visible in small telescopes as a faint patch if it can be observed high enough in a good sky.

Through the 8-inch under good skies, the galaxy had a bright, fuzzy nucleus surrounded by the soft glow of the unresolved spiral arms. Even from my Hawaii observing sites, NGC 1365 is never high in the sky, and for more northerly observers, it is too low to offer any chance of detecting detail even in large telescopes. Southern Hemisphere observers with large telescopes and good skies may be able to see the spiral arms.

VISUAL ASTRONOMY OF THE DEEP SKY

Photograph of NGC 1365. South is up. (Courtesy Wayne C. Annala, Copyright University of Hawaii, Institute for Astronomy.)

Drawing of NGC 1365.

Scale: 1.2 arc-min/cm	Viewing Distance (cm)	
8-inch f/11.5 Cassegrain		
20mm Erfle (117×)	25×:115	200×:14
12.4mm Erfle (188×)	50×: 57	300×:10
	100×: 29	400×: 7

air mass: 1.97, faintest star: 14.2 at zenith, 188×
no tracking
10/8/83 14:00–14:10 UT at Waianae ranch,
Hawaii; R. Clark

M45, THE PLEIADES OPEN CLUSTER IN TAURUS

R.A. $03^h\ 46.9^m$, Dec. $24°\ 07'$ (2000.0)

Technical. The Pleiades have been familiar the world over since ancient times. The cluster is very close, relatively speaking, at a distance of only 400 light-years. It is also unusually large, 30 light-years in diameter, so it is easily resolved by the unaided eye.

This unique grouping has had special significance in all cultures for which records can be found. The earliest recorded observation is found in Chinese annals from 2357 BC. The cluster had particular meaning in agricultural societies, since its rising and setting near sunrise and sunset marked important times in the growing season. Even Halloween is tied to the Pleiades. In the Middle Ages the cluster culminated around midnight on the "Witch's Sabbath", which had its origin in ancient Druids' rites. (Since then the midnight culmination of the Pleiades has shifted to November 21st owing to precession.)

References to the Pleiades are found throughout literature, music, and religion. The great pyramid at Teotihuacan, 28 miles northwest of Mexico City, has its west face directed to the setting of the Pleiades (14° north of west there), and all the east–west streets of the ancient city are oriented in the same direction. Other cultures that gave the Pleiades special significance include American Indian, Maya, Aztec, Australian Aborigine, Egyptian, Greek, Roman and Persian. Robert Burnham devotes many pages to Pleiades lore in his *Burnham's Celestial Handbook*.

The Pleiades are often called the Seven Sisters, though nine of the stars now have names. Seven of the names date back at least as far as the Greek poet Aratus in the 3rd century BC. (Aratus took his constellation lore from writings by Eudoxus that were already about a century old, but are now lost.) Aratus spoke of a lost Pleiad, and indeed, many people can see only six stars here. Perhaps, it has been speculated, a star did fade sometime within the historical memory of the ancient Greeks – possibly Pleione, since it is variable today. References to the lost Pleiad are also found in Japanese literature and legends from several other cultures.

M45 contains about 300 to 500 stars. Since all are at essentially the same distance, their apparent brightnesses reflect their intrinsic luminosities. The brightest Pleiades are all blue–white; fainter ones are yellow and reddish. When the stars' brightnesses are graphed as a function of their color (the so-called color–magnitude diagram), they all fall on the curve known as the main sequence. This is where a star resides for most of its active lifetime while it converts hydrogen to helium. None of the Pleiades visible today has yet evolved off the main sequence to the red giant stage.

All the brightest Pleiades except Maia are spinning very fast, with rotation periods of about two days. They are also quite young, about 80 million years old, and are rapidly consuming their hydrogen fuel. There is one known white dwarf in the cluster. Astronomer Alan Sandage has theorized that in the distant past, the Pleiades probably contained two stars even brighter than those we see now, which have since become white dwarfs.

On photographs the brighter Pleiades are surrounded by a delicate web of nebulosity. This is a fine example of a reflection nebula, one visible because of starlight reflected and scattered by small dust grains. Color photography shows the Pleiades nebulosity to be quite blue. The reason is not just that the stars illuminating the dust are bluish. The grains tend to be smaller than the wavelength of light and therefore scatter blue light preferentially – a process similar to the Rayleigh scattering that makes our sky blue. Because the light is reflected starlight, its spectrum is continuous, rather than concentrated in emission lines as in emission nebulae like M42 (the Great Nebula in Orion) or M8 (the Lagoon Nebula).

Throughout the Pleiades, the nebulosity appears streaked in long filaments almost like cirrus clouds. Magnetic fields may be responsible for aligning this dust. Some of the filaments appear only a few arc-seconds in width and many arc-minutes long.

The brightest patch of nebulosity, NGC 1435, appears to surround and extend south of the star Merope, and hence is known as the Merope Nebula. It has also been called the Thumbprint from its appearance on an early drawing. In 1965 the astronomer F. O'Dell showed that the nebula is actually behind Merope, not enveloping it or in front of it.

Visual. M45 is a beautiful sight to the unaided eye, in binoculars, or in a telescope. Its stars have a total magnitude of 1.4 and are spread across 100 arc-minutes. Because the cluster is easily resolved by the normal unaided eye, its mean surface brightness (20.0 magnitudes per square arc-second) has little meaning except for comparison to other objects in this book.

Most people can see six or eight Pleiades with the naked eye, though people with good visual acuity can often see 10 to 12, and as many as 18 have been claimed. The number visible to the naked eye is not only a test of how faint one can see, but also a test of visual acuity and the sky's freedom from haze. The table of bright Pleiades on page 94 shows that 10 stars are brighter than magnitude 5.65. Only because they are so close together are the stars of this magnitude and fainter hard to see. I have never been able to see more than six Pleiades while wearing glasses, even under excellent skies, and without glasses the cluster appears as a blob. However, even though my nearsightedness is severe, the sensitivity of my retina to the faintest of light appears to be normal.

Detection of faint telescopic members of the Pleiades is also confused by the glare of the brighter stars. Even under excellent skies, stars of only about magnitude 14.4 can just be seen with an 8-inch telescope in the Pleiades. Elsewhere the limit is about 15.2. That is a loss of nearly a magnitude.

The Pleiades are often confused with the Little Dipper (Little Bear) by beginning amateurs and laypeople. The cluster does have the shape of a small dipper about twice the size of the Moon, and it is easier to see than the Little Dipper of Ursa Minor because the main stars are brighter.

M45 is certainly a favorite object among amateur astronomers for its beauty in virtually any optical instrument with an adequate field of view. Binoculars provide a pretty sight, but a 3- to 8-inch telescope at about 30 power probably shows the most spectacular view. A wide-field eyepiece at 60× will still show all the bright stars at once, including the dipper. At higher powers only smaller portions of the cluster are seen, though these higher powers can be useful and are discussed below.

The Pleiades nebulosity is a prized trophy for visual observers. Dust- and dew-free optics are essential for the pursuit, since light scattered from the bright stars will hide the nebulosity. The easiest portion is NGC 1435 south of Merope. Other nebulosity appears to surround Alcyone, Maia, and Electra. There is little or none around Celaeno, Taygeta, and Asterope, and definitely none around Atlas and Pleione.

The Merope Nebula was discovered by W. Tempel in 1859 with a 4-inch refractor. In 1874 Lewis Swift found it detectable with a 2-inch refractor at 25×. Modern observers seem to be having more difficulty, and it is often stated that a 6- to 8-inch telescope is required. This discrepancy is no doubt due to increased light pollution, and probably also to increased atmospheric dust particles, which scatter starlight. The mean surface brightness of NGC 1435 is 21.6 magnitudes per square arc-second and the total magnitude is 6.8, only slightly fainter than the Helix Nebula, NGC 7293.

Modern observers have reported viewing the nebulae around other stars. John Mallas in *The Messier Album* reported nebulosity, including fine streaks, around Maia, Taygeta, Alcyone and Celaeno with his 4-inch refractor. At least some of this must be spurious, because there is little or no nebulosity around Celaeno and Taygeta. Walter Scott Houston saw an unusual sight through an 8-inch telescope on an exceptional night in the southern Arizona mountains: "When I looked into the eyepiece, expecting to see a few faint wisps, the field was laced from edge to edge with bright wreaths of delicately structured nebulosity...".

The brightest of the linear streaks are near Electra. One about 20 arc-seconds wide extends from just south of that star about a third of the way to Alcyone. A second streak about half as long runs parallel to it one arc-minute farther south. They are easily confused with artificial streaks caused by diffraction or reflections in the telescope.

The drawing shows what was seen through the 8-inch under good skies. The lowest magnification with the 1.25-inch eyepiece holder was 82×, giving a field of view 40 arc-minutes in diameter. This field encompassed the bowl of the dipper quite nicely. I had seen the Merope Nebula many times before, at low power under excellent skies. This

South is to the left. Photograph of M45. (Courtesy Mount Wilson and Las Campanas Observatories, Carnegie Institution of Washington.)

A VISUAL ATLAS OF DEEP-SKY OBJECTS

Drawing of M45

Scale: 4.4 arc-min/cm	Viewing Distance (cm)		air mass: 1.05, faintest star: 14.2 at zenith, 188×, no tracking
8-inch f/11.5 Cassegrain			
28mm Kellner (82×)	25×:31	200×:4	10/08/83 11:40–12:15 UT at Waianae ranch,
20mm Erfle (117×, best	50×:16	300×:2.6	Hawaii; R. Clark
for nebula)	100×: 8	400×:2	

Brightest stars of the Pleiades

Star	Name	Magnitude	Star	Name	Magnitude
25 Tauri (Eta)	Alcyone	2.87	19 Tauri	Taygeta	4.29
27	Atlas	3.62	28	Pleione	5.09
17	Electra	3.70	16	Celaeno	5.44
20	Maia	3.86	21	Asterope	5.64
23	Merope	4.17	18	—	5.65
			22	—	6.41

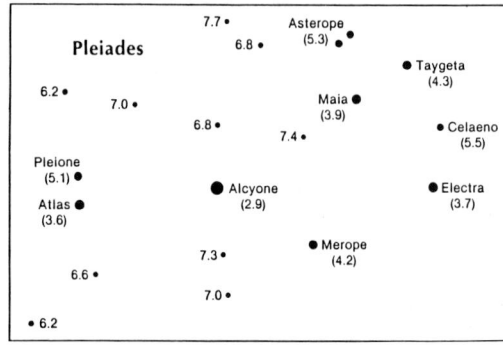

observation proved different, however, because the nebula could be seen well at as high as 82×. And when the magnification was increased to 117×, the nebula became easier! Under excellent conditions, the Merope Nebula is easily seen in the 3-inch finder at 31×.

The Pleiades' many and varied stars and delicate nebulosity provide a beautiful view to the amateur. It is no wonder the cluster is called the most studied and photographed of astronomical objects.

Figure 7.1. The brightest Pleiades, with their visual magnitudes. To the naked eye Asterope appears single; the magnitude is the combined light of both stars. The 6.2-magnitude star at upper left is also a naked-eye blend. From *Sky & Telescope*, November, 1985.

M1 (NGC 1952), THE CRAB NEBULA: SUPERNOVA REMNANT IN TAURUS

R.A. $05^h 34.5^m$, Dec. $22° 01'$ (2000.0)

Technical. Messier 1 has been known as the Crab Nebula ever since the third Earl of Rosse observed it with his 36-inch telescope in 1844. In the drawing he made, the nebula's filaments suggest the legs of a crab. Photographs show a beautiful network of red filaments throughout a diffuse green oval.

M1 is one of the youngest objects in the sky and certainly the youngest Messier object. It is the expanding remnant of a brilliant naked-eye supernova that was seen in July, 1054. Now 5 by 3 arc-minutes in size, M1 is growing by nearly a half arc-minute per century. Changes have been photographed in only a couple of decades.

M1 is about 6000 light-years distant and 6 light-years across. It is one of the strongest radio sources in the sky and also emits strong X-rays. Near the center of the nebula is a 16th-magnitude star that flashes 30 times a second in visible light, radio, and X-rays. This is a pulsar, the superdense neutron-star core of an old supernova. It is thought to be only 20 kilometers in diameter, and so dense that a teaspoonful of its matter would have a mass of several million tons! The flash rate corresponds to its period of rotation. Its very strong magnetic field spins with the star 30 times a second; energetic electrons trapped in the field produce the radiation. The pulsar is actually pumping energy into the nebula, so the expansion rate is increasing.

Visual. M1 can be seen in binoculars and small telescopes as a small, faint patch. With a visual magnitude of 9 and a size of about 5 by 3 arc-minutes, it has a mean surface brightness of 20.6 magnitudes per square arc-second. Large amateur telescopes under dark skies will show some of the filaments, though they are often difficult.

At low powers (60× or less) few details are visible. At higher powers the nebula's outline begins to depart from a smooth oval. In the 8-inch at 188×, the "bay" to the east is visible, and the whole thing takes on the appearance of two oblong nebulae side by side. No stars could be seen inside the nebula with the 8-inch, but many were around it.

Photograph of M1. South is up. (Courtesy Evered Kreimer, The Messier Album, J.H. Mallas & E. Kreimer.)

Drawing of M1.

Scale: 1.2 arc-min/cm	Viewing Distance (cm)
8-inch f/11.5 Cassegrain	
12.4mm Erfle (188×)	25×:115 200×:14
	50×: 57 300×:10
	100×: 29 400×: 7

air mass: 1.01, faintest star: 13.7 at zenith, 188×, no tracking
11/14/82 12:48–13:10 UT at Barbers Point, Hawaii; R. Clark

M42 (NGC 1976), M43 (NGC 1982), THE GREAT NEBULA IN ORION

M42: R.A. $05^h 35.4^m$ Dec. $-05° 23'$ (2000.0)
M43: R.A. $05^h 35.6^m$ Dec. $-05° 16'$

Technical. Messier 42 and 43 are probably the brightest and most spectacular nebulae in the sky, rivaled only by the Eta Carinae Nebula. Often referred to as the Great Nebula in Orion, M42 and M43 are a beautiful example of an H II region: an emission nebula containing mostly hydrogen, fluorescing in the ultraviolet light of very hot, newborn stars in its midst. The light of the Orion Nebula consists primarily of green emission lines of oxygen, with other colors from blue to red coming from emission of hydrogen, helium, nitrogen, and neon.

The distance of the Orion Nebula is usually given as around 1300 light-years and its size as about 30 light-years. The bright region is surrounded by vastly larger, dark clouds of gas and dust, which in fact fill much of Orion itself. The dust is thought to be primarily silicate (rock) particles only a micron in diameter. The composition of the glowing gas has been given as: hydrogen 90.8%, helium 9.08%, carbon 0.05%, oxygen 0.02%, nitrogen 0.02%, sulfur 0.003%, neon 0.0009%, chlorine 0.0002%, argon 0.0001%, and fluorine 0.00001%.

Deep within the cloud many stars are forming. One sign that the dark cloud is much bigger than the portion we see is that the region is full of bright infrared sources – stars whose visible light is blocked by dust. The bright nebula appears to be a hole blown in the dark cloud's wall, allowing us to see part way in.

In the brightest part of the hole are the four bright Trapezium stars, which are responsible for illuminating the gas. These very young, hot stars are estimated to be only 100 000 years old. They are among the youngest stars that amateurs can see.

Visual. The Orion Nebula is easily seen in the middle of the "Sword of Orion" as a fuzzy star. Curiously, however, its haziness is not mentioned in ancient records. The first known discovery of the nebula was by Nicholas Pieresc in 1611.

It is very pretty in any instrument from binoculars to the largest telescopes. M42 is 65 arc-minutes in diameter (twice the size of the full Moon) with a total visual magnitude of 4. M43 is about 7 by 5 arc-minutes and has a total magnitude of 8. The average surface brightness of the Orion Nebula is 21.7 magnitudes per square arc-second – but the faintest parts are much dimmer and the Trapezium region very much brighter, in the range of 17 magnitudes per square arc-second. This large range in brightness is difficult to photograph, and most pictures overexpose the Trapezium region. The eye has a larger dynamic range than film and, given good skies, can see the faint and bright portions of the nebula at the same time.

The detail visible in the Orion Nebula is truly spectacular. Here is one of the few sights in the sky that, when seen through modest amateur telescopes, impresses even those not excited by astronomy. The intracacy is beyond description, and even after hundreds of hours of viewing this nebula, it still remains a beautiful and wondrous sight.

Its very complexity makes the nebula difficult to draw. The drawing made with the 8-inch on page 101 took two nights (January 17 and 18, 1983) and nearly six hours of work (2 hours of observations and 4 hours for the final drawing). This short a time was only possible because many hours had been spent viewing and sketching the nebula on other nights. In fact a drawing done a year before, shown on page 103, shows considerably less detail. At that time I had less experience making detailed observations, used lower magnifications, and spent less time at the telescope.

The Trapezium region contains the brightest patch of the nebula, about 4 by 3 arc-minutes in size. The surface brightness here is so high that the amount of detail is often limited by atmospheric turbulence (seeing) rather than the limitations of the eye – a situation almost unknown in visual deep-sky work. With so much light, very high powers can be used.

Many stars dot this region. The four brightest are the famous multiple star system θ^1 Orionis. They form a trapezoidal figure that inspired the group's name. The four stars are easily resolved in small telescopes, as their separations range from 8.7 to 19.2 arc-seconds and their magnitudes from 5 to 8. Two additional components, shining at 11th magnitude, can be made out with a 6-inch telescope in steady air. The rest of the nebula is full of fainter stars: over 300 brighter than magnitude 17 are within just 5 arc-minutes of the Trapezium. The nebula contains more than 50 variable stars with maxima greater than magnitude 14.

The fainter regions of the nebula contain many loops of nebulosity familiar from photographs. They are visible in moderate sized amateur telescopes. The brightest arc, extending south on the east side, is visible in telescopes as small as about 2 inches. Since M42 is over one degree in diameter, low powers are needed to view it all at once. However, only at moderate to high powers do most of the elegant arcs become visible. In the 8-inch, they were best seen at powers near 200×.

The faintest portions of the nebula form a loop at the southern end, opposite the Trapezium, giving the nebula a circular, almost bubble-like shape. At low power this outer loop is seen only on very dark nights. Surprisingly, though, at medium to high powers it can be seen on moderate nights such as those on which the drawing of January 1983 was made. The outer loop appears continuous at low powers, but at 117× in the 8-inch, the individual sections seen on photographs could be resolved.

Powers of 100× to 200× show a dark patch just south of the Trapezium. This patch is next to a star 2 arc-minutes south and slightly west of the southern tip of the bright zone containing the Trapezium stars. This bright corner of the nebula is a nearly perfect right angle, and its edges appear remarkably straight. It looks almost unnatural.

To most observers the Orion Nebula appears pale green. The visibility of other colors is somewhat controversial. A few observers report a faint reddish color, especially on the southern edge of the Trapezium region – which is indeed quite red in photographs. On extremely dark nights, I have seen the Trapezium and faint outer regions to appear green while most of the nebula appeared a vivid pastel pink, and the bright arc extending south on the east side appeared pastel blue. Blocking the bright area around the Trapezium makes the detection of red in the outer regions easier.

Photograph of M42 and M43. South is up. (Courtesy Mount Wilson and Las Campanas Observatories, Carnegie Institution of Washington.)

Drawing of M42 and M43 made in January, 1983 on previous page. The main loops of M42 and detail in the Trapezium region were best seen at 188×. The faint outer loop was first detected at 82×, then seen better at 117×, and could also be made out at 188×. The final drawing took about four hours.

Scale: 2.5 arc-min/cm
8-inch f/11.5 Cassegrain

	Viewing Distance (cm)	
28mm Kellner (82×)	25×:55	200×:7
20mm Erfle (117×)	50×:28	300×:5
12.4mm Erfle (188×)	100×:14	400×:3

Drawing Method 2 (see Chapter 5) on two nights:
air mass: 1.12, faintest star: 14.0 at zenith, 188×, no tracking
1/16/83 8:00–8:40 UT at Barbers Point, Hawaii

air mass: 1.12, faintest star: 13.8 at zenith, 188×, no tracking
1/18/83 8:40–9:48 UT at Hawaii Kai, Hawaii;
R. Clark

Drawing of M42 and M43 made in January, 1982 on next page. Compare this drawing with the one made in January, 1983. The greater detail in the latter is due to the use of several magnifications and more time at the telescope.

Scale: 2.5 arc-min/cm
8-inch f/11.5 Cassegrain

	Viewing Distance (cm)	
28mm Kellner (82×)	25×:55	200×:7
20mm Erfle (117×)	50×:28	300×:5
	100×:14	400×:3

air mass: 1.24, faintest star: 13.4 at zenith, 117×; no tracking
1/24/82 9:07–9:55 UT at Ewa Beach, Hawaii;
R. Clark

Photograph of the inner region of M42, the Trapezium area. South is up. (Courtesy Lick Observatory.)

A VISUAL ATLAS OF DEEP-SKY OBJECTS

Drawing of M42 Trapezium area.

Scale: 0.5 arc-min/cm
8-inch f/11.5 Cassegrain
12.4mm Erfle (188×)

Viewing Distance (cm)
100×:69   300×:23
200×:34   400×:17

air mass: 1.17, faintest star: 13.5 at zenith, 188×; no tracking
1/13/83 9:30–9:55 UT at Hawaii Kai, Hawaii; R. Clark

NGC 2023
NGC 2024, IC 434 (THE HORSEHEAD NEBULA) NEBULAE IN ORION

IC 434:   R.A. $05^h 41.1^m$ Dec. $-02°24'$
NGC 2023: R.A. $05^h 41.7^m$, Dec. $-02°13'$
NGC 2024: R.A. $05^h 41.9^m$, Dec. $-01°50'$
(2000.0)

Technical. The 2nd-magnitude star Zeta Orionis, the easternmost one of Orion's belt, is surrounded by bright and dark nebulae. The region is famous for the photographically spectacular Horsehead Nebula, B33. This is a dark dust cloud remarkably like the head of a ghostly horse rearing up in silhouette against the bright background of the emission nebula IC 434. The energy source illuminating IC 434 is the star Sigma Orionis, about 1° southwest of Zeta.

IC 434 appears to be colliding with, or perhaps burning its way into, a very large, dark cloud. The sharp, nearly straight boundary between them extends about 1° south of Zeta Orionis. The Horsehead, about halfway down the boundary's length, is a protuberance of the dark cloud. The boundary extends about 18 light-years from north to south, and the Horsehead is about 1 light-year across. The dark cloud covers the entire eastern part of the drawing and photograph, as can be seen by the relative absence of stars on the right, and in fact it extends many degrees still farther east. The whole complex has been placed at a distance of about 1200 light-years.

NGC 2024 is a complex nebulous patch about 15 arc-minutes east-northeast of Zeta Orionis. It is an emission nebula, probably excited by Zeta, crossed by a large, dark lane running north to south.

NGC 2023 is a bluish reflection nebula surrounding an 8th-magnitude star a little less than half a degree farther south, east-northeast of the Horsehead.

Visual. NGC 2024 is the brightest and easiest nebula in this fascinating region. It is very close to 1.8-magnitude Zeta Orionis, so placing the bright star out of the field of view reduces glare. NGC 2024 shows considerable detail in medium size telescopes, and under good skies it is visible through the 3-inch finder at 31×. Through the 8-inch the main dark lane could be seen, along with several smaller ones, as shown in the drawing. In a 13-inch telescope, many more dark lanes appeared.

NGC 2023, a diffuse patch around an 8th-magnitude star, is quite easy to see if the optics are clean and free of dew. The best way to check whether the nebula is real, rather than starlight scattered in the atmosphere, telescope, or eye, is to examine other stars of similar brightness to make sure they are free of "nebulosity".

IC 434 is very faint and considered one of the most difficult deep-sky objects to observe visually. It is often said that excellent skies and an 8- to 12-inch telescope at low power are required. The Horsehead, B33, is even more difficult, being only 5 arc-minutes across. Many amateurs find the dark lane in NGC 2024 and believe they've found the Horsehead.

Glare from Zeta Orionis is one reason for failing to detect IC 434, especially with dirty or dewed optics. Low power is another. Medium powers not only make IC 434 and B33 larger but also increase their apparent distance from Zeta Orionis. As can be seen in the drawing with the 8-inch telescope, IC 434 and B33 were detected under only moderate to good skies. The drawing by Ray Fabre using a 13.1-inch Dobsonian telescope at nearly the same powers shows a much better view of both. This view of the Horsehead is similar to that in an 8-inch under excellent skies. If good to excellent skies can be found, the Horsehead Nebula should be fairly easy in a 6-inch telescope. I have also seen it through the 3-inch at 50× under the excellent skies of the Colorado Rockies.

Unfortunately, there are no magnitude or surface brightness measures of either NGC 2024, NGC 2023, or IC 434. However, IC 434 is slightly harder to detect than the Merope Nebula in the Pleiades. The average surface brightness of that nebula is 21.6 magnitudes per square arc-second, so IC 434 is probably 22 or slightly fainter.

Drawing of NGC 2023, NGC 2024, and IC 434 by Ray Fabre with a 13-inch Dobsonian reflector at 60× and 90×. Note the increased detail seen with the larger aperture compared with the 8-inch drawing, made at the same time and site. The scale and viewing distances are the same as that for the 8-inch drawing.

air mass: 1.1, faintest star: 15.2 at zenith, 90×; no tracking
2/12/83 7:15–8:00 UT at Waianae Ranch, Hawaii; R. Fabre

*Photograph of NGC 2023, NGC 2024, and IC 434.
South is up. (Courtesy James E. Gunn.)*

A VISUAL ATLAS OF DEEP-SKY OBJECTS

Drawing of NGC 2023, NGC 2024, and IC 434 with an 8-inch telescope. The Horsehead was barely seen with the 8-inch telescope. A power of 117× was better for detecting it than 82×; the higher power may have helped reduce the glare from Zeta Orionis.

Scale: 4.6 arc-min/cm
8-inch f/11.5 Cassegrain
28mm Kellner (82×)
20mm Erfle (117×)

Viewing Distance (cm)

10×:75	100×:7
25×:30	200×:4
50×:15	300×:2

air mass: 1.12, faintest star: 14.2 at zenith, 117×; no tracking
2/12/83 7:00–8:00 UT at Waianae Ranch, Hawaii; R. Clark

M78 (NGC 2068), NGC 2071, DIFFUSE NEBULAE IN ORION

M78: R.A. $05^h 46.8^m$, Dec. $00°\ 03'$
NGC 2071: R.A. $05^h 47.2^m$, Dec. $00°\ 18'$,
(2000.0)

Technical. M78 is a small reflection nebula a little more than 2° northeast of Zeta Orionis. It was discovered by Pierre Mechain in 1780. The nebula is believed to be about 1600 light-years distant, roughly the same as Zeta Orionis. It is 2 or 3 light-years across. The light it scatters is from the 10th-magnitude star HD 38563.

Just northeast of M78 is NGC 2071, another reflection nebula. The entire region is enveloped in a dark nebula, so stars are few. The total magnitude of M78 is 8, and with a size of 8 by 6 arc-minutes, its average surface brightness is 20.8 magnitudes per square arc-second. NGC 2071 is 4 by 3 arc-minutes, and its average surface brightness is somewhat dimmer than M78's.

Photograph of M78 and NGC 2071. South is up. (Courtesy Evered Kreimer, The Messier Album.)

A VISUAL ATLAS OF DEEP-SKY OBJECTS

Visual. M78 often looks like a small comet. With a visual magnitude of 8 and a size of 8 by 6 arc-minutes, it has a mean surface brightness of 20.8. Its northern border is sharp, while the southern part dims gradually, starting near the northern edge. The nebula contains two 10th-magnitude stars, one near the northern edge, the other 53 arc-seconds to the south. A third and fainter star is near the southern limit of the nebula, slightly east of the other two.

NGC 2071 appears smaller and fainter than M78. It has a size of 4 by 3 arc-minutes, but there are no magnitude estimates.

Drawing of M78 and NGC 2071.

Scale: 2.0 arc-min/cm	Viewing Distance (cm)	air mass: 1.09, faintest star 13.7 at zenith, 188×, no tracking
8-inch f/11.5 Cassegrain		
12.4mm Erfle (188×)	25×:69 200×:9	11/14/82 11:51–12:15 UT at Barbers Point, Hawaii; R. Clark
	50×:34 300×:6	
	100×:17 400×:4	

NGC 2261, HUBBLE'S VARIABLE NEBULA IN MONOCEROS

R.A. 06^h 39.1^m, Dec. 08° 43' (2000.0)

Technical. NGC 2261 was discovered by Sir William Herschel in 1783, but it has been known as Hubble's Variable Nebula ever since Edwin Hubble found, on photographs taken between 1900 and 1916, that it had changed shape. NGC 2261 is a small reflection nebula that appears rather like a comet. The variable star R Monocerotis is at its apex, and appears as if it were the "comet's" nucleus. R Mon varies by up to 4 magnitudes in an irregular way. Its variations are responsible for the changing appearance of the nebula, which it illuminates. However, the nebula and star do not vary together. It seems that dust clouds orbit close to the star, blocking its light to a greater or lesser degree and casting their moving shadows on the nebula.

Some changes have been seen to occur in as little as one day, and shadows have moved by up to 1 arc-second in four days. Changes that might be seen in large amateur telescopes can occur on timescales as short as a month. On nights of very good seeing, with large telescopes R Mon does not appear quite stellar. Apparently dust close to the star forms a very small, brilliant reflection nebula.

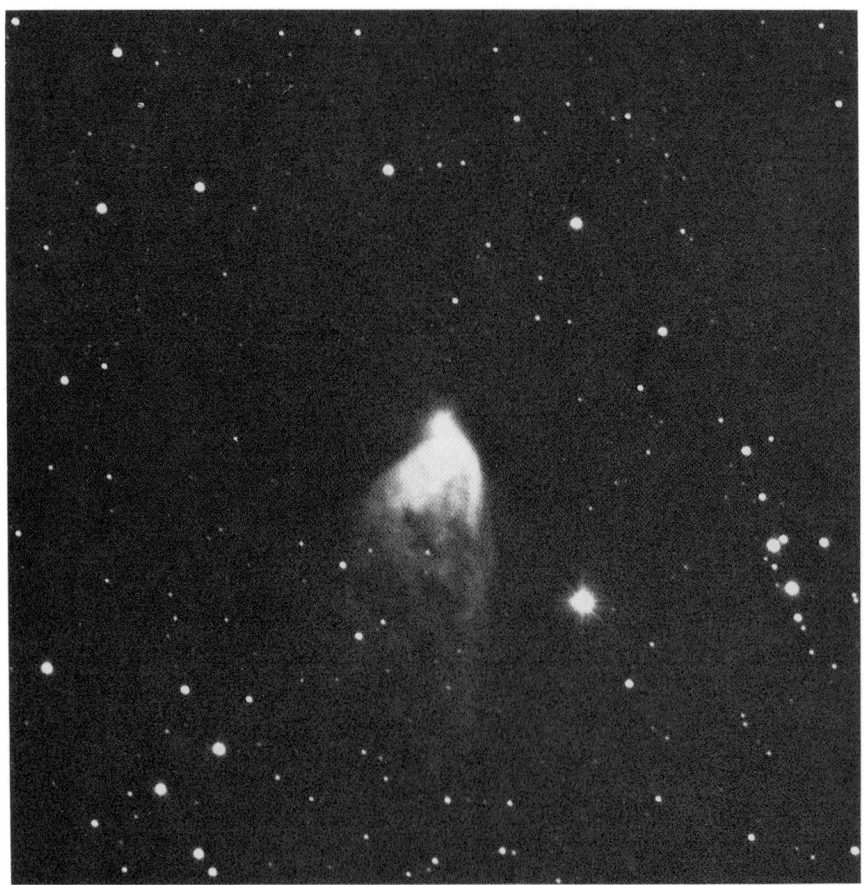

Photograph of NGC 2261. South is up. (Courtesy National Optical Astronomy Observatories.)

Visual. NGC 2261 is quite easy in medium size telescopes even under only moderate skies. I first found it while randomly sweeping the winter skies and immediately thought it was a comet. A check of several star charts and *Burnham's Celestial Handbook* set me straight. At magnitude 10 with a size of 2 by 2 arc-minutes, it has a mean surface brightness of 20.1 magnitudes per square arc-second. This is quite bright compared to most nebulae. Although no detail was seen in the 8-inch under moderate skies, it should be possible to see detail under excellent skies if the shadows at the time are high in contrast. Since the surface brightness is high, and the nebula is a reflection from dust grains, it may appear bluish in large telescopes under dark skies.

Drawing of NGC 2261.

Scale: 0.8 arc-min/cm	Viewing Distance (cm)		air mass: 1.07, faintest star: 13.5 at zenith, 188×;
8-inch f/11.5 Cassegrain			no tracking
12.4mm Erfle (188×)	25×:172	200×:21	1/13/83 10:35–11:05 UT at Hawaii Kai, Hawaii;
9mm Kellner (260×)	50×: 86	300×:14	R. Clark
	100×: 43	400×:11	

M46 (NGC 2437), OPEN CLUSTER IN PUPPIS WITH NGC 2438, A PLANETARY NEBULA

M46:   R.A. $07^h 41.9^m$, Dec. $-14° 49'$
NGC 2438: R.A. $07^h 41.9^m$, Dec. $-14° 43'$
            (2000.0)

Technical. M46 was discovered in 1771 by Charles Messier. The cluster is about 5400 light-years distant and 30 light-years in diameter. It contains between 200 and 500 stars, with about 150 between magnitudes 10 to 13.

M46 is unusual in that it appears to contain a planetary nebula, NGC 2438. But the nebula is not a member of the star cluster, since it is moving through space at a different velocity. The nebula appears 65 arc-seconds in diameter. It is probably in the foreground about 3000 light years from us, and has a diameter of about 1 light-year. Its central star is 16th magnitude.

Visual. M46 is a pretty cluster at any magnification, where it all fits within the field of view. The cluster is round, with a diameter of 25 arc-minutes. The total visual magnitude is 8, and in small instruments and low power, where the cluster is not resolved, the mean surface brightness is 23.6 magnitudes per square arc-second.

The planetary nebula is best seen at medium to high powers. At 150× or 200× with a wide-field eyepiece, the whole cluster and the planetary can be seen simultaneously. The nebula is 11th magnitude, with a mean surface brightness of 19.8. In medium-sized telescopes, a star slightly northwest of center is often mistaken for the central star, which is fainter. In the 8-inch under moderate skies, the nebula appeared as a doughnut noticeably fainter on the west side.

*Photograph of M46 and NGC 2438. South is up.
(Courtesy Evered Kreimer, The Messier Album.)*

Drawing of M46 and NGC 2438.

Scale: 3.0 arc-min/cm
8-inch f/11.5 Cassegrain
12.4mm Erfle (188×)
9mm Kellner (260×)

Viewing Distance (cm)

10×: 115   100×: 11
25×: 46   200×: 6
50×: 23   300×: 4

air mass: 1.39, faintest star: 13.5 at zenith, 188×,
no tracking
1/13/83 8:45–9:18 UT at Hawaii Kai, Hawaii;
R. Clark

M67 (NGC 2682), OPEN CLUSTER IN CANCER

R.A. $08^h\ 51.0^m$, Dec. $11°\ 49'$ (2000.0)

Technical. This open star cluster is unusual in that it is one of the oldest known (estimates range from 3 to 10 billion years) and is very far from the plane of our Galaxy. Most open clusters, also known as galactic clusters, lie close to the galactic disk, whereas M67 lies approximately 1500 light-years above it. The cluster is about 2500 light-years distant and 12 light-years in diameter. It includes some 500 stars brighter than magnitude 16; the brightest is magnitude 10.

The stellar population is unlike that of most open clusters, but rather similar to the very old globular clusters. The only other open cluster known to have a similar pattern is NGC 188, in Cepheus. The most evolved stars in M67 have only about half the luminosity of similar stars in globulars, however. This difference is attributed to chemical composition: the typical globular is made almost entirely of hydrogen and helium, whereas M67 has a heavy-element ratio similar to our Sun.

Photograph of M67. South is up. (Courtesy Optical Astronomy Observatories.)

A VISUAL ATLAS OF DEEP-SKY OBJECTS

Visual. M67 has a total visual magnitude of 7, and it is 15 arc-minutes in diameter. When unresolved in very small instruments such as binoculars, its mean surface brightness is 21.5 magnitudes per square arc-second. The drawing shows an interesting visual effect of closely spaced stars. The cluster contains many faint stars not resolved in an 8-inch telescope. With averted vision, the eye can detect their light without resolving them. This gives an appearance of a faint background nebulosity. The drawing is a representation of this effect. With direct vision the "nebulosity" vanishes.

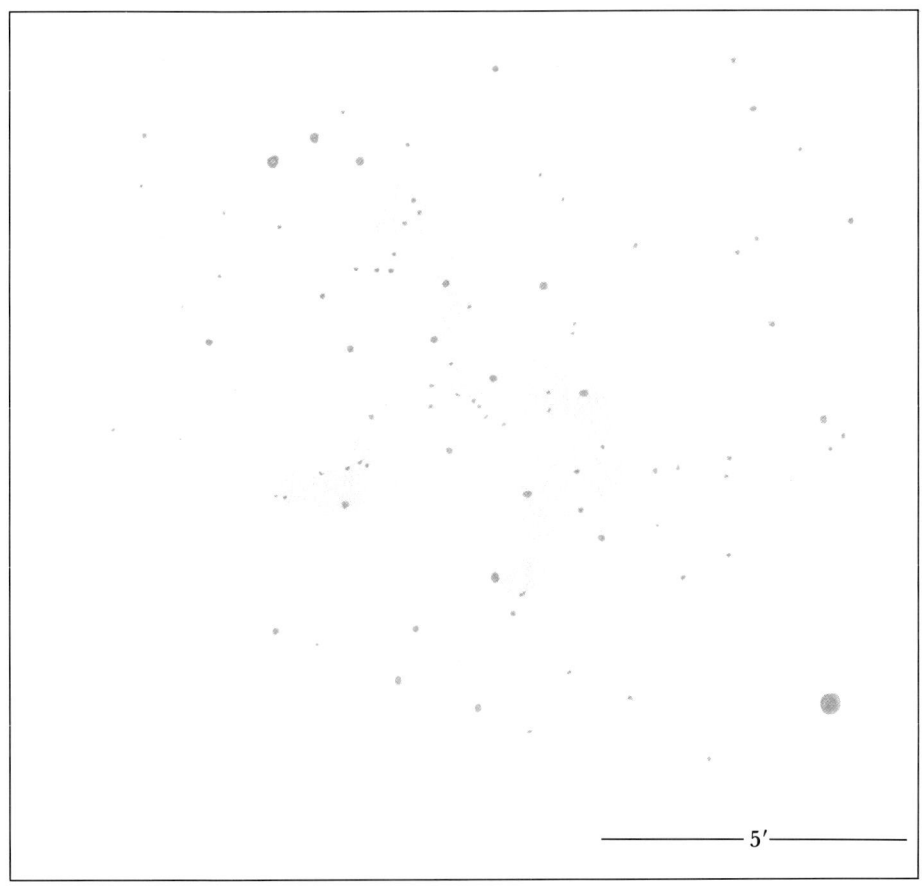

Drawing of M67.

Scale: 1.2 arc-min/cm	Viewing Distance (cm)		air mass: 1.05, faintest star: 14.0 at zenith, 188×, no tracking
8-inch f/11.5 Cassegrain			
20mm Erfle (117×)	25×: 115	200×: 14	1/16/83 10:30–10:45 UT at Barbers Point, Hawaii;
12.4mm Erfle (188×)	50×: 67	300×: 10	R. Clark
	100×: 29	400×: 7	

NGC 2903, SPIRAL GALAXY IN LEO

R.A. 09^h 32.1^m, Dec. 21° 31' (2000.0)

Technical. NGC 2903 is a large, many-armed type-Sb spiral galaxy, one of the brightest galaxies in Leo. It shows complex detail on deep-sky photographs. The only other galaxy of comparable brightness in Leo is M66.

Visual. In angular size NGC 2903 is 11.0 by 4.7 arc-minutes, and its total visual magnitude is 9.7. The mean surface brightness is 22.6 magnitudes per square arc-second. In the 8-inch under moderate skies, NGC 2903 was seen as a diffuse glow surrounding a brighter nucleus. Under excellent skies it should reveal some detail in an 8-inch or larger telescope.

Photograph of NGC 2903. South is up. (Courtesy Palomar Observatory.)

Drawing of NGC 2903.

Scale: 1.1 arc-min/cm
8-inch f/11.5 Cassegrain
12.4 mm Erfle (188×)

Viewing Distance (cm)	
25×:125	200×:16
50×: 63	300×:10
100×: 31	400×: 8

air mass: 1.00, faintest star: 13.5 at zenith, 188×; no tracking
1/13/83 11:40–12:40 UT at Hawaii Kai, Hawaii; R. Clark

Notes: Intermittent clouds increased the time needed to make the observation.

M81 (NGC 3031), SPIRAL GALAXY IN URSA MAJOR

R.A. 09^h 55.6^m, Dec. $69°$ $04'$ (2000.0)

Technical. M81 is sometimes known as Bode's Nebula, after its discovery by J. E. Bode in 1774. It is a spiral galaxy tilted 32° from edge-on. Relatively close to us at about 8 or 10 million light-years, it has a diameter of 40 000 to 50 000 light-years. M81 is the brightest member of a group of about a dozen galaxies; its nearest neighbor and closest rival in brightness is M82 (NGC 3034). Radio observations show hot gas extending twice the length of the spiral arms seen on photographs. This hot gas has been traced all the way across a "bridge" to M82. Presumably the bridge is the result of a close encounter between the two galaxies about 200 million years ago.

Visual. M81 appears as a diffuse oval with a bright center. It has a size of 18 by 10 arc-minutes, and its total visual magnitude is 8.0. The mean surface brightness is 22.3 magnitudes per square arc-second. Through the 8-inch under good skies, only the slightest hint of the spiral arms was seen. The outer spiral arms are of very low contrast. On short-exposure photographs through large telescopes, spiral arms with thin dust lanes can be traced to within half an arc-minute of the nucleus, but none could be seen through the 8-inch.

The easiest part of the outer arms to detect is west and slightly north of the nucleus. The arm here consists of several bright knots, which in large telescopes look like a string of faint stars.

*Photograph of M81. South is up.
(Courtesy National Optical Astronomy Observatories.)*

Drawing of M81.

Scale: 1.3 arc-min/cm	Viewing Distance (cm)		air mass: 1.51, faintest star: 14.3 at zenith, 188×;
8-inch f/11.5 Cassegrain			no tracking
12.4mm Erfle (188×,	25×:106	200×:13	2/12/83 9:55–10:05 UT at Waianae ranch,
best view)	50×: 53	300×: 9	Hawaii; R. Clark
7mm Erfle (334×)	100×: 26	400×: 7	

M82 (NGC 3034), PECULIAR GALAXY IN URSA MAJOR

R.A. 09^h 56.1^m, Dec. $69°$ $42'$ (2000.0)

Technical. M82 is an irregular galaxy discovered, along with M81 about a degree south, by J. E. Bode in 1774. This galaxy is unusual because its central region is very dusty and seems to be violently disrupted. It was long thought that the core of the galaxy had undergone an explosion a few million years ago. Now it is thought that the inner region is merely in vigorous turmoil, probably owing to great bursts of star formation in the dusty areas. Many of the abundant gas clouds are moving outward at speeds of up to 500 km/sec. The violent gas motions have produced many shock waves, which have compressed the remaining gas and dust to trigger the formation of yet more stars.

This thick interstellar matter provides spectacular detail on photographs. Much of the gas is heated by the newly formed stars to a temperature of 100 kelvin (–170°C). This is very hot compared with typical interstellar dust, which is only a few degrees above absolute (zero kelvin). The stars that do this heating are in many large clusters, each with as many as 10 000 members. The heated dust makes M82 the brightest galaxy in the sky at thermal infrared wavelengths, aside from the Milky Way. M82 is also a strong radio source.

Recent reconstructions of the history of M82 may explain what happened to it. About 200 million years ago M81 passed close to M82, and the gravitational tidal effects perturbed the orbits of billions of stars and great masses of gas. The result was collisions between gas clouds, causing many of them to collapse and produce new stars, while other clouds were thrown out of the disk plane of M82. The ejected material eventually slowed and fell back into the plane, causing more cloud collisions and the burst of star formation we see today.

M82 is about 8 or 10 million light-years distant and 16 000 light-years across. Its mass is estimated at only one-tenth that of M81.

Visual. M82 shows considerable detail in amateur telescopes. At a total magnitude of 9.2 and with a size of 8 by 3 arc-minutes, the mean surface brightness is 21.3 magnitudes per square arc-second. This is a full magnitude more surface brightness than the mean of neighbor M81 (though M81's central region is admittedly much brighter). M82's high surface brightness allows high magnifications to be used even with small telescopes.

In the 8-inch, the main irregular dark lane in the center of the galaxy was very easy to see at medium powers under good skies. The dark lane, the galaxy's most prominent feature, is difficult in 4-inch or smaller telescopes. Other structure was visible in the 8-inch, such as the sharp drop in intensity about 1 arc-minute east of the nucleus. The dark lane gave the appearance of an arrow pointing west and slightly south. A bright region was seen to the west of the nucleus. Powers near 200× seemed to give the best overall view. More detail might have been visible if the seeing was not so poor (star images were 2 to 3 arc-seconds in size).

A VISUAL ATLAS OF DEEP-SKY OBJECTS

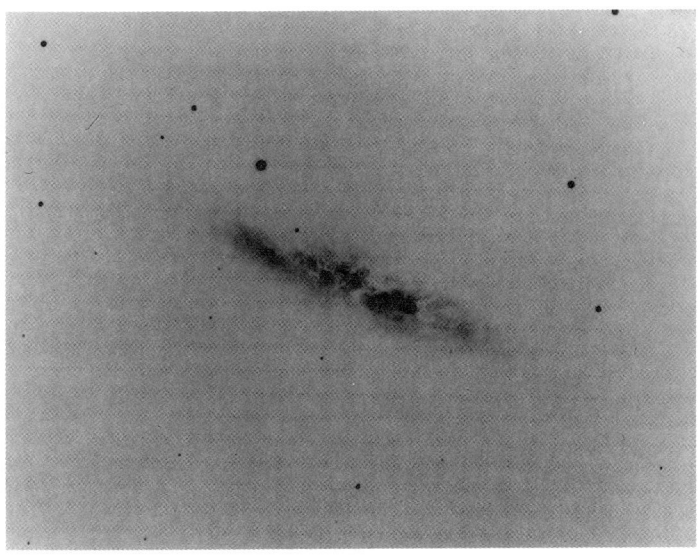

Photograph of M82. South is up. (Courtesy Canada–France–Hawaii Telescope Corporation.)

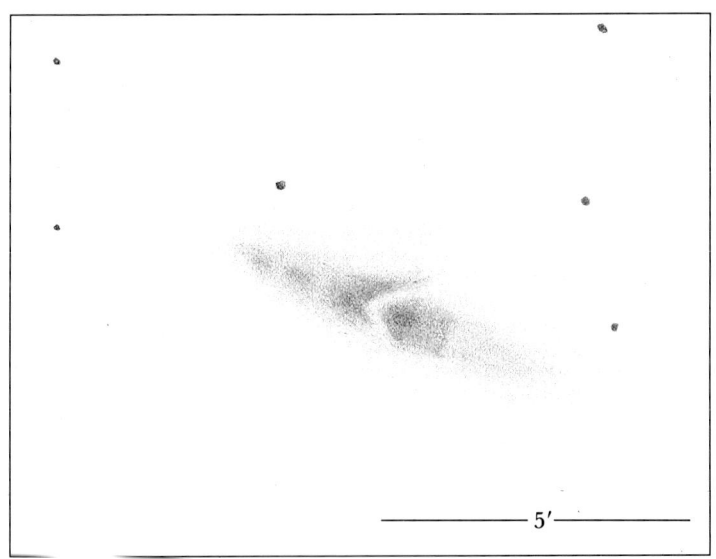

Drawing of M82.

Scale: 1.2 arc-min/cm
8-inch f/11.5 Cassegrain
20mm Erfle (117×)
12.4 mm Erfle (188×, best view)
7mm Erfle (334×)

Viewing Distance (cm)

25×: 115   200×: 14
50×: 57   300×: 10
100×: 29   400×: 7

air mass: 1.55, faintest star: 14.3 at zenith, 188×; no tracking
2/12/83 9:15–9:55 UT at Waianae ranch, Hawaii; R. Clark

M96 (NGC 3368), GALAXY IN LEO
R.A. 10^h 46.8^m, Dec. 11° 49' (2000.0)

Technical. M96 is an Sa-type spiral galaxy in the Leo Cluster of Galaxies, which includes M95, M65, M66, and M105. The group is around 30 million light-years distant. As with all type -Sa galaxies, a large central bulge dominates tightly wound spiral arms. M96 has a faint outer ring about 6 arc-minutes in diameter.

Visual. M96 has a total magnitude of 10.2 and a diameter of 6 by 4 arc-minutes, for a mean surface brightness of 22.3 magnitudes per square arc-second. Most of the light is concentrated in the central bulge, however, and the surface brightness there is much higher. Through the 8-inch in good skies, a dark lane was easily seen at medium powers. I did not know it existed before making the observation. The lane is not visible in telescopes smaller than 4 inches.

A VISUAL ATLAS OF DEEP-SKY OBJECTS

Photograph of M96. South is up. (Courtesy Evered Kreimer, The Messier Album.)

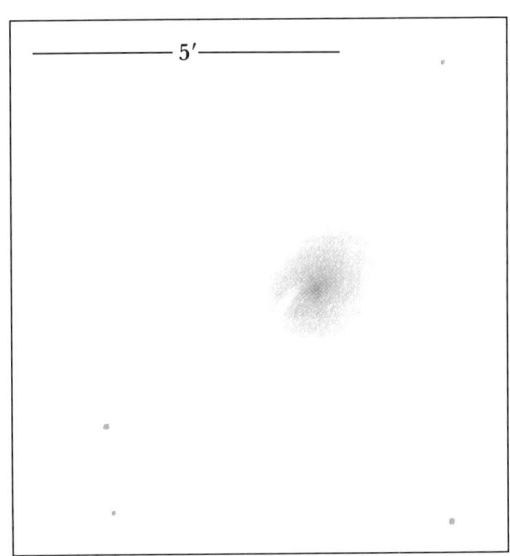

Drawing of M96.

Scale: 1.2 arc-min/cm
8-inch f/11.5 Cassegrain
12.4mm Erfle (188×)

Viewing Distance (cm)	
25×:115	200×:14
50×:57	300×:10
100×:29	400×:7

air mass: 1.09, faintest star: 14.3 at zenith, 188×; no tracking
2/12/83 10:10–10:30 UT at Waianae ranch, Hawaii; R. Clark

M105 (NGC 3379) NGC 3384, NGC 3389, GALAXIES IN LEO

M105:   R.A. 10^h 47.8^m, Dec. 12° 35'
NGC 3384: R.A. 10^h 48.3^m, Dec. 12° 38'
NGC 3389: R.A. 10^h 48.4^m, Dec. 12° 32'
(2000.0)

Technical. M105 and NGC 3384 are elliptical galaxies and NGC 3389 is a type-Sc spiral. All are part of the Leo Cluster of Galaxies, which includes M95, M96, M65 and M66 at a distance of about 30 million light-years. NGC 3384 is only 8 arc-minutes from M105, and NGC 3389 is 10 arc-minutes from it.

Visual. M105 is the brightest member of the trio, with a total magnitude of 10.6. Being 2 arc-minutes in diameter, its mean surface brightness is 20.8 magnitudes per square arc-second – but, like most galaxies, it is far brighter in the middle than at the edges.

NGC 3384 is only slightly fainter at magnitude 11.0. It is larger, 4 by 2 arc-minutes, yielding a mean surface brightness of 21.9 magnitudes per square arc-second. Most of the light is concentrated in the central 1 arc-minute.

NGC 3389 is the most difficult: it has a total magnitude of 12.2 and appears 2 by 1 arc-minutes. This size corresponds to a mean

Photograph of M105 (at left), NGC 3384 (at lower right), and NGC 3389 (at upper right). South is up. (Courtesy Evered Kreimer, The Messier Album.)

surface brightness of 21.7 magnitudes per square arc-second. The light is distributed fairly uniformly, so the peak surface brightness is not much greater, unlike with M105 and NGC 3384.

All three appeared featureless through the 8-inch under moderate skies. The elliptical galaxies showed brighter centers, and NGC 3384 appeared slightly elongated, while M105 was round. NGC 3389 appeared as a very faint oval of uniform brightness. At medium powers with wide-field eyepieces, all three galaxies fit into the field of view and provide a nice view.

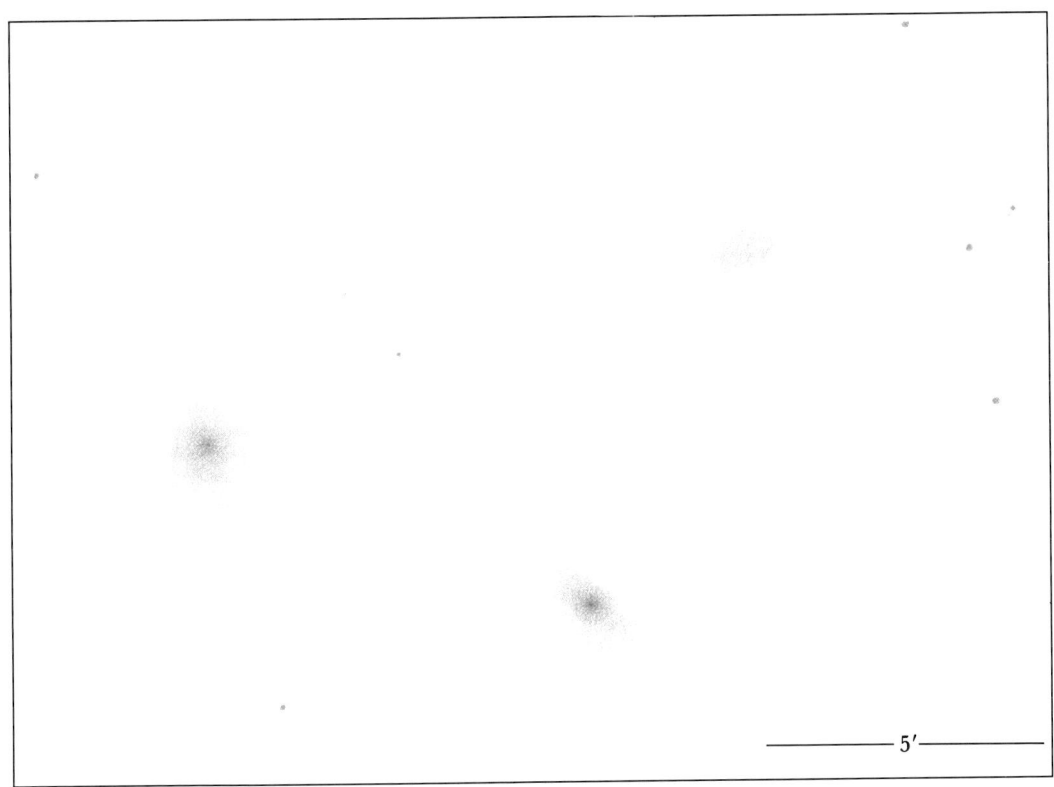

Drawing of M105, NGC 3384, and NGC 3389.

Scale: 1.3 arc-min/cm
8-inch f/11.5 Cassegrain
20mm Erfle (117×)
12.4mm Erfle (188×, best view)
9mm Kellner (260×)

Viewing Distance (cm)
25×: 106   200×: 13
50×: 53   300×: 9
100×: 26   400×: 7

air mass: 1.03, faintest star: 13.5 at zenith, 188×; no tracking
1/13/83 12:45–13:20 UT at Hawaii Kai, Hawaii; R. Clark

M108 (NGC 3556), GALAXY IN URSA MAJOR

R.A. 11^h 11.6^m, Dec. 55° 41' (2000.0)

Technical. M108 is a nearly edge-on spiral galaxy of type Sc. First discovered by Pierre Mechain in 1781, it was not commonly included in Messier's catalog until the early 1970s. M108 is about 25 million light-years away and has many obscuring dust lanes and no pronounced central bulge.

Visual. M108 appears 7.8 by 1.4 arc-minutes in angular size, with a total brightness of magnitude 10.8 and a mean surface brightness of 22.0 magnitudes per square arc-second. It is 48 arc-minutes northwest of the Owl Nebula, M97 (NGC 3587). Through the 8-inch under moderate skies, no detail corresponding to dark lanes was seen. The nucleus looked like a star, and an actual star is superimposed on the galaxy's western side. The galaxy appeared as a uniform, elongated glow, fading at the edges into the sky background.

Photograph of M108. South is up. (Courtesy Evered Kreimer, The Messier Album.)

Drawing of M108.

Scale: 1.2 arc-min/cm
8-inch f/11.5 Cassegrain
20mm Erfle (117×)
12.4mm Erfle (188×, best view)
7mm Erfle (334×)

Viewing Distance (cm)
25×: 115   200×: 14
50×: 57   300×: 10
100×: 29   400×: 7

air mass: 1.41, faintest star: 13.8 at zenith, 188×; no tracking
5/15/83 8:41–8:52 UT at Barbers Point, Hawaii; R. Clark

M97 (NGC 3587) THE OWL NEBULA: PLANETARY NEBULA IN URSA MAJOR

R.A. 11^h 14.9^m, Dec. 55° 02' (2000.0)

Technical. M97 was discovered by Charles Messier's compatriot Pierre Mechain in 1781. When William Parsons, Lord Rosse, observed it with his 72-inch speculum-metal reflector in 1848, he made a drawing remarkably like an owl's face, and M97 has been the Owl Nebula ever since. It is thought to be about 1600 light-years distant and about 1.5 light-years in diameter. Its gas is only 1/10 to 1/100 as dense as that of a typical planetary nebula, suggesting that M97 is relatively old and has been expanding and thinning for a long time. The 14th-magnitude central star has an effective temperature of 85 000 kelvin.

Visual. The two dark "owl's eyes" are hard to detect in small to medium telescopes. With a diameter of 2.5 arc-minutes and a total magnitude of 11.0, the nebula has a mean surface brightness of 22.6 magnitudes per square arc-second – very low for a planetary nebula. By comparison, the famous Ring Nebula, M57 in Lyra, has a mean surface brightness of 17.9 magnitudes per square arc-second. With the 8-inch under moderate to good skies, the eyes were barely seen. If I had not known of their existence I probably would not have detected them at all. When making the drawing I did not know the eyes' true orientation, so the fact that they are placed correctly indicates that they really were seen – unless blind luck was at work. This uncertainty shows the extreme difficulty in observing the owl's eyes. Better skies or a larger telescope are needed.

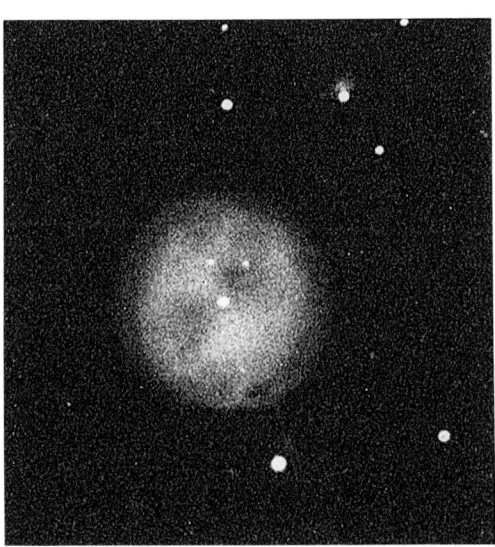

Photograph of M97. South is up. (Courtesy Evered Kreimer, The Messier Album.)

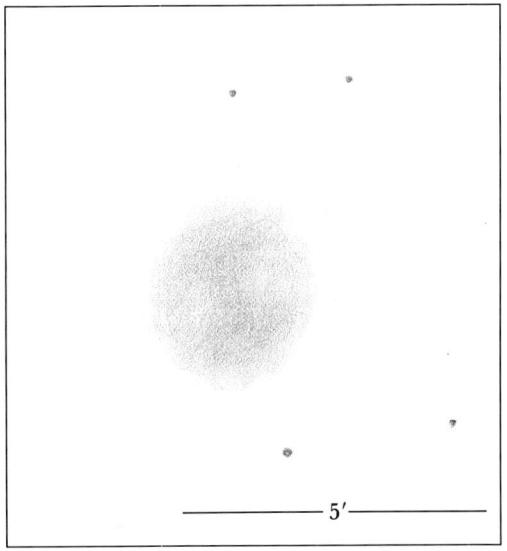

Drawing of M97.

Scale: 1.2 arc-min/cm	Viewing Distance (cm)	
8-inch f/11.5 Cassegrain		
20mm Erfle (117×)	25×:115	200×:46
12.4mm Erfle (188×,	50×: 57	300×:10
best view)	100×: 29	400×: 7

air mass: 1.36, faintest star: 13.8 at zenith, 188×;
no tracking
5/15/83 8:25–8:40 UT at Barbers Point, Hawaii;
R. Clark

M66 (NGC 3627) NGC 3628, M65 (NGC 3623), GALAXIES IN LEO

M65: R.A. $11^h 18.9^m$, Dec. $13° 07'$
M66: R.A. $11^h 20.2^m$, Dec. $13° 01'$
NGC 3628: R.A. $11^h 20.3^m$, Dec. $13° 37'$
(2000.0)

Technical. This trio of galaxies fits into a low-power field of view. They are members of the Leo Cluster of Galaxies, about 30 million light-years away, which also includes M95, M96, and M105.

All three galaxies are type-Sb spirals, with NGC 3628 being nearly edge-on. M66 has thick spiral arms and dust lanes. M65 appears like a miniature Andromeda Galaxy (M31); it has a dark dust lane on the east side, and the spiral arms are wound much like M31s. M66 has a diameter of 50 000 light-years; M65 60 000 light-years; and NGC 3628, about 90 000 light-years.

Visual. M65 and M66 can be detected in binoculars, while NGC 3628 requires a small telescope under good skies. M66 has a total magnitude of 9.7 and an angular size of 8.0 by 2.5 arc-minutes, yielding a mean surface brightness of 21.6 magnitudes per square arc-second. M65 has a total magnitude of 10.3, a size of 7.8 by 1.6 arc-minutes, and a mean surface brightness essentially the same, 21.7. NGC 3628 is more difficult, with a total magnitude of 10.3 and a size of 12 by 2 arc-minutes, yielding a fainter mean surface brightness, 22.4 magnitudes per square arc-second.

In the 8-inch, M66 shows an oval central region surrounded by the oval glow of the fainter spiral arms. The arms appear dimmer than the impression given by most photographs, no doubt because the central region is usually overexposed. The oval corresponding to the arms is oriented generally north – south, while that of the bright central region is more northwest – southeast. The southern arm looked pointed, an appearance familiar from photographs.

M65 showed a bright central region surrounded by the fainter spiral arms. The dark lane could not be seen in the 8-inch. NGC 3628 appeared considerably fainter than either M65 or M66 and was only slightly brighter in the center. The spiral arms faded gradually into the sky background, and no dark lanes could be seen during this observation. However, under excellent skies the dark lane so evident in the photograph is visible in an 8-inch telescope.

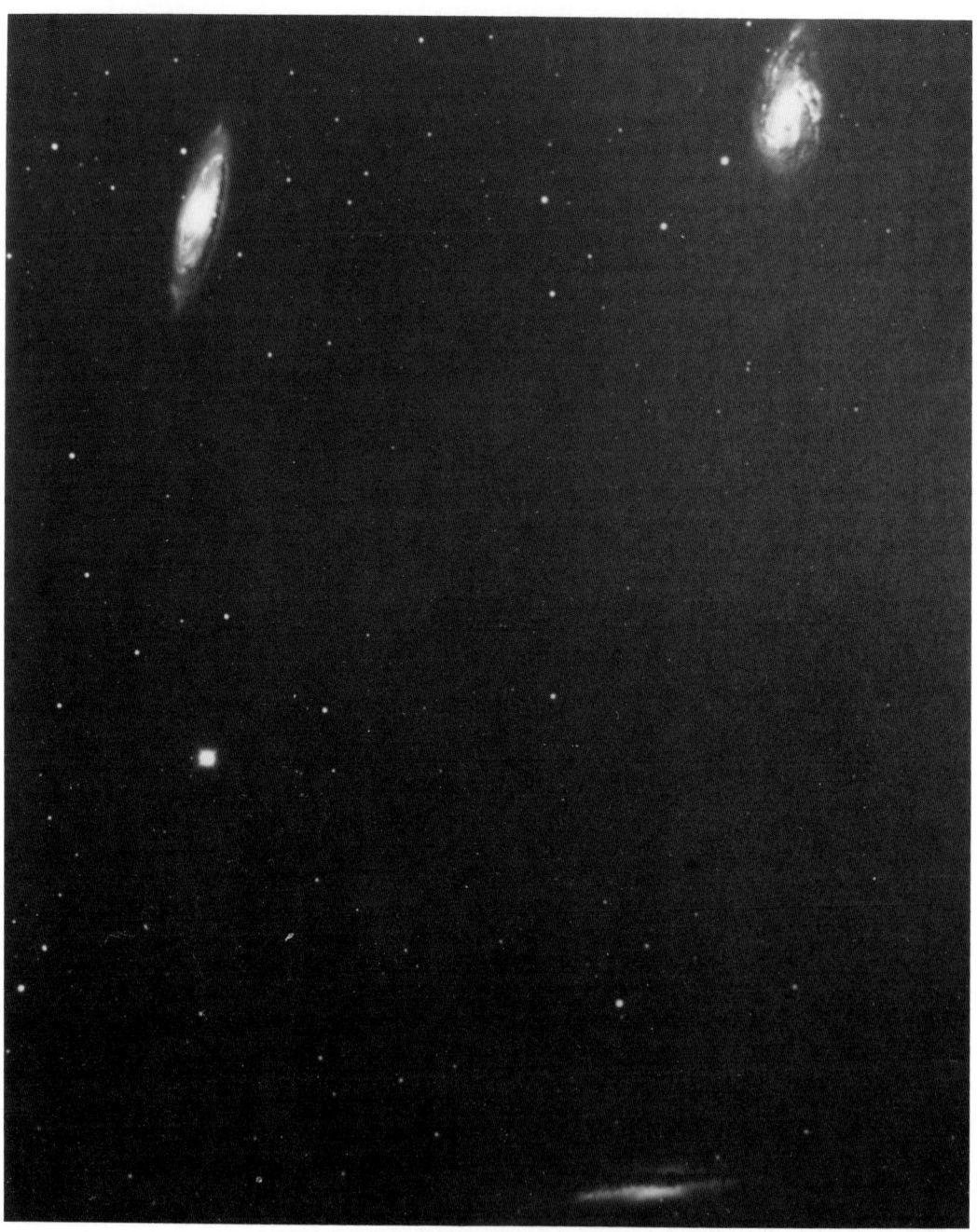

Photograph of M66 (at upper right), NGC 3628 (at bottom), and M65 (at upper left). South is up (Courtesy Canada–France–Hawaii Telescope Corporation.)

Drawing of M66 (at upper right), NGC 3628 (at bottom), and M65 (at upper left).

Scale: 2.3 arc-min/cm	Viewing Distance (cm)		air mass: 1.10, faintest star: 14.3 at zenith, 188×;
8-inch f/11.5 Cassegrain			no tracking
20mm Erfle (117×)	10×:115	100×:11	2/12/83 10:30–10:56 UT at Waianae ranch,
12.4mm Erfle	25×: 46	200×: 6	Hawaii; R. Clark
	50×: 23	300×: 4	

M109 (NGC 3992), GALAXY IN URSA MAJOR

R.A. 11^h 57.6^m, Dec. 53° 22' (2000.0)

Technical. NGC 3992 was added to Messier's catalog as "M109" as recently as 1953, by Owen Gingerich. It is a barred spiral galaxy of type SBb. A short bar extends through the central hub. At its ends are long, thin spiral arms that can be traced for about three-quarters of a turn.

Visual. M109 is magnitude 10.9 and 6.4 by 3.5 arc-minutes in size; the mean surface brightness comes to 22.9 magnitudes per square arc-second. Through the 8-inch under moderate skies, M109 appeared as a diffuse round glow. The central region appeared stellar. There was no hint of spiral structure.

A VISUAL ATLAS OF DEEP-SKY OBJECTS

Photograph of M109. South is up. (Courtesy Evered Kreimer, The Messier Album.)

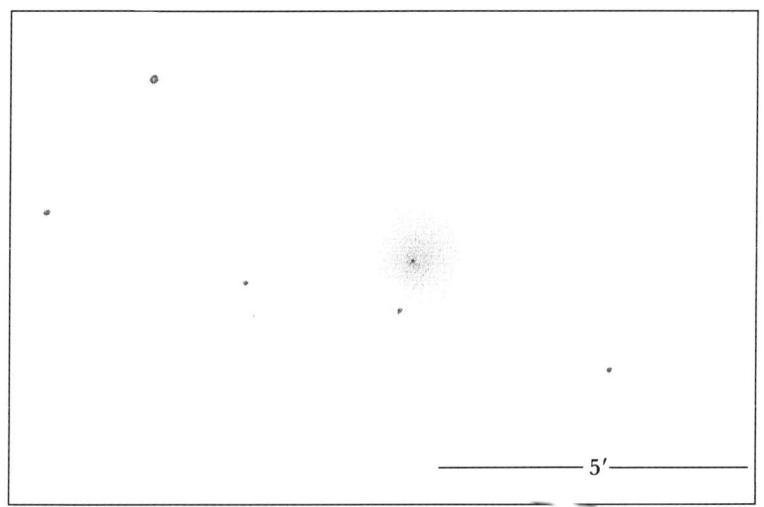

Drawing of M109.

Scale: 1.2 arc-min/cm	Viewing Distance (cm)	
8-inch f/11.5 Cassegrain		
20mm Erfle (117×)	25×:115	200×:14
12.4mm Erfle (188×,	50×:57	300×:10
best view)	100×:29	400×:7

air mass: 1.32, faintest star: 13.8 at zenith, 188×;
no tracking
5/15/83 9:05–9:16 UT at Barbers Point, Hawaii;
R. Clark

NGC 4038, NGC 4039, THE RINGTAIL GALAXY IN CORVUS

NGC 4038: R.A. $12^h 01.9^m$, Dec. $-18°52'$
NGC 4039: R.A. $12^h 01.9^m$, Dec. $-18°51'$
(2000.0)

Technical. NGC 4038 and 4039 are interacting galaxies. The pair is known as the Ringtail Galaxy or the Antennae because of two very long, thin "tails" that extend roughly north – south from the small central part. The distance to the pair is about 50 million light-years. Recent computer models indicate the two galaxies were quite normal before their encounter, and that the tails are stars and gas flung from the outlying regions of the original galaxies. The gas, enough to form about 1.5 billion suns, has been detected at radio wavelengths, being primarily hydrogen. The two galaxies have probably taken several hundred million years to reach their present state.

The galaxies' main portions show intricate detail. Both lack a nuclear region, and measurements of velocities show complex motions that at first seemed difficult to explain. However, each galaxy is rotating about its own axis, and the two are orbiting each other. Computer models can account for the motions when this combination of orbits is considered.

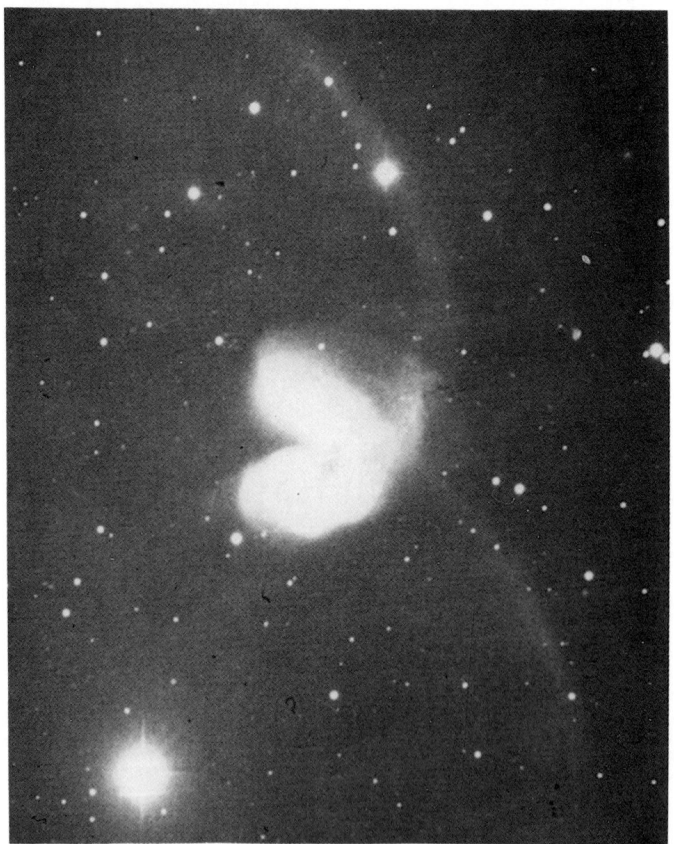

Photograph of NGC 4038 (the bottom portion of the irregularly shaped U) and NGC 4039 (the top portion). South is up. (Courtesy Palomar Observatory.)

A VISUAL ATLAS OF DEEP-SKY OBJECTS

Visual. NGC 4038-39 *look* unusual, quite unlike normal galaxies. They appear as a "U" shaped object with the open end to the west. The northern leg of the U appears larger. This is NGC 4038, magnitude 11.0, about 2.5 by 2.5 arc-minutes in size not including the tails. NGC 4039 is magnitude 12 and 2.5 by 2.0 arc-minutes not including tails. The mean surface brightness is 21.6 magnitudes per square arc-second for NGC 4038 and 22.4 for NGC 4039.

The parts seen in the 8-inch were somewhat smaller than the sizes above. Two "stars" were seen, one in each galaxy. They may be two of the many bright knots that appear on photographs. If so, better skies should reveal much of the intricate detail with an 8-inch. The tails are extremely faint, and it is questionable if they can be seen at all in medium size telescopes.

Drawing of NGC 4038 (the larger object) and NGC 4039 (the smaller, thin extension above it).

Scale: 1.2 arc-min/cm	Viewing Distance (cm)		air mass: 1.31, faintest star: 13.8 at zenith, 188×;
8-inch f/11.5 Cassegrain			no tracking
20mm Erfle (117×)	25×:115	200×:14	5/15/83 6:45–7:04 UT at Barbers Point, Hawaii;
12.4mm Erfle (188×,	50×: 57	300×:10	R. Clark
best view)	100×: 29	400×: 7	

M99 (NGC 4254), GALAXY IN COMA BERENICES

R.A. 12^h 18.8^m, Dec. 14° 25' (2000.0)

Technical. M99 is a type -Sc spiral in the great Virgo Cluster of Galaxies. It is probably 50 million light-years distant and has one of the greatest redshifts in the Virgo Cluster: 2300 kilometers per second. M99 was discovered in 1781 by Pierre Mechain. It is nearly face-on and has a well-defined spiral pattern. The bright southern arm extends westward unusually far from the nucleus. This was the second galaxy to be recognized as spiral, by Lord Rosse in 1848.

Visual. M99 appears at magnitude 10.4, with an angular size of 4.5 by 4.0 arc-minutes, yielding a low mean surface brightness of 22.2 magnitudes per square arc-second. In the 8-inch, the central portion appeared as a small, bright diffuse area, while the spiral arms formed a soft, uniform glow around it. No hint of spiral structure was seen under moderate to good skies. Detection of the bright southern spiral arm has been reported by observers under good to excellent skies with 8-inch telescopes.

Photograph of M99. South is up. (Courtesy Evered Kreimer, The Messier Album)

Drawing of M99.

Scale: 1.4 arc-min/cm
8-inch f/11.5 Cassegrain
20mm Erfle (117×)
12.4mm Erfle (188×, best view)
7mm Erfle (334×)

Viewing Distance (cm)	
25×:98	200×:12
50×:49	300×: 8
100×:25	400×: 6

air mass: 1.38, faintest star: 13.8 at zenith, 188×; no tracking
5/15/83 10:13–10:28 UT at Barbers Point, Hawaii; R. Clark

M106 (NGC 4258), GALAXY IN CANES VENATICI

R.A. 12^h 19.0^m, Dec. 47° 18' (2000.0)

Technical. NGC 4258 was added to Messier's catalog as "M106" by Helen Sawyer Hogg in 1947. The galaxy appears to have erupted about 20 million years ago. Its nucleus is unusually bright at both visible and radio wavelengths. The spiral arms terminate in knots of bright young stars. "Ghost" radio arms, unseen optically, trail behind the visible arms. Two clouds of material, amounting to several tens of millions of solar masses, appear to have been ejected from the nucleus in the galaxy's plane.

Visual. M106, magnitude 9.0, is 20 by 6.5 arc-minutes in size, though the brighter part is only 8 by 3 arc-minutes. The overall mean surface brightness is 22.9 magnitudes per square arc-second.

Through the 8-inch under moderate to good skies, this galaxy was quite a surprise. I was quite familiar with photographs of it and had seen it many times before, sometimes under good to excellent skies. But these observations were done at low powers before the research for this book was completed. The surprise was a distinct dark lane easily detected next to the nucleus on the west side. It is rarely seen on photographs because the central region is usually overexposed. The lane was best seen at powers near 200×. The spiral arm to the north was also seen, though its contrast was low. The arm ended in a point about 3 arc-minutes from the nucleus. Excellent skies or larger telescopes would surely show beautiful detail in this galaxy.

Photograph of M106. South is up. (Courtesy Palomar Observatory.)

Drawing of M106.

Scale: 2.3 arc-min/cm
8-inch f/11.5 Cassegrain
20mm Erfle (117×)
12.4mm Erfle (188×, best view)

Viewing Distance (cm)	
25×:60	200×:7
50×:30	300×:5
100×:15	400×:4

air mass: 1.24, faintest star: 13.8 at zenith, 188×; no tracking
5/15/83 9:18–9:31 UT at Barbers Point, Hawaii; R. Clark

M100 (NGC 4321), GALAXY IN COMA BERENICES

R. A. 12^h 22.9^m, Dec. 15° 49' (2000.0)

Technical. M100 is a spiral galaxy of class Sc seen nearly face-on. Its distance is estimated to be about 40 million light-years and its linear diameter slightly more than 100 000 light-years. The galaxy's total luminosity is about 20 billion times that of our Sun, its mass about 160 billion suns. There are two main spiral arms and complex dust lanes. The arms can be followed all the way in to the nucleus.

Photograph of M100. South is up. (Courtesy Evered Kreimer, The Messier Album.)

A VISUAL ATLAS OF DEEP-SKY OBJECTS

Visual. In small to medium telescopes, M100 appears as a bright central spot surrounded by a uniform, diffuse glow. The total magnitude is 10.4 and the diameter 5 arc-minutes. This corresponds to a mean surface brightness of 22.6 magnitudes per square arc-second. The main spiral arms are only 3 arc-minutes across, which is often as large as the galaxy appears. The arms beyond 3 arc-minutes are quite faint and low in contrast even under excellent skies.

In the 8-inch, no spiral structure could be detected under moderate to good skies. Some observers have reported hints of spiral arms in medium size telescopes.

Drawing of M100.

Scale: 1.2 arc-min/cm	Viewing Distance (cm)	air mass: 1.28, faintest star: 13.8 at zenith, 188×; no tracking
8-inch f/11.5 Cassegrain		
20mm Erfle (117×)	25×:115 200×:14	5/15/83 9:55–10:12 UT at Barbers Point, Hawaii; R. Clark
12.4mm Erfle (188×, best view)	50×: 57 300×:10	
	100×: 29 400×: 7	

M84 (NGC 4374), M86 (NGC 4406), AND 13 OTHER GALAXIES IN VIRGO

M84:   R.A. $12^h 25.1^m$, Dec. 12° 53'
NGC 4387: R.A. $12^h 25.7^m$, Dec. 12° 49'
NGC 4388: R.A. $12^h 25.8^m$, Dec. 12° 39'
NGC 4402: R.A. $12^h 26.1^m$, Dec. 13° 07'
M86:   R.A. $12^h 26.2^m$, Dec. 12° 56'
NGC 4413: R.A. $12^h 26.5^m$, Dec. 12° 36'
NGC 4425: R.A. $12^h 27.2^m$, Dec. 12° 44'
NGC 4435: R.A. $12^h 27.7^m$, Dec. 13° 04'
NGC 4438: R.A. $12^h 27.8^m$, Dec. 13° 00'
NGC 4458: R.A. $12^h 29.0^m$, Dec. 13° 15'
NGC 4459: R.A. $12^h 29.0^m$, Dec. 13° 58'
NGC 4461: R.A. $12^h 29.1^m$, Dec. 13° 11'
NGC 4473: R.A. $12^h 29.8^m$, Dec. 13° 25'
NGC 4477: R.A. $12^h 30.0^m$, Dec. 13° 38'
NGC 4479: R.A. $12^h 30.3^m$, Dec. 13° 35'
(2000.0)

Technical. These galaxies are near the core of the Virgo Cluster of Galaxies, which in turn is the center of the gigantic Local Supercluster. Our Galaxy and the Local Group of galaxies is near the edge of the supercluster. The Virgo Cluster is about 20 million light-years across and roughly 50 million light-years distant. It contains about 250 large galaxies and more than a thousand small ones.

M84 is an elliptical system about 25 000 light-years across with a mass of about 500 billion suns. M86 is another elliptical galaxy, but it seems somewhat out of place in the cluster. Its spectral lines show no red shift, but rather a blue shift! While M84 has a red shift indicating that it is receding from us at nearly 900 kilometers per second, M86 is approaching us at 400 kilometers per second. M86 may be escaping from the Virgo Cluster, or it could be close to us and not a cluster member at all.

Two galaxies very near M84 and M86 on the sky are the nearly edge-on spirals NGC 4388 and 4402. Attributes of all the galaxies observed are listed in Table 7.1.

Visual. Almost all these galaxies appear just as small fuzzy patches, either round or elliptical in the 8-inch telescope. Only NGC 4388 and 4402, the edge-on spirals, appeared very elongated. M84 and M86 appeared noticeably brighter toward their centers, an effect most pronounced in M86. NGC 4413 is a barred spiral; it showed a bright nucleus surrounded by a uniform glow corresponding to the spiral arms.

This collection of galaxies is quite interesting because several at a time can be seen at low to medium powers. The only difficulty is finding the region and identifying them, because there are few bright stars and many other galaxies to confuse the observer.

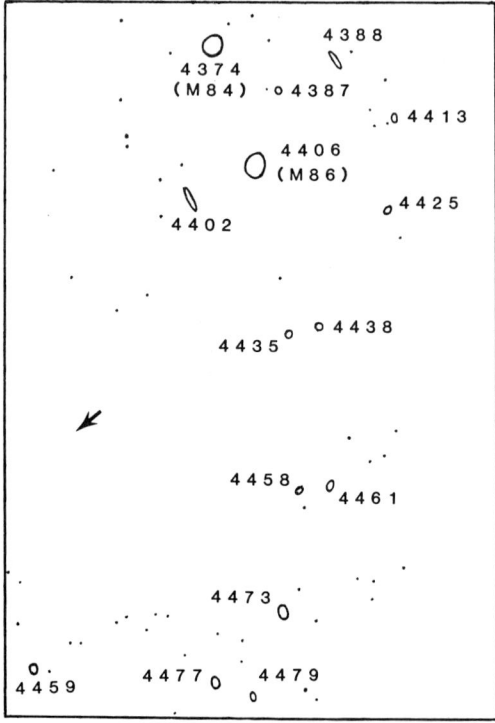

Figure 7.2. Finder chart for Virgo galaxies. The arrow indicates the direction of north.

A VISUAL ATLAS OF DEEP-SKY OBJECTS

Table 7.1. *Characteristics of some Virgo Galaxies*

Object	Galaxy type	Visual magnitude	Size arc-min	Surface brightness
NGC 4374 (M84)	E1	10.5	2.0 × 1.8	20.5
NGC 4387	E5	12.8	1.9 × 1.1	21.4
NGC 4388	SBc	12.0	5.0 × 1.0	22.4
NGC 4402	Sb	13.0	2.0 × 0.8	22.1
NGC 4406 (M86)	E3	10.5	3.0 × 2.8	21.4
NGC 4413	SBa	13.2	1.1 × 0.7	21.5
NGC 4425	S0	12.9	2.0 × 1.5	21.5
NGC 4435	E4	11.8	1.4 × 0.9	20.7
NGC 4438	Sa	11.0	4.0 × 1.5	21.6
NGC 4458	E0	12.0	1.9 × 1.8	22.0
NGC 4459	S0	11.7	1.5 × 1.0	20.8
NGC 4461	S0	12.2	2.0 × 1.0	21.6
NGC 4473	E4	11.3	2.0 × 1.0	20.7
NGC 4477	SBa	11.8	4.0 × 3.5	21.9
NGC 4479	SB0	12.6	1.5 × 1.5	22.0

Surface brightnesses are in magnitudes per square arc-second.

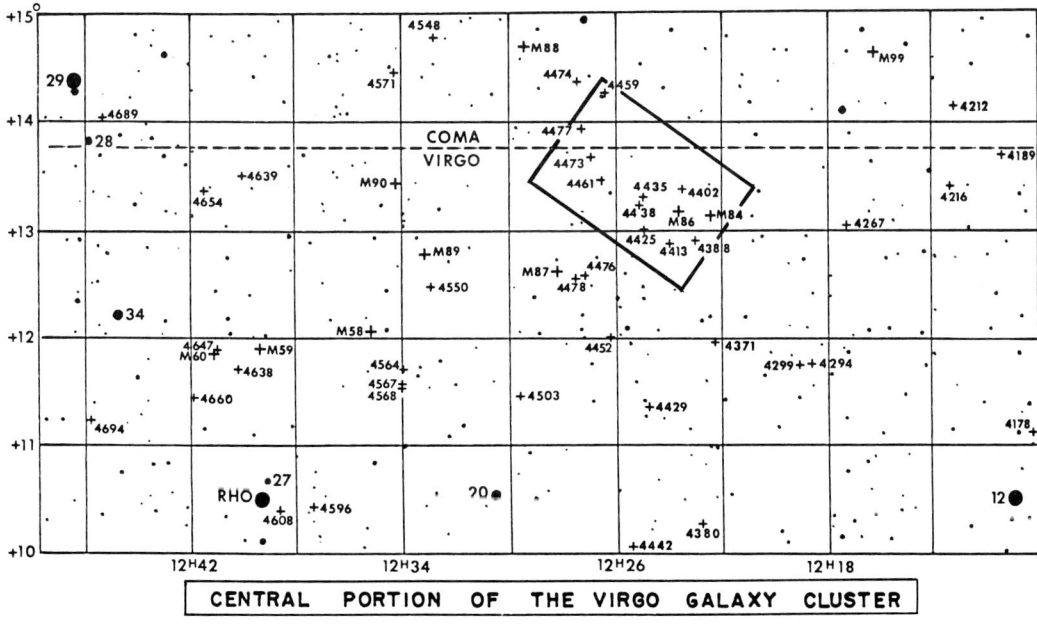

Figure 7.3. A wide-field finder chart for the Virgo Cluster region. The diagonal box indicates the area covered by the drawing and finder chart on the previous page. Adapted from *Burnham's Celestial Handbook* with permission.

Photograph of M84 (at top, just left of center) and M86 (the large object 3 centimeters below and slightly to the right of M84). North is to the lower left. A finder chart for the other galaxies is on page 146, and one for the whole inner Virgo Cluster region is on page 147. (Courtesy Akita Fujii.)

A VISUAL ATLAS OF DEEP-SKY OBJECTS

Drawing of M84, M86, and many other galaxies in Virgo. A finder chart is on page 146.

Scale: 5.4 arc-min/cm	Viewing Distance (cm)		air mass: 1.14, faintest star: 14.2 at zenith, 188×; no tracking
8-inch f/11.5 Cassegrain			
20mm Erfle (117×)	10×:64	100×:6	2/12/83 11:10–11:58 UT at Waianae ranch, Hawaii; R. Clark
	25×:25	200×:3	
	50×:13	300×:2	

NGC 4449, GALAXY IN CANES VENATICI
R.A. $12^h\ 28.2^m$, Dec. $44°\ 05'$ (2000.0)

Technical. NGC 4449 is an irregular galaxy in the approximate shape of a rectangle. On the northern corner is a small hook that points east, composed mostly of H II regions.

Visual. This galaxy appears 4.2 by 3.0 arc-minutes in size and has a total magnitude of 10.5. The mean surface brightness is 21.9 magnitudes per square arc-second. Through the 8-inch, the galaxy appeared as an elongated glow with a faint extension on the north side that corresponds to the hook in the photograph. Under some conditions at low powers with medium size telescopes, the galaxy may appear more rectangular.

Photograph of NGC 4449. South is up. (Courtesy of K.A. Brownlee and Deep Sky Magazine.)

A VISUAL ATLAS OF DEEP-SKY OBJECTS

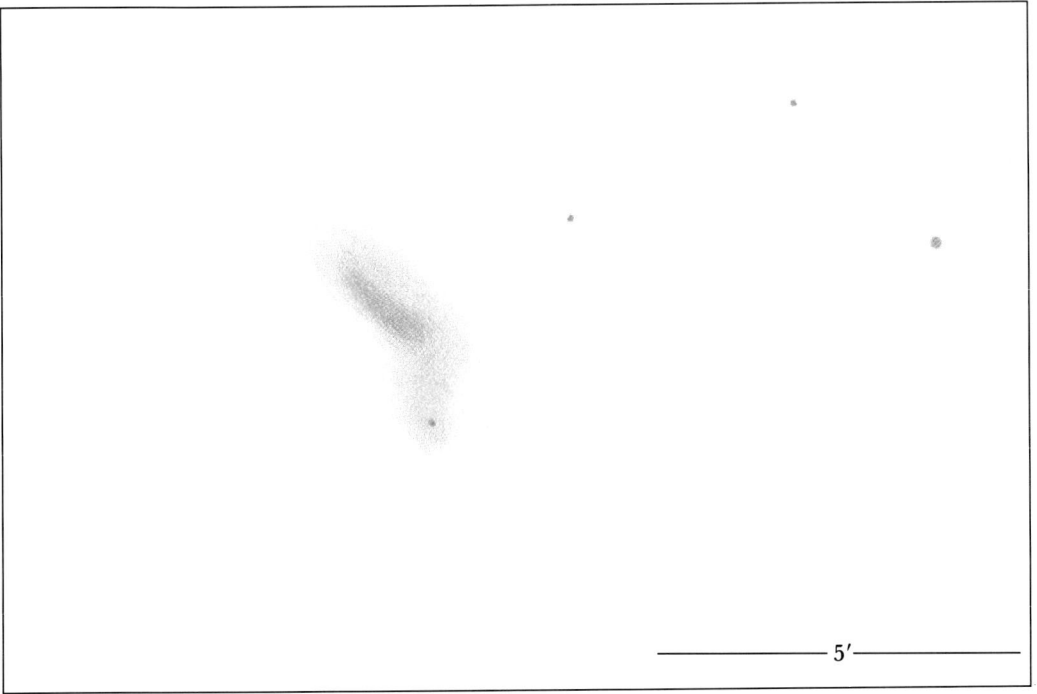

Drawing of NGC 4449.

Scale: 1.0 arc-min/cm
8-inch f/11.5 Cassegrain
12.4mm Erfle (188×)

Viewing Distance (cm)

25×:138   200×:17
50×: 68   300×:11
100×: 34 400×: 9

air mass: 1.16, faintest star: 14.3 at zenith, 188×;
no tracking
5/11/83 9:17–9:30 UT at Waianae ranch, Hawaii;
R. Clark

M87 (NGC 4486) NGC 4476, NGC 4478, GALAXIES IN VIRGO

NGC 4476: R.A. $12^h 30.0^m$, Dec. $12°20'$
NGC 4478: R.A. $12^h 30.3^m$, Dec. $12°19'$
M87:   R.A. $12^h 30.8^m$, Dec. $12°23'$
(2000.0)

Technical. M87 is a giant elliptical galaxy, the brightest member of the Virgo Cluster and one of the intrinsically brightest galaxies in the sky. It is also one of the most massive. By one estimate, M87 contains 3 trillion stars and dozens of times the mass of our Galaxy! An estimated 10 000 globular clusters surround this galaxy, over 800 of which have been detected by large observatory telescopes. In contrast, our Galaxy has only a couple hundred known globular clusters.

M87 is about 120 000 light-years across, not much more than many spirals. But since it is spherical, while spirals are flattened, M87 may have the largest volume of any known galaxy.

M87 also has a very violent nucleus. A high-velocity jet of matter is emerging from the nucleus toward the northwest. It consists of a line of at least six blobs of very hot gas (20 million kelvin) and has a strong magnetic field. The jet extends along the galaxy's axis of rotation. It is estimated to be 5000 light-years long and only 15 000 years old. A second, fainter jet was recently found emerging from the nucleus in the opposite direction. A third, broad fan also extends to one side. The whole area emits considerable radiation, from radio to X-ray.

Near M87's innermost core the stars are orbiting very rapidly, indicating that within 300 light-years of the center there is a very massive object that emits little or no light. This object, with an estimated mass of 5 billion suns, is usually presumed to be a black hole. Gas falling toward it would form an extremely hot and energetic "accretion disk" orbiting just outside the hole as it spiraled in. The accretion disk could be responsible for the creation of the jets.

The galaxies to the west of M87 are NGC 4478 and NGC 4476, both ellipticals that are small only in comparison to their giant neighbor.

Visual. M87 is magnitude 8.6 and about 3 arc-minutes in diameter. In this area the mean surface brightness is high, 19.6 magnitudes per square arc-second. The very faintest outer portions of the galaxy extend to 7 arc-minutes, and the globular clusters to 9 arc-minutes, though these outer parts cannot be detected visually through amateur telescopes.

The jet of M87 can be photographed with medium to large amateur telescopes but cannot be detected visually. The jet's total magnitude is 15, and it extends only about 20 arc-seconds from the bright center. Moreover, the jet is superimposed on essentially the brightest part of the galaxy, so the contrast is very low. A telescope a couple of meters in aperture is probably needed to view the jet! The only recorded visual observation of it is by Otto Struve with the 100-inch telescope on Mount Wilson, California.

In the 8-inch, M87 appears quite bright compared with other ellipticals in the area. It has a bright center and the intensity decreases uniformly to the edge, with no features.

NGC 4476 has a visual magnitude of 13.3, an apparent size of 0.7 by 0.4 arc-minutes, and a mean surface brightness of 20.5 magnitudes per square arc-second. NGC 4478 has a visual magnitude of 12.4, a size of 1.0 by 1.8 arc-minutes, and a mean surface brightness of 21.7 magnitudes per square arc-second.

In the 8-inch, NGC 4476 appeared brighter and smaller than NGC 4478, even though the latter is nearly a magnitude brighter. This is because the light of the latter is spread over a larger area. The mean surface brightness values indicate this effect, illustrating their value for detecting objects. Both galaxies appeared as small, faint, fuzzy spots. Medium to high powers are needed to see them.

Photograph of M87, the large galaxy at right. NGC 4478 is in the upper left, and NGC 4476 is not shown, being just off the left edge. South is up. (Courtesy Evered Kreimer, The Messier Album.)

A VISUAL ATLAS OF DEEP-SKY OBJECTS

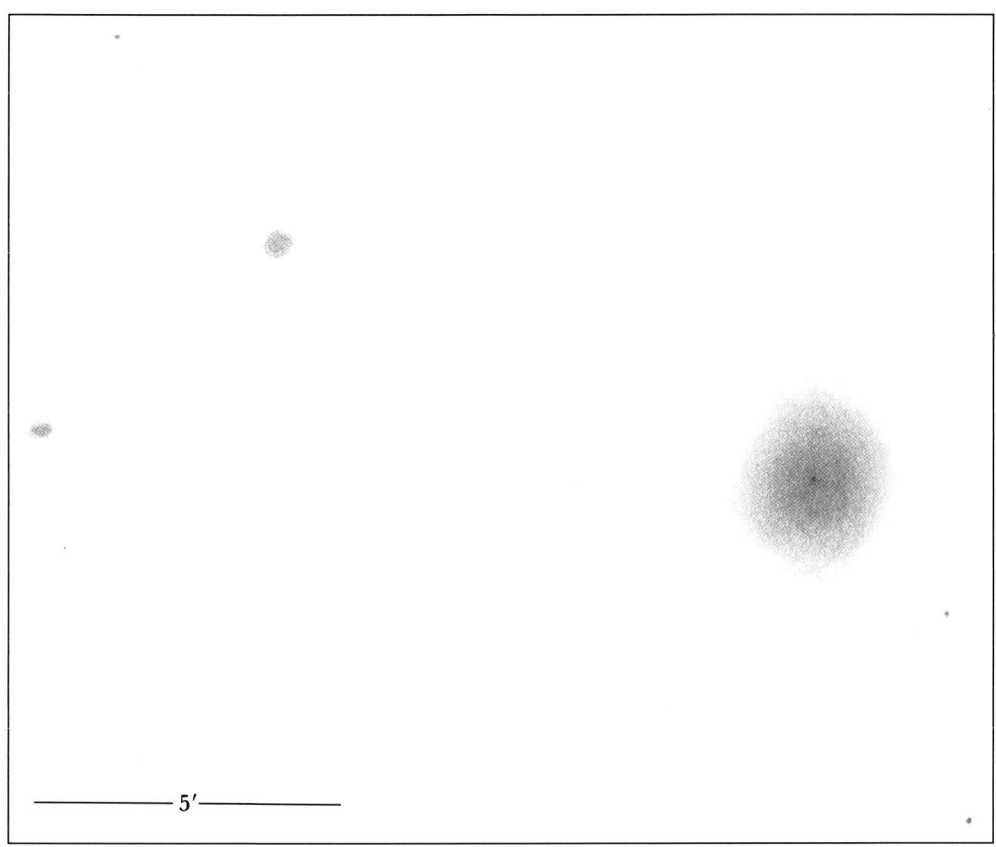

Drawing of M87 (the large galaxy at right), NGC 4478 (upper left), and NGC 4476 (near the left-center edge). The positions of NGC 4476 and the star at upper left are only approximate.

Scale: 1.2 arc-min/cm	Viewing Distance (cm)	
8-inch f/11.5 Cassegrain		
20mm Erfle (117×)	25×:115	200×:14
12.4mm Erfle (188×)	50×: 57	300×:10
	100×: 29	400×: 7

air mass: 1.46, faintest star: 14.3 at zenith, 188×;
no tracking
5/11/83 10:50–11:05 UT at Waianae ranch,
Hawaii; R. Clark

NGC 4565, EDGE-ON GALAXY IN COMA BERENICES

R. A. 12^h 36.4^m, Dec. 25° 59' (2000.0)

Technical. NGC 4565 is a marvelous galaxy that is oriented only 4° from perfectly edge-on. It is probably an outlying member of the Virgo Cluster. Its distance is thought to be about 20 million light-years, and its apparent length at that distance corresponds to about 90 000 light-years. Its total light is estimated to be about 3 billion times that of our Sun.

The entire length of the galaxy exhibits dark dust lanes. High-resolution photographs show beautiful, intricate detail in the dust lanes reminiscent of the festoons seen on Jupiter. The gravitational and magnetic fields in a spiral galaxy tug at the dust clouds, density waves compress them, and clouds collide with one another. This continual energy input causes upwellings that often reach hundreds of light-years above the galactic plane. As clouds fall back to the plane of the rotating galaxy, their paths form arcs that produce the festoon features. This is a common occurrence in spiral galaxies, but here we can view it especially clearly. The features in NGC 4565 are quite small in angular size, appearing no more than 5 arc-seconds wide and 10 to 15 arc-seconds long.

Visual. NGC 4565 is a spectacular sight in medium and large amateur telescopes. This galaxy is often overlooked by amateurs, but in my opinion it is one of the most beautiful in the sky, being more spectacular than M104, the commonly observed, nearly edge-on galaxy 37° to the south. The visual magnitude of NGC 4565 is 10.5, its apparent size is 15 by 1.1 arc-minutes, and its mean surface brightness is 22.2 magnitudes per square arc-second. Through the 8-inch in moderate to good skies, the dark lane was quite easily seen without averted vision by several observers at powers from 117 to 188×.

I had viewed this galaxy several times in poorer skies at lower powers without spotting the dark lane. At 188× in a 12.4-mm Erfle eyepiece, the galaxy stretched about halfway across the 23 arc-minute field of view. The nuclear region showed detail rarely seen in photographs, because they usually overexpose the center in order to show the faint outer parts; the nucleus itself appeared like a bright star. The intensity dropped fairly rapidly for the first 10 arc-seconds or so, then about 25 arc-seconds on each side of the nucleus, in the plane of the galaxy, two bright spots were seen that also rarely appear in photographs. Apparently, dust clouds darken the region between the nucleus and the spots. When this observation was made, none of the observers had seen photographs of the nuclear region, so the details were a pleasant discovery. More details are in the description on page 55 (Chapter 5).

A VISUAL ATLAS OF DEEP-SKY OBJECTS

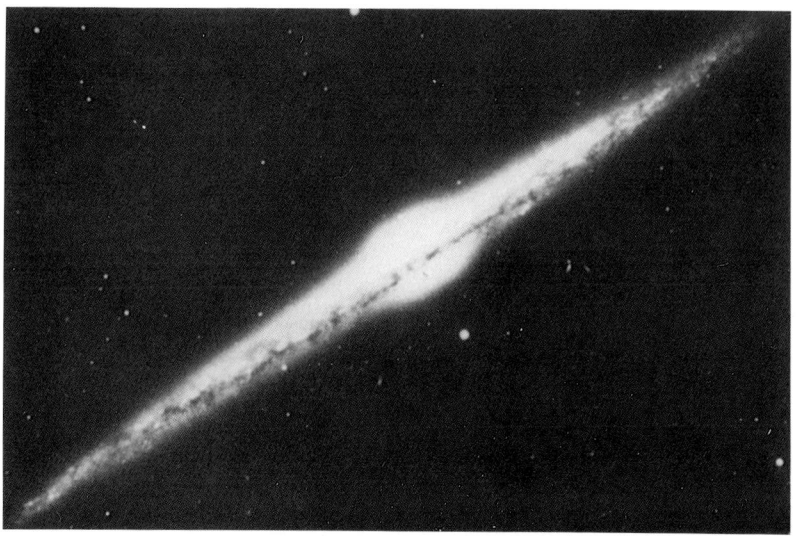

Photograph of NGC 4565. South is up. (Courtesy Palomar Observatory.)

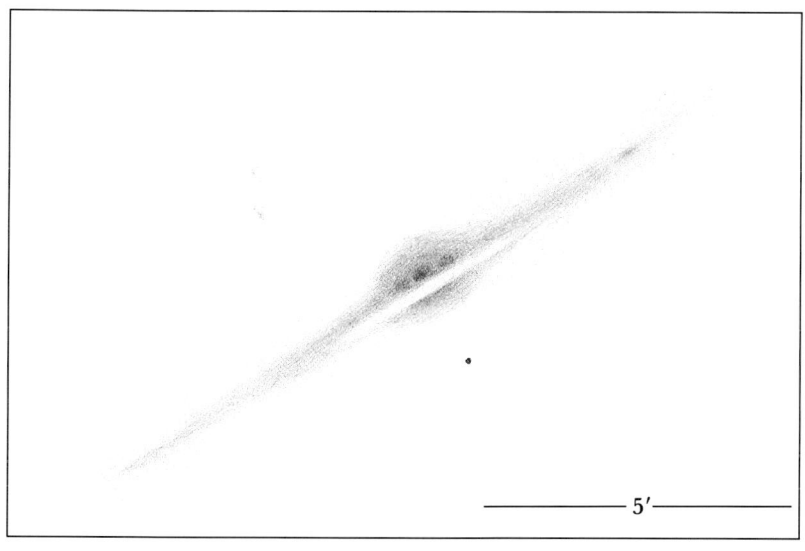

Drawing of NGC 4565.

Scale: 1.2 arc-min/cm	Viewing Distance (cm)	
8-inch f/11.5 Cassegrain		
20mm Erfle (117×)	25×: 115	200×: 14
12.4mm Erfle (188×,	50×: 57	300×: 10
best view)	100×: 29	400×: 7

air mass: 1.21, faintest star: 14.3 at zenith, 188×;
no tracking
5/11/83 10:00–10:42 UT at Waianae ranch,
Hawaii; R. Clark

M90 (NGC 4569), GALAXY IN VIRGO
R.A. 12^h 36.8^m, Dec. 13° 09' (2000.0)

Technical. M90 is a spiral galaxy in the Virgo Cluster. Its distance is about 40 million light-years and its diameter about 80 000 light-years. The galaxy has been estimated to weigh about 80 billion solar masses. It was discovered by Charles Messier in 1781.

Visual. M90 appears at magnitude 9.0, with an apparent diameter of 7 by 2.5 arc-minutes. The mean surface brightness is 20.7 magnitudes per square arc-second. The nucleus appears like an 11th-magnitude star, surrounded by the galaxy's faint oval glow. No details were seen in the 8-inch under moderate skies, though numerous dark dust lanes appear near the bright central region on

Photograph of M90. (Courtesy Evered Kreimer, The Messier Album.)

photographs. A large amateur telescope or better skies may begin to show these. Only the innermost 4.5 by 0.9 arc-minutes of the galaxy was seen; the outer spiral arms are very faint.

Drawing of M90.

Scale: 1.2 arc-min/cm
8-inch f/11.5 Cassegrain
20mm Erfle (117×)
12.4mm Erfle (188×, best view)

Viewing Distance (cm)

25×: 115   200×: 14
50×: 57   300×: 10
100×: 29   400×: 7

air mass: 1.38, faintest star: 13.8 at zenith, 188×; no tracking
5/15/83 10:31–10:41 UT at Barbers Point, Hawaii; R. Clark

M104 (NGC 4594), THE SOMBRERO GALAXY IN VIRGO

R.A. 12^h 39.9^m, Dec. $-11°$ $37'$ (2000.0)

Technical. M104 is a beautiful spiral galaxy tipped only 6° from edge-on. It is probably at a distance of 40 million light-years and a member of the Virgo Cluster. This is thought to be an especially massive galaxy, with about 1.3 trillion solar masses. The diameter of the main part is 80 000 light-years, though faint outer portions extend to 130 000 light-years.

M104 played a key role in support of the "Island Universe" theory. Early in this century its red shift was measured at 1100 kilometers per second. That was the highest then known anywhere in the universe, and it indicated that M104 was unlikely to be a small object within our Galaxy. Subsequent studies of other "spiral nebulae" showed their true extragalactic nature.

Visual. M104 appears at magnitude 8.2, with an angular size of 7.0 by 1.5 arc-minutes. The mean surface brightness is high, 19.4 magnitudes per square arc-second, suggesting that this galaxy is a good target under poor skies. Through the 8-inch telescope under moderate skies, the dark lane was easily seen at medium to high powers. M104 appears very similar visually to its photographs, but this is partly an illusion. Compare the size of the drawing with the photo. The central portion of the galaxy appears like a miniature of the whole thing. Only the brighter central part can be seen visually, while the fainter outer portions are brought out by long-exposure photography. A similar effect is seen with many globular clusters.

A VISUAL ATLAS OF DEEP-SKY OBJECTS

Photograph of M104. South is up. (Courtesy National Optical Astronomy Observatories.)

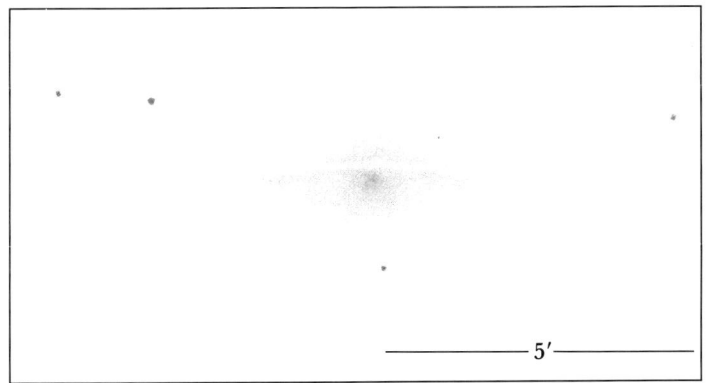

Drawing of M104.

Scale: 1.2 arc-min/cm
8-inch f/11.5 Cassegrain
20mm Erfle (117×)
12.4mm Erfle (188×, best view)

Viewing Distance (cm)

25×: 115 200×: 14
50×: 57 300×: 10
100×: 29 400×: 7

air mass: 1.25, faintest star: 13.5 at zenith, 188×;
no tracking
1/13/83 14:20–14:43 UT at Hawaii Kai, Hawaii;
R. Clark

M94 (NGC 4736), GALAXY IN CANES VENATICI

R.A. 12^h 51.0^m, Dec. 41° 07' (2000.0)

Technical. M94 is a violently active galaxy with tightly wound spiral arms. Although it looks rather normal, there is evidence of an explosion that may have occurred as recently as 10 million years ago. Some astronomers speculate that all spiral galaxies have periodic violent outbursts. Thus the so-called violent galaxies may actually be quite normal. (See also the description of M106, another of this class.)

There are two sets of spiral arms. The inner set contains many irregular dust lanes, which are usually lost in deep-sky photographs owing to overexposure. The outer arms are much fainter and rich in dust. A ring of material beyond them extends 15 arc-minutes from the nucleus.

Visual. M94 appears at magnitude 8.9 and has a size of 5.0 by 3.5 arc-minutes, yielding a mean surface brightness of 20.6 magnitudes per square arc-second. That size corresponds to the faint outer arms, which may be un-

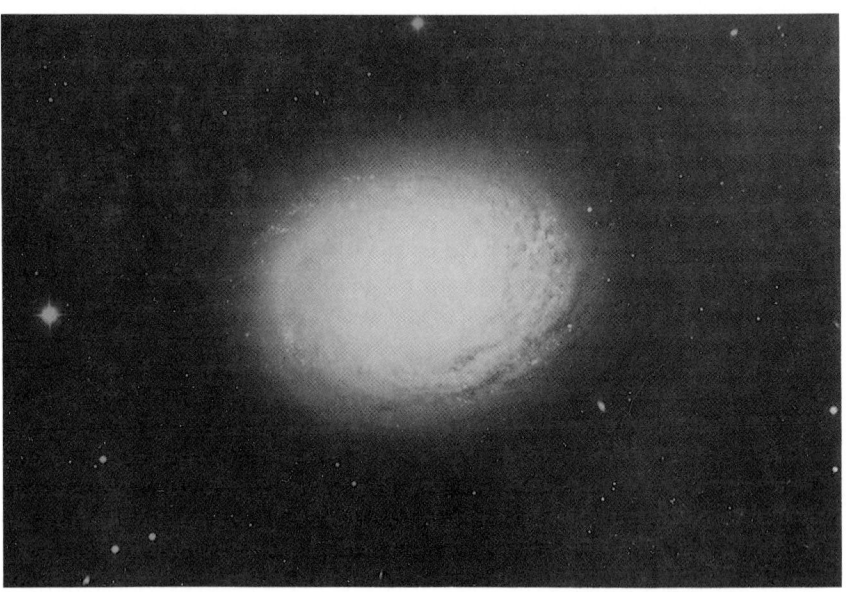

Photograph of M94. South is up. (Courtesy Palomar Observatory.)

A VISUAL ATLAS OF DEEP-SKY OBJECTS

observable under most conditions. The brighter inner arms are about 2 by 1.5 arc-minutes, and the mean surface brightness in this region is a magnitude or more brighter.

The high surface brightness helps in the detection of detail in this galaxy. Through the 8-inch under only moderate skies and at powers near 200×, the inner spiral arms were seen as a ring surrounding the nucleus. The nucleus is very bright compared to the arms, but does not appear stellar, being 30 arc-seconds across.

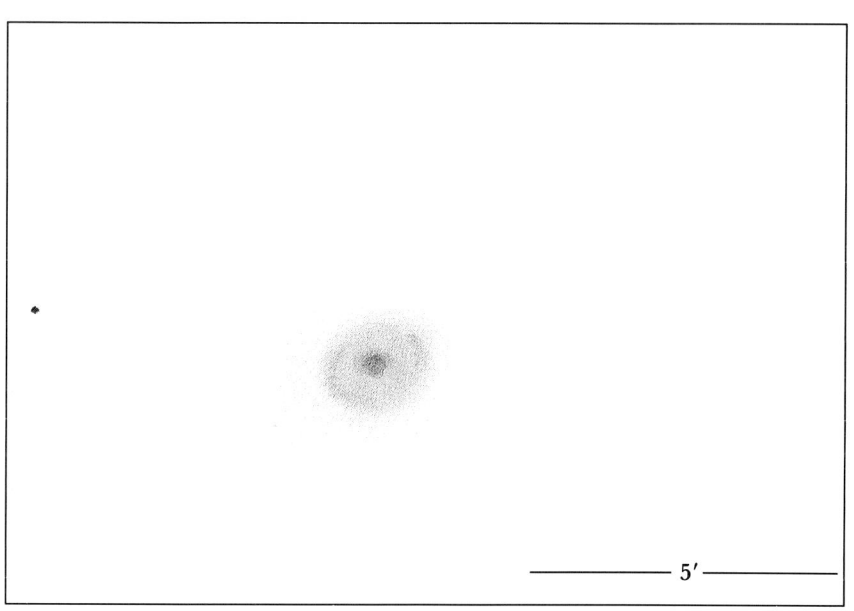

Drawing of M94.

Scale: 1.2 arc-min/cm
8-inch f/11.5 Cassegrain
12.5mm Orthoscopic
(187×)

Viewing Distance (cm)

25×:115 200×:14
50×: 57 300×:10
100×: 29 400×: 7

air mass: 1.31, faintest star: 13.3 at zenith, 188×;
no tracking
6/22/82 8:00–8:15 UT at Hawaii Kai, Hawaii; R. Clark

M64 (NGC 4826), THE BLACK EYE GALAXY IN COMA BERENICES

R.A. 12^h 56.8^m, Dec. 21° 41' (2000.0)

Technical. M64 is an unusual spiral galaxy. The outer spiral arms are very smooth in appearance on the finest observatory photographs, indicating a lack of large dust clouds or bright star-forming regions. Nor are there any concentrations of stars in the arms. Galaxies with this characteristic are called anemic. What makes the lack of dust in the outer arms particularly strange is that near the nucleus, about one-quarter of the way to the outer edge, is an extremely large and thick dark dust lane. This is the "Black Eye".

M64, discovered by Johann Bode in 1779, is one of the 12 brightest galaxies in the sky. It is somewhat closer than the Virgo Cluster of Galaxies and is about 65 000 light-years in diameter.

Visual. M64 is an interesting object, but many amateurs have found detection of the black-eye feature difficult. Nevertheless, amateur astronomer John Mallas saw the eye in 4-inch and 2.4-inch telescopes. Failure to detect the dark spot is probably due most often to – once again – inadequate magnification. M64 has a visual magnitude of 8.6 and a size of 7.5 by 3.5 arc-minutes. The mean surface brightness is 20.8 magnitudes per square arc-second.

Through the 8-inch in only moderate skies, the spot was visible at powers over 100×. Closer to 50× it was hard to see. The galaxy's nucleus is nearly stellar, and the arms fade smoothly into the sky background.

A VISUAL ATLAS OF DEEP-SKY OBJECTS

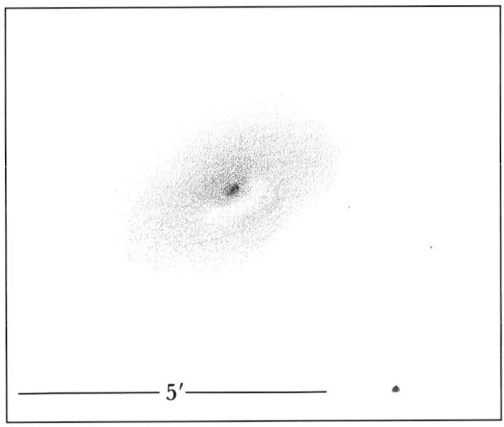

Photograph of M64. South is up. (Courtesy Martin Germano.)

Drawing of M64.

Scale: 1.2 arc-min/cm
8-inch f/11.5 Cassegrain
20mm Erfle (117×)

Viewing Distance (cm)

25×:115   200×:14
50×: 57   300×:10
100×: 29 400×: 7

air mass: 1.16, faintest star: 13.3 at zenith, 188×; no tracking
6/22/82 7:30–7:40 UT at Hawaii Kai, Hawaii; R. Clark

M63 (NGC 5055), GALAXY IN CANES VENATICI

R.A. 13^h 15.7^m, Dec. 42° 01' (2000.0)

Technical. M63 has been called the Sunflower Galaxy. Its tightly coiled spiral arms, tilted about 30° from edge-on, enclose a nucleus about 6 arc-seconds in diameter. The spiral arms appear as two parts: bright and tightly coiled out to a distance of 50 arc-seconds, and fainter, looser arms out to nearly 4.5 arc-minutes. The outer arms contain many bright knots of stars intermingling with dust lanes. These arms have been described as sparks thrown out by a fiery pinwheel. M63 is estimated to be 30 or 40 million light-years away. It is in the class of multiple-arm spirals, which have many more than the typical two to four arms.

A supernova was discovered in M63 by J. Golly in May 1971, and was found independently by myself four days later. I was using an 8-inch Newtonian telescope, and the supernova, a Type I, was magnitude 12.1. Having observed the galaxy before, I thought a star 3 arc-minutes south of the nucleus seemed out of place. This casual familiarity with the field paid off; the supernova was confirmed by the Smithsonian Astrophysical Observatory a few days later. Making drawings such as those in this book is an asset to finding novae and supernovae, since the observer gains experience with the field as well as a record of "permanent" stars in the area.

Visual. M63 is magnitude 9.8 and 7 by 3.5 arc-minutes in size, with the faintest portions reaching out to 9 by 4 arc-minutes. The mean surface brightness is 22.3 magnitudes per square arc-second, though the inner portions are much brighter. The nuclear region appears quite small. No detail could be seen through the 8-inch under good skies. The inner spiral arms appeared as a region nearly uniform in brightness; the outer-arm region faded uniformly into the sky background, with a maximum extent of about 5.5 by 2.0 arc-minutes. Through a 16-inch telescope in excellent skies, many bright knots could be seen at 250× and the whole galaxy was as big as 7 by 3.5 arc-minutes.

A VISUAL ATLAS OF DEEP-SKY OBJECTS

Photograph of M63. South is up. (Courtesy Evered Kreimer, The Messier Album.)

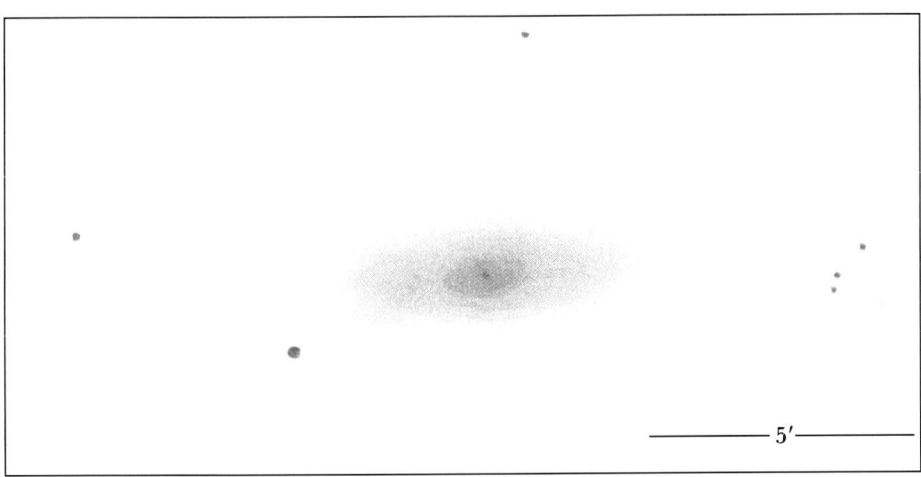

Drawing of M63.

Scale: 1.4 arc-min/cm	Viewing Distance (cm)	
8-inch f/11.5 Cassegrain		
20mm Erfle (117×)	25×:98	200×:12
12.4mm Erfle (188×,	50×:49	300×: 8
best view)	100×:25	400×: 6

air mass: 1.08, faintest star: 14.3 at zenith, 188×; no tracking
5/11/83 9:00–9:15 UT at Waianae ranch, Hawaii; R. Clark

NGC 5128, PECULIAR GALAXY IN CENTAURUS

R.A. 13^h 25.3^m, Dec. $-43°$ $01'$ (2000.0)

Technical. NGC 5128 is a very unusual giant elliptical galaxy; it has an enormous, complex dust band across its center. Recently astronomers pieced together clear evidence of what they have long suspected: NGC 5128 is the result of the collision and merger of two galaxies, an elliptical and a spiral.

This object is one of the strongest radio sources in the sky; radio astronomers know it as Centaurus A. The radio emissions come from a huge area, 6° by 10°. At the estimated distance of 16 million light-years, this corresponds to a width of 3 million light-years! High-resolution radio telescopes have yielded dramatic images of this source: two great, smoky-looking lobes of gas blown into intergalactic space by two thin jets coming from opposite sides of the galaxy's nucleus.

One of the jets has been found optically; its visible portion is 130 000 light-years long, consisting of blobs and streaks. It is very much like the jet from the nucleus of M87 (see page 153), and like it may be streaming from the face of an accretion disk surrounding a supermassive black hole. The gas upon which the hole is "feeding" may have been diverted in the course of the two-galaxy collision. NGC 5128 is also a source of X- and gamma rays.

Visual. NGC 5128 is extremely fascinating. It is big and bright, as galaxies go, with a total visual magnitude of 7.2 and a size of 10 by 8 arc-minutes. The mean surface brightness is 20.6 magnitudes per square arc-second. Unfortunately for Northern-Hemisphere observers, it is far south at declination $-43°$. It rises only 7° above the horizon as seen from latitude 40° north; for a decent view of its details it should be at least at 20° altitude. So the observer ought to be at least as far south as 27° latitude (Florida, Texas) and should plan to catch the galaxy when it is on the meridian.

The observations through the 8-inch were made in Hawaii at latitude 21° north, and even then the air mass was a very thick 2.41. Nevertheless the galaxy was awe-inspiring. It appeared about 5 arc-minutes in diameter, and the dark band was obvious at all powers. The band should be easy in small amateur telescopes, since it is a full arc-minute across at its narrowest point. Through the 8-inch, a bright patch could be seen within the dark lane on the west side. This galaxy must be even more spectacular from the Southern Hemisphere, where it can be seen overhead.

Photograph of NGC 5128. South is up. (Courtesy David Malin, Anglo-Australian Telescope Board.)

A VISUAL ATLAS OF DEEP-SKY OBJECTS

Drawing of NGC 5128.

Scale: 1.2 arc-min/cm
8-inch f/11.5 Cassegrain
20mm Erfle (117×)
12.4mm Erfle (188×, best view)
7mm Erfle (334×)

Viewing Distance (cm)

25×:115   200×:14
50×: 57   300×:10
100×: 29 400×: 7

air mass: 2.41, faintest star: 13.8 at zenith, 188×;
no tracking
5/15/83 7:15–7:40 UT at Barbers Point, Hawaii;
R. Clark

NGC 5139 (OMEGA CENTAURI), THE GREAT GLOBULAR CLUSTER IN CENTAURUS

R.A. $13^h 26.8^m$, Dec. $-47° 29'$ (2000.0)

Technical. Omega Centauri is without a doubt the grandest globular cluster in the sky. It has the largest true diameter of any yet measured (620 light-years) and also happens to be one of the closest known (16 500 light-years). It is about 15 billion years old, roughly the age of our entire Galaxy. Omega Centauri contains an estimated one million stars. It is one of the most massive clusters known, with about 500 000 times the mass of our Sun. In addition, it is one of the fastest rotating globulars. As a result it is decidedly elliptical, about 25 percent longer on one axis. We are fortunate to have it as such a nearby neighbor.

Like other globular clusters, Omega Centauri is deficient in heavy elements and contains no measurable gas or dust. Since the cluster's stars are nearly as old as the Universe, they formed out of material relatively fresh from the Big Bang: hydrogen and helium, with only small traces of heavy elements that were synthesized inside previous generations of stars.

The lack of gas and dust is explained by the cluster's orbit. A detailed study of the motions of individual stars in the cluster allowed an accurate determination of the entire cluster's motion and hence its orbit around and through our Galaxy. Omega Centauri follows a highly elliptical orbit around the center of our Galaxy that brings it to within 6 200 light-years of the galactic nucleus and as far away as 21 000 light-years. Its many passes through the plane of our Galaxy have swept it clean of gas. This orbit is typical of other globulars. Since they have no interstellar gas, they have not been able to produce new stars for billions of years.

Omega Centauri was catalogued by Claudius Ptolemy over 1,800 years ago. In 1603 it was again listed as a star, this time by Johann Bayer, who gave it the Greek letter Omega in his sky atlas *Uranometria*. Edmond Halley seems to have been the first to recognize the object as a cluster, in 1677. When high in the sky, it definitely appears larger than a star and quite fuzzy to the naked eye.

Visual. Omega Centauri is beyond compare. Its total visual magnitude of 3.65 makes it the brightest globular in the sky. With a diameter of 30 arc-minutes, the mean surface brightness is 19.6 magnitudes per square arc-second. Only the cluster 47 Tucanae far to the south can come close to its visual splendor.

Through the 8-inch telescope under moderate skies, Omega Centauri was awe-inspiring even when very low in the sky. At powers of more than 100× it was resolved to the center. Countless stars filled the field of medium-power Erfle eyepieces.

The star density increases only slowly toward the center. The cluster's bright central portion is 6 to 7 arc-minutes across, and the maximum extent observed through the 8-inch was about 13 arc-minutes. On deep-sky photographs Omega Centauri is well over one degree in diameter, though visually when high in the sky it appears about 30 arc-minutes across.

East of the center are two U-shaped strings of stars with the Us connected at their bottoms. Farther east, and above and below the double U, are two strings of stars that extend farther eastward still. The southernmost is an arc about 3.5 arc-minutes long. In the accompanying drawing, only a few of the brightest stars are in their geometrically correct positions. Plotting thousands of stars by eye is hardly feasible!

The brightest stars in Omega Centauri are of 11th magnitude, so the cluster can be at least partially resolved with telescopes as small as 2 inches, when it is high in the sky. These bright stars are red giants, and their color may be detectable in large amateur telescopes. Omega Centauri contains over 160 known variable stars, second only to M3.

The well-known globular M13 (NGC 6205) in Hercules appears miniscule in comparison with Omega Centauri. From low northern latitudes, such as in Hawaii, M13 and Omega Centauri are above the horizon at the same time, so the two can be compared. Such a comparison can be made here: the drawing of Omega Centauri and that of M13 on page 186 are at the same scale. Note that M13 appeared only half the size of Omega Centauri through the 8-inch – even though the latter was very low in the sky, while the former was observed nearly overhead. This is not intended to denigrate the splendor of M13, but rather to put it in perspective for northern observers for whom Omega Centauri is one of those tantalizing, legendary objects on or below the southern horizon. Since M13 is called the Great Globular Cluster in Hercules, perhaps Omega Centauri should be named the "Super Great Cluster in Centaurus".

Photograph of NGC 5139 (Omega Centauri). South is up. (Courtesy National Optical Astronomy Observatories.)

A VISUAL ATLAS OF DEEP-SKY OBJECTS

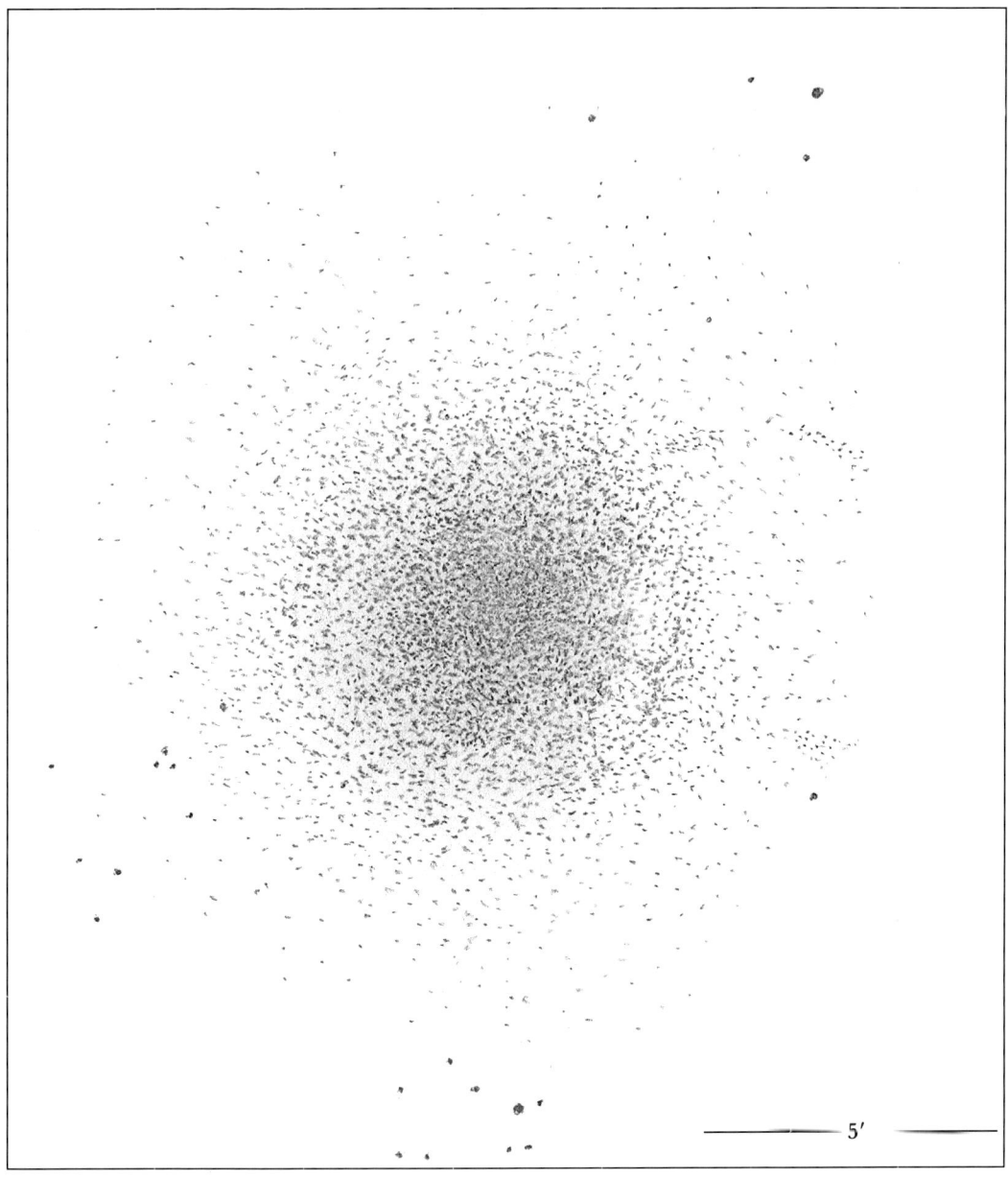

Drawing of NGC 5139 (Omega Centauri).

Scale: 1.2 arc-min/cm	Viewing Distance (cm)	
8-inch f/11.5 Cassegrain		
20mm Erfle (117×)	25×: 115	200×: 14
12.4mm Erfle (188×, best view)	50×: 57	300×: 10
	100×: 29	400×: 7
7mm Erfle (334×)		

air mass: 2.78, faintest star: 13.8 at zenith, 188×;
no tracking
5/15/83 7:40–8:20 UT at Barbers Point, Hawaii;
R. Clark

M51 (NGC 5194), THE WHIRLPOOL GALAXY IN CANES VENATICI NGC 5195

M51:   R.A. $13^h 29.9^m$, Dec. 47° 12'
NGC 5195: R.A. $13^h 30.0^m$, Dec. 47° 16'
    (2000.0)

Technical. M51 was discovered by Charles Messier in 1773, and it was the first galaxy recognized as having spiral characteristics, by Lord Rosse in 1845. It and NGC 5195 are a pair of interacting galaxies. M51 is estimated to be three times as massive as its companion, or about 100 billion times as massive as our Sun. This pair of galaxies is only about 13 million light-years away.

M51 is tilted 60° from edge on, but is somewhat distorted because NGC 5195 has stretched it into an elliptical shape. The small galaxy is passing the larger one, with tidal forces modifying the appearance of both. Computer simulations have been able to reproduce the structure of each galaxy. NGC 5195 passed M51 many millions of years ago and has bent one spiral arm away from its normal spiral pattern; this is the arm that appears to trail after the smaller galaxy.

In photographs the spiral pattern of M51 is nothing less than spectacular, highlighted by complex star clouds and bright and dark nebulosity. Some of the highest-resolution photographs even show a few individual supergiant stars. On the inner side of the spiral arms are long, thin, dark dust lanes showing great complexity. The arms can be traced for about one-and-one-half turns, and can be detected as close as 15 arc-seconds from the nucleus. The nucleus appears about 2.7 arc-seconds, or approximately 500 light-years, in diameter.

NGC 5195 has a bright nucleus no more than 140 light-years across, and radio waves are emitted from a region several times larger than this. Apparently many hot young stars populate the nucleus, intermingled with gas heated to 30 000 °C and dust. This galaxy is classed as an irregular, though it is believed to be evolving into a barred spiral.

Visual. The M51 system can be an excellent sight in amateur telescopes, and it may show its spiral structure. M51 is visual magnitude 8.1 and 10 by 5.5 arc-minutes in diameter, while NGC 5195 is magnitude 11.0 and 2.0 by 1.5 arc-minutes in size. Here again, mean surface brightness is a better indication of the visibility of the two objects. M51 has a mean surface brightness of 21.1, and NGC 5195 20.8 magnitudes per square arc-second. Although NGC 5195 is 3 magnitudes fainter, its slightly higher mean surface brightness makes it seem about the same brightness and size as the central region of M51.

The visibility of M51's spiral structure in amateur telescopes is analyzed in detail in Chapter 6. The smallest telescope required is about 6 inches. Amateur John Mallas reported hints of spiral structure in a 4-inch refractor, though he remarked that the observation may be spurious since he was quite familiar with photographs of the object.

The view through the 8-inch telescope is shown here for three different observing conditions and levels of observer experience. The first drawing, made in 1983, was with good skies after the research for this book was complete. The detail that could be seen with careful observation was phenomenal. Some details have never been reported before with medium-size amateur telescopes so far as I know. The two spiral arms were reasonably easy at all powers. At high power (334×) a hub was visible around the central core and was slightly brighter at its east side, appearing like a rim. After observing this galaxy many times without detecting detail in the companion, it came as a surprise to see structure in NGC 5195. It had a bright, starlike nucleus, but at 188× and higher, there was a decrease in brightness and then an *increase* away from the nucleus. Apparently a dark dust lane surrounds the nucleus. The next day I searched observatory photographs for this feature. This central part of the galaxy, as well as the hub of M51, are commonly overexposed. When a properly exposed photograph was found, the dust lane was quite apparent. It is quite satisfying to find something in the sky that one had not seen before, and then prove the observation was not spurious.

Compare that drawing with the one made in 1971. Less was seen then even though the sky was better (the limiting magnitude was at least a half magnitude fainter) and my observing experience was already good. But the concepts in this book had not been realized at that time. First notice that the highest power used in 1971 is the *lowest* used in 1983. That had a large effect on what was seen. The observing time was substantial in both cases, but in 1971 no details were detected in the companion galaxy or the central hub. The two main spiral arms were seen in 1971 but were more distinct in the 1983 observation.

Under moderate or poorer skies, a medium-size telescope may not show even a hint of spiral structure. Such a view is shown in the third drawing, made on June 20, 1982. The contrast in such skies is so low that no matter what power is used, virtually no detail can be seen. A hint of spiral structure was possibly seen, but higher powers showed the same or less, even in the brighter central regions. A larger aperture in such skies *would* show more detail – but a small telescope in the country often outperforms a large one in the city. Sky quality, not telescope size, is the most important factor for seeing detail in deep-sky objects.

Under good to excellent skies, large amateur telescopes will begin to show dark lanes as well as bright knots in the spiral arms.

Photograph of M51 (the main galaxy), with its smaller companion NGC 5195. South is up. (Courtesy National Optical Astronomy Observatories.)

A VISUAL ATLAS OF DEEP-SKY OBJECTS

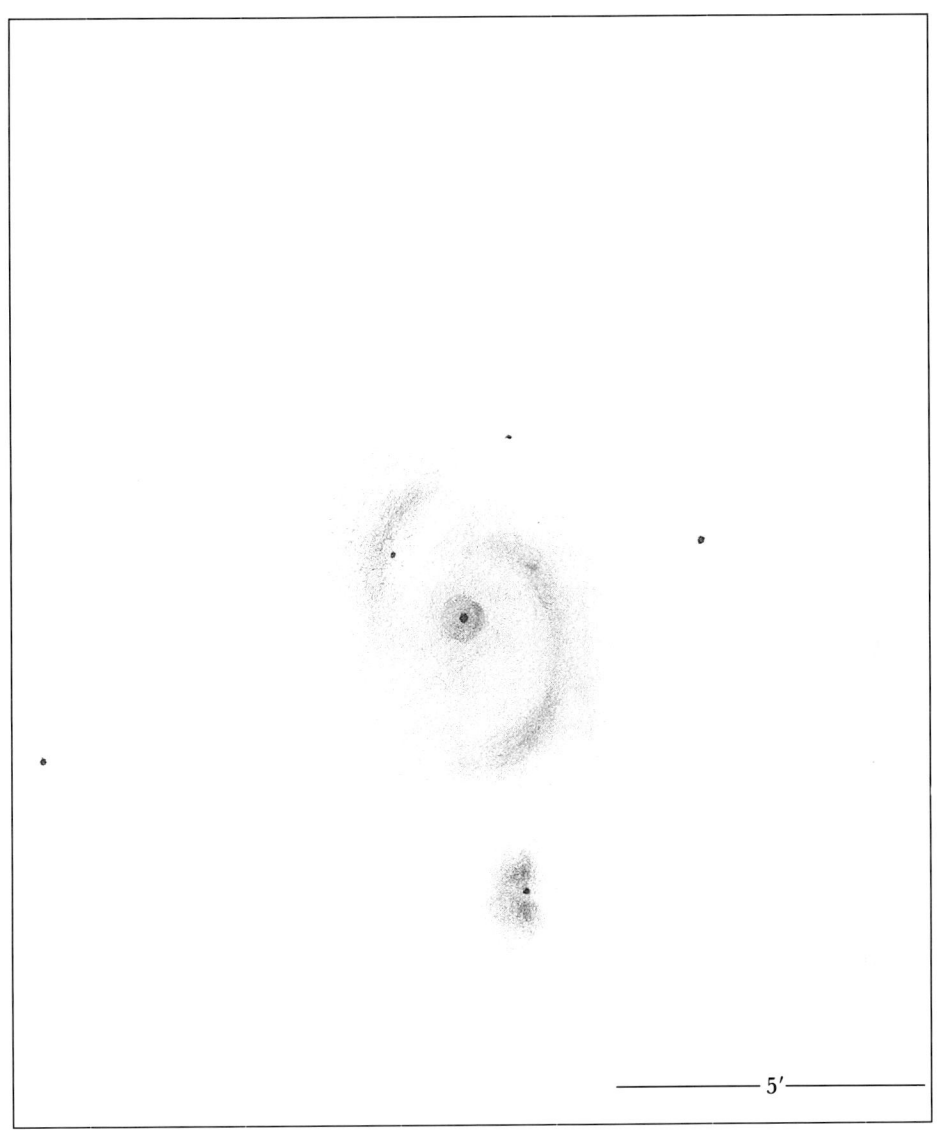

Drawing of M51 and NGC 5195, its smaller companion. Compare with the drawings on the next two pages. The reasons for the differences are described in the text.

Scale: 1.2 arc-min/cm	Viewing Distance (cm)
8-inch f/11.5 Cassegrain	
20mm Erfle (117×)	25×:115 200×:14
12.4mm Erfle (188×)	50×: 57 300×:10
7mm Erfle (334×)	100×: 29 400×: 7

Drawing Method 2
air mass: 1.11, faintest star: 14.3 at zenith, 188×;
no tracking
5/11/83 8:15–8:54 UT at Waianae ranch, Hawaii;
R. Clark

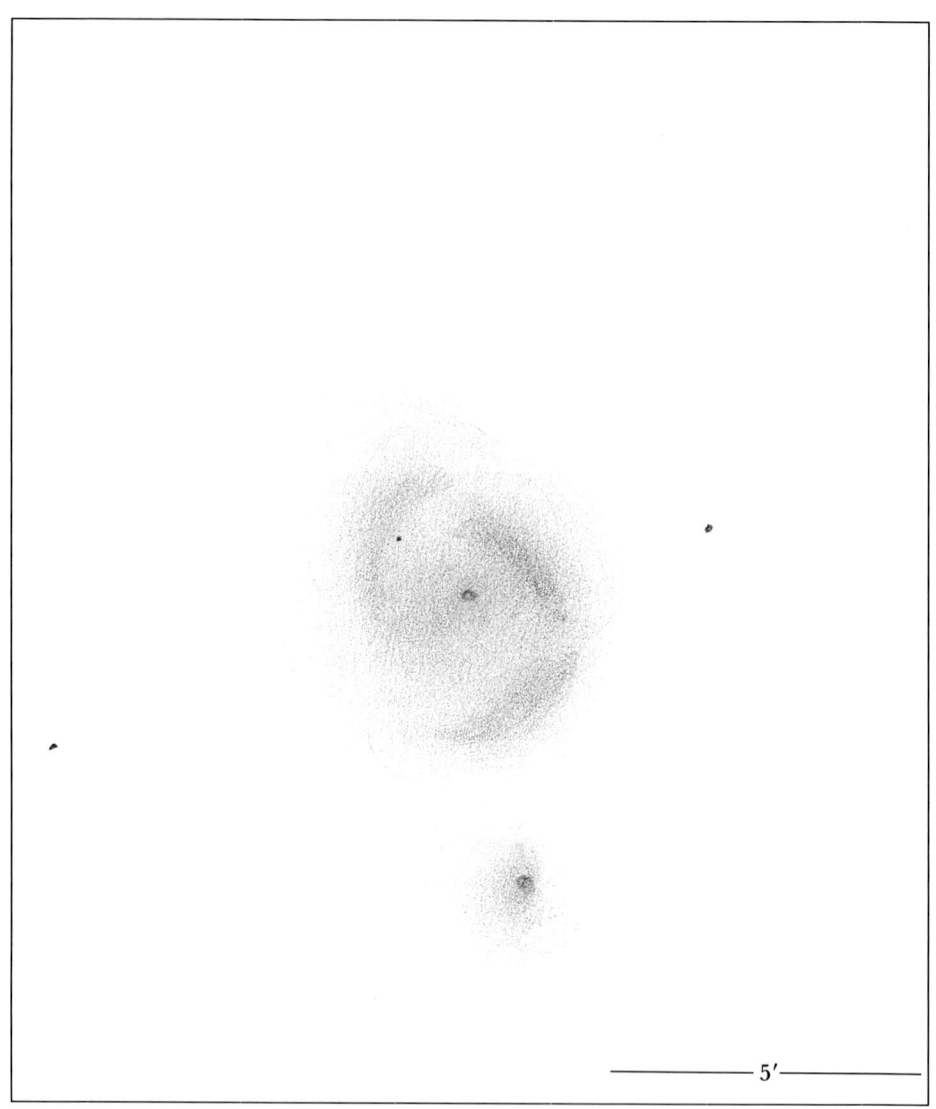

Drawing made May 22, 1971, of M51 and NGC 5195. Compare the detail with the previous drawing, made in 1983. The sky conditions were much better for the one above, but better observing techniques led to more being seen in 1983.

Scale: 1.2 arc-min/cm	Viewing Distance (cm)	
8-inch f/7.25 Newtonian		
28mm Kellner (52×)	25×:115	200×:14
12.5mm Orthoscopic	50×: 57	300×:10
(120×)	100×: 29	400×: 7

air mass: 1.02, faintest star: 14.8 at zenith, 188×; tracking
5/22/71 6:37–6:55 UT at Manastash Ridge, Washington; R. Clark

Drawing made June 20, 1982, of M51 and NGC 5195. Compare this drawing with the previous M51 drawings, and note how much less detail is visible under poor sky conditions. Maybe a hint of spiral structure was seen. Higher powers could not be used because of the low contrast.

Scale: 1.2 arc-min/cm
8 inch f/11.5 Cassegrain
20mm Erfle (117×)

Viewing Distance (cm)

25×: 115   200×: 14
50×: 57   300×: 10
100×: 29   400×: 7

air mass: 1.25, faintest star: 13.3 at zenith, 117×; no tracking
6/20/82 8:15–8:25 UT at Hawaii Kai, Hawaii; R. Clark

M83 (NGC 5236), GALAXY IN HYDRA
R.A. 13^h 37.1^m, Dec. $-29°$ $52'$ (2000.0)

Technical. M83 is a beautiful face-on spiral galaxy. Although usually classed as an Sc type, it shows characteristics of a barred spiral. M83 is about 30 million light-years away and has a mass estimated at almost one trillion times that of our Sun. The total light amounts to about 100 billion times that of the Sun. Thus M83 is comparable with our Galaxy in light output.

On color photographs, the spiral arms show all the signs of vigorous star formation: many knots of young blue star clusters and pink emission nebulae, bordered by dark lanes of dust. The arms have been rich in supernovae: one every 10 or 15 years. The galaxy's central region, on the other hand, displays the characteristic yellow of an older stellar population.

While most spiral galaxies have flat disks, that of M83 is warped. The arms are bent toward us on one side and away on the other. Left behind in the plane of the galaxy is a vast amount of hydrogen gas. The reason for this

Photograph of M83. South is up. (Courtesy Evered Kreimer, The Messier Album.)

strange warping is unclear. No other galaxies of substantial mass are nearby. NGC 5253, two degrees away, might be a possibility, but it seems to have only a tenth as much mass and would have had to practically collide with M83 to cause the warping.

Visual. M83 is magnitude 8.0 and 10 by 8 arc-minutes across; the mean surface brightness is 21.4 magnitudes per square arc-second. M83 is among the 25 brightest galaxies in the sky and a showpiece for Southern Hemisphere observers.

Through the 8-inch under moderate to good skies, M83 began to show its spiral structure even when low in the sky (air mass 2.1). In fact it appeared like a barred spiral: the arms extended straight out from the nucleus, then began to curve. The detection of spiral structure was quite a surprise, as I had observed the galaxy several times before with no sign of it. But that was using powers below 60×. The arms were evident at near 200×. This galaxy must be a grand sight in medium and large amateur telescopes when high in a dark sky.

Drawing of M83.

Scale: 1.2 arc-min/cm 8-inch f/11.5 Cassegrain 20mm Erfle (117×) 12.4mm Erfle (188×, best view)	Viewing Distance (cm) 25×:115 200×:14 50×: 57 300×:10 100×: 29 400×: 7	Drawing Method 2 air mass: 2.14, faintest star: 14.3 at zenith, 188×; no tracking 5/11/83 11:07–11:30 UT at Waianae ranch, Hawaii; R. Clark

M4 (NGC 6121), GLOBULAR STAR CLUSTER IN SCORPIUS

R.A. $16^h 23.7^m$, Dec. $-26° 31'$ (2000.0)

Technical. M4 is a beautiful globular that is easily found near Antares. It is about 6000 light-years distant and at present considered the nearest globular cluster. Its mass is about 60 000 times that of our Sun, quite small compared with giants such as Omega Centauri. M4 is dimmed by the foreground dust in the Antares region. Its stars are only loosely compacted, as globular clusters go, and a nearly straight line of stars appears to cut through its middle. The cluster was discovered in 1746 by P. L. de Cheseaux.

Photograph of M4. South is up. (Courtesy Martin Germano.)

Visual. M4 has a total magnitude of 7.4 and a maximum diameter of 20 arc-minutes. This works out to a mean surface brightness of 22.5 magnitudes per square arc-second, though, of course, the central region is far brighter. Since M4 is a loose globular cluster, it can be partially resolved into stars with telescopes somewhat smaller than 4 inches, under good skies. Medium-size amateur telescopes partially resolve it under most sky conditions.

Through the 8-inch, the line of stars in the center appeared like a brilliant knife edge slicing through the cluster. The line is also visible in small telescopes. With large apertures, the individual stars making up the "knife" can be discerned, making for a unique and beautiful view. Whereas most globular clusters are quite similar, the "knife" makes M4 distinctive.

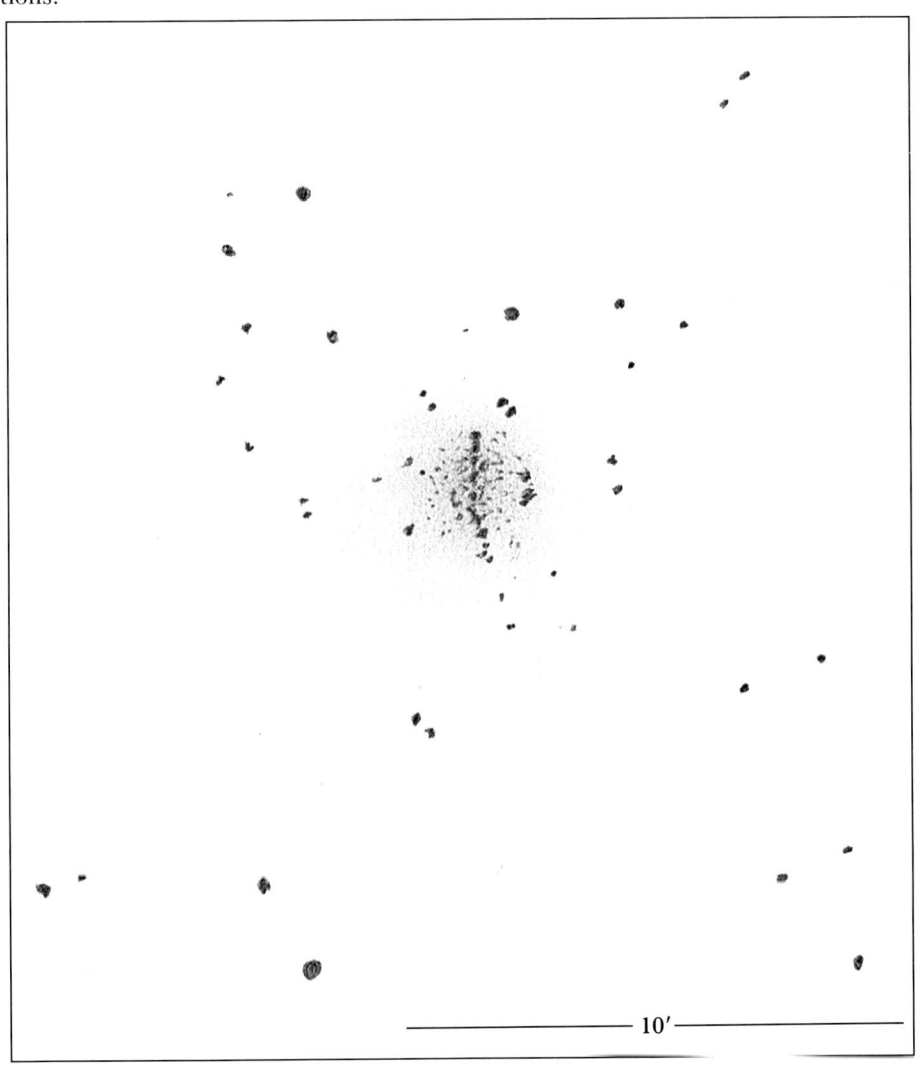

Drawing of M4.

Scale: 1.5 arc-min/cm
8-inch f/11.5 Cassegrain
20mm Erfle (117×)

Viewing Distance (cm)

25×:92   200×:11
50×:46   300×: 8
100×:23 400×: 6

air mass: 1.49, faintest star: 13.3 at zenith, 188×; no tracking
6/20/82 9:10–9:25 UT at Hawaii Kai, Hawaii; R. Clark

M13 (NGC 6205), GLOBULAR CLUSTER IN HERCULES

R.A. 16^h 41.7^m, Dec. 36° 27' (2000.0)

Technical. M13, the Great Globular Cluster in Hercules, was discovered by Edmond Halley in 1714. At a distance of 20 000 to 25 000 light-years, it has a diameter of around 175 light-years, though some of its stars lie outside that diameter. Estimates of the total stellar population range from a little over 500 000 to more than a million, and the mass is about 500 000 times the Sun's. The cluster's brightest stars are 11th-magnitude red giants; the faintest ones are below magnitude 22. Stars in this cluster that are as luminous as the Sun are magnitude 19.

Visual. At magnitude 5.7, M13 can be seen by the unaided eye under moderate to good skies, a fact noted by Halley. This is the brightest and most spectacular globular in the celestial northern hemisphere (though it has several close rivals). The cluster has a maximum diameter of 23 arc-minutes, yielding a mean surface brightness of 21.2 magnitudes per square arc-second. Because most of its stars are concentrated in the center, however, the surface brightness there is several magnitudes brighter.

Photograph of M13. South is up. (Courtesy Palomar Observatory.)

A VISUAL ATLAS OF DEEP-SKY OBJECTS

Some stars are resolved in telescopes as small as 3 inches, under good skies. In medium-size amateur telescopes many stars can be resolved. Through the 8-inch under moderate skies, the cluster was mostly resolved at magnifications near 200×. Notice that M13 is only half the size in the drawing that it is in the photograph. The bright central region in the photograph is overexposed and washed out, but to the eye it is quite spectacular. M13 is certainly the globular showpiece for Northern Hemisphere observers who cannot see the larger Omega Centauri (NGC 5139).

The visual appearance of M13 seems to vary considerably, depending on sky quality, telescope aperture, magnification, and observer experience. The central region has several small star-poor areas, which may be more or less apparent, depending on the observing conditions. There are also many star chains, some of which appear in the drawing. Their visibility too depends on the observing conditions.

Half a degree northeast of M13 is a small galaxy, NGC 6207. It is a 12.3-magnitude, type-Sc spiral 2 by 1 arc-minutes in size. Its mean surface brightness is 21.7 magnitudes per square arc-second.

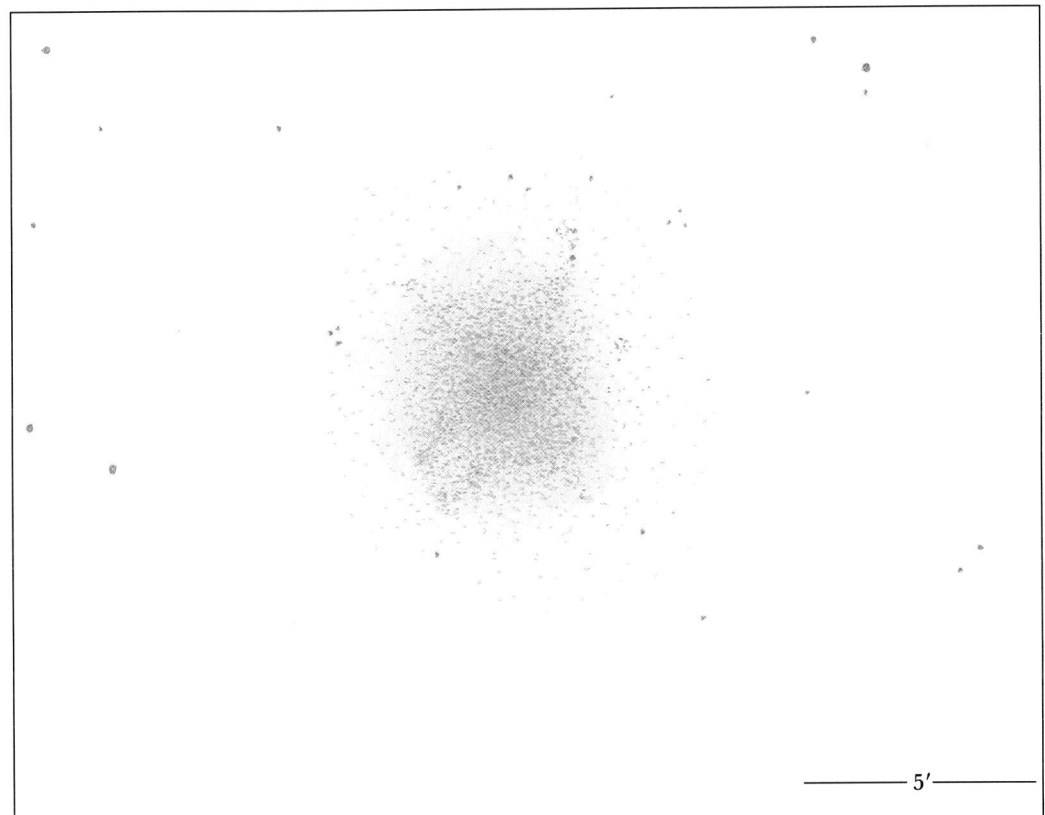

Drawing of M13. Compare this drawing with that of Omega Centauri on page 175.

Scale: 1.6 arc-min/cm	Viewing Distance (cm)		air mass: 1.05, faintest star: 13.8 at zenith, 188×;
8-inch f/11.5 Cassegrain			no tracking
20mm Erfle (117×)	25×:86	200×:11	5/15/83 12:05–12:36 UT at Barbers point, Hawaii;
12.4mm Erfle (188×,	50×:43	300×: 7	R. Clark
best view)	100×:21	400×: 5	

M20 (NGC 6514), THE TRIFID NEBULA IN SAGITTARIUS

R.A. 18^h 01.9^m, Dec. $-23°$ $02'$ (2000.0)

Technical. The Trifid Nebula is a spectacular mass of glowing hydrogen gas with dust clouds silhouetted in front of it. Discovered by LeGentil in 1747, it gained its popular name from the three main dark lanes that divide it into three bright regions. Several stars near the nebula's center provide the ultraviolet light that makes the hydrogen fluoresce. Two bright stars can be seen near the intersection of the dark lanes. Each is a spectroscopic binary, a pair of stars rotating closely around each other that betray their duplicity by the Doppler shifts in their spectral lines.

The main nebula is glowing hydrogen. South of the main nebula is another cloud that is not glowing, but reflecting starlight from the bright star near its center. On color photographs the hydrogen emission nebula appears red, the reflection nebula blue. The light of the reflection nebula is scattered from very small dust grains, as small or smaller than the wavelength of light. This type of scattering affects blue light more than red, which is why a reflection nebula can appear bluer than its illuminating star.

The distance to M20 is estimated at only 4500 light years, about the same as the Lagoon Nebula, M8. The stars that excite the Trifid's hydrogen are thought to be only about 7 million years old. The nebula's distinctive structure has been explained in the following way. Radiation pressure and possibly a stellar wind from the young stars cause the hot hydrogen to expand. As it does so it encounters an intricate network of cold, dark material. It divides and flows around dense lumps in this material, leaving them sharply outlined. Such "bright rims" are common in emission nebulae and give them much of their beauty and drama on photographs. In the case of the Trifid we seem to be viewing the dark masses from behind. On the other hand, they may be much closer than the nebula and have nothing to do with it at all.

Visual. M20 can be a stunning object, but this depends critically on the sky conditions. Magnitude estimates for the nebula are hard to come by, perhaps because of the difficulty posed by the many field stars. Recent estimates put it at magnitude 8.5 and its size is 29 by 27 arc-minutes; however, the brightest two spots are about 9 and 5 arc-minutes in diameter. If the magnitude estimate is accurate, the mean surface brightness is 24.4 magnitudes per square arc-second, although the two main parts are a couple of magnitudes brighter. Even so, this value appears too low when compared with the surface brightnesses of other nebulae. Spot measurements with a photoelectric photometer are needed to better determine the nebula's brightness.

M20 seems visible to the unaided eye under good skies, but what's seen is probably not the nebula but the many stars and clusters in its immediate vicinity.

M20 can show a wealth of detail under some skies, or virtually nothing at all. In my experience, this nebula presents a wider range of views than any other object in the sky. In a dark environment, it is easily visible in 7 × 50 binoculars and the dark lanes can be seen through a 2.4-inch telescope at powers of 20× to 40×. Under good skies I have easily seen the dark lanes in the 3-inch finder at 31×. Under moderate to poor skies, the nebula is difficult or impossible in any telescope. But under conditions that are just a little better, the nebula becomes easily visible in most instruments.

Through the 8-inch under moderate to good skies, the dark lanes are easily seen and structure within them becomes evident at medium powers. M20 benefits greatly from high power, which many observers tend not to use on it. At 117× through the 8-inch with a wide-field Erfle eyepiece, the nebula just filled the field of view. This magnification was best for viewing the faint outer regions. A power of 188× was better for defining the detail in the dark lanes. The drawing was hastened by the onset of clouds; even more detail probably could have been seen if more time had been available.

As noted previously, on photographs the northern portion of M20 appears red, the southern part blue. I have observed it many times without detecting color. But one occasion was different. This was an excellent night in the Cascade Mountains of Washington state, when several observers easily saw magnitude 14.8 stars in 8-inch telescopes (the actual faintest star was probably right at the theoretical limit of 15.2). When I turned an 8-inch onto M20, it came as quite a surprise to see beautiful blue and red, just like in color photographs but more pastel! What a magnificent object in dark skies.

Photograph of M20. South is up. (Courtesy National Optical Astronomy Observatories.)

A VISUAL ATLAS OF DEEP-SKY OBJECTS

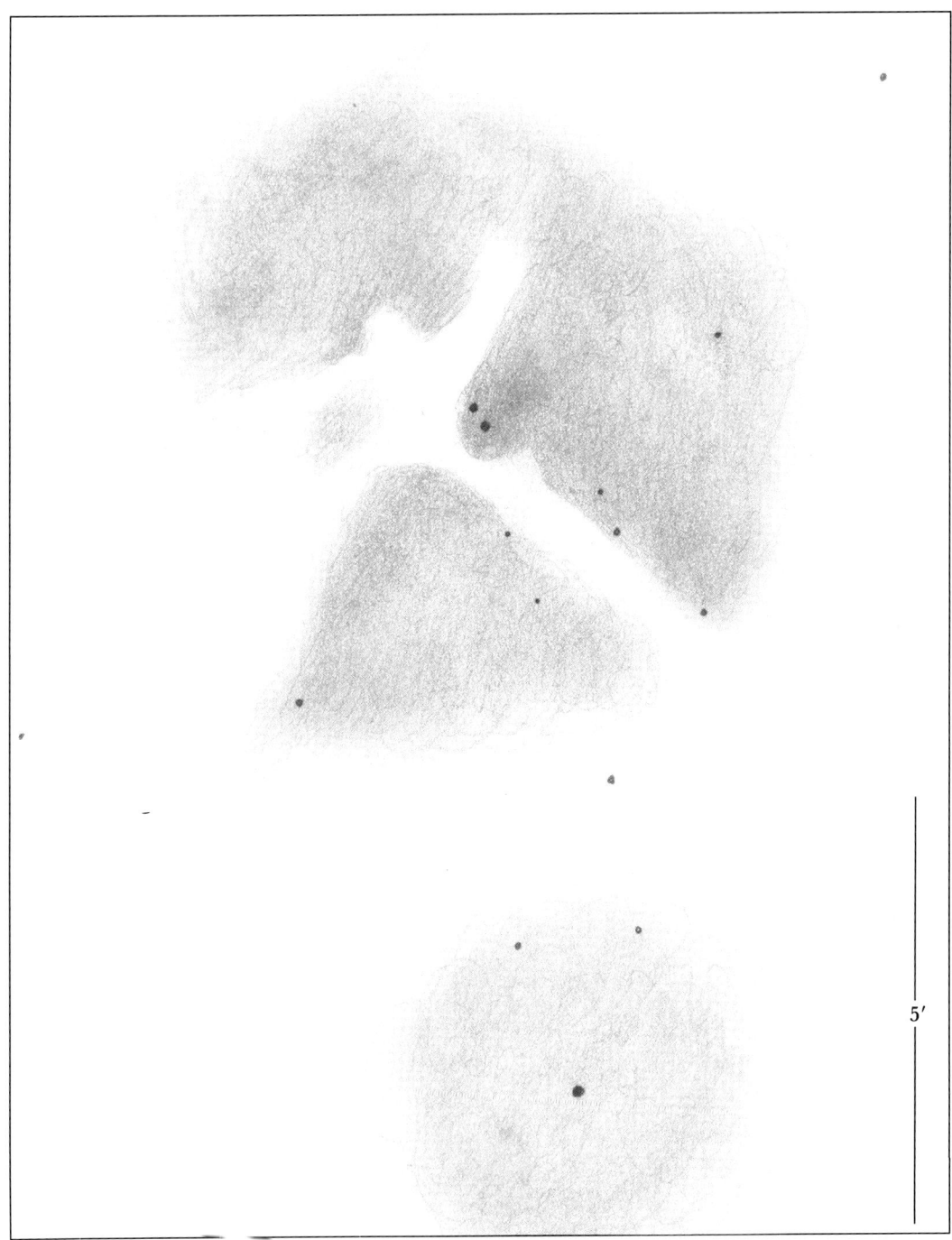

Drawing of M20.

Scale: 0.8 arc-min/cm	Viewing Distance (cm)		air mass: 1.40, faintest star: 13.8 at zenith, 188×;
8-inch f/11.5 Cassegrain			no tracking
20mm Erfle (117×)	25×: 172	200×: 21	5/15/83 12:54–13:14 UT at Barbers Point, Hawaii;
12.4mm Erfle (188×)	50×: 86	300×: 14	R. Clark
	100×: 43	400×: 11	

M8 (NGC 6523), THE LAGOON NEBULA IN SAGITTARIUS, NGC 6530

R.A. 18^h 04.7^m, Dec. $-24°$ $20'$ (2000.0)

Technical. The Lagoon Nebula, M8, is a beautiful and complex region of dust and glowing gas. The main nebula appears cut in two by a dark lane, which the late-19th-century astronomy writer Agnes Clerke claimed looked like a lagoon. The nebula was discovered in 1747 by Le Gentil, but a star cluster, which is located near the center of the nebula and designated NGC 6530, was discovered by John Flamsteed as early as 1680.

The nebula is estimated to be about 4500 light-years away, and the age of the star cluster to be about 2 million years, depending on the authority cited. This distance would make the nebula's outlying portions well over 100 light-years across.

One of the stars that irradiates the hydrogen gas into glowing is Herschel 36, magnitude 9.5, near the nebula's brightest part. Other stars also illuminate the nebula, including 6th-magnitude 9 Sagittarii, and probably several more embedded in the nebula that we cannot see. These young stars are pushing gas away from the nebula's center with their intense radiation and stellar winds. As the hot hydrogen collides with cold, dark, denser patches, it flows around them and outlines them with bright rims. Such rims and dark patches appear in many photographs of emission nebulae.

The small, very bright patch next to Herschel 36 has been named the Hourglass Nebula for its shape. It is extremely dense, as nebulae go, and only a half light-year across. The heat of Herschel 36 will soon cause it to dissipate. Astronomers have calculated that its present situation next to the bright star is so unstable that the star must be much younger than 10 000 years. Herschel 36 is certainly one of the youngest stars known.

Visual. M8 can be seen with the unaided eye. The total magnitude of the nebula is estimated as about 5, with a total size of 80 by 40 arc-minutes. This yields a mean surface brightness of only 22.4 magnitudes per square arc-second. But the brightest portions are many magnitudes brighter. It has been said that the Hourglass Nebula can be seen in a telescope when the full Moon is nearby. Its surface brightness must be similar to that of the Trapezium region in M42.

On the west side of the Hourglass, with Herschel 36, are three dark lanes like a miniature Trifid Nebula. Even after observing M8 many times, I never saw them because I never tried powers above about 60×. However, 80× or more will bring them into view in medium amateur telescopes. After "discovering" this dark marking, I surveyed many amateur observers and had a hard time finding anyone else who had seen it. The lesson is clear: a beautiful and easy object has been missed because M8 seems only to be explored at the lowest possible magnifications.

The drawing shows the appearance of M8 through the 8-inch telescope under moderate skies. The star Herschel 36 was seen next to the small dark lane, but the bright Hourglass outline was not recorded. Probably more time should have been spent at higher magnifications. When the observation was made, I did not know of the Hourglass or the adjacent dark "Trifid".

The Hourglass Nebula has a size of about 0.8 by 0.7 arc-minutes, with a visual magnitude of about 8. The surface brightness is close to 16 magnitudes per square arc-second. After I knew of the existence of the Hourglass I searched for it under the excellent skies of the Colorado Rockies. In seeing of 1.5 arc-seconds the hourglass shape could be seen in the 8-inch at 334×. The dark lanes could be seen in the 3-inch at 266× but only with difficulty. The Hourglass region was also seen, but it appeared only as a bright patch separated from the star Herschel 36. Even considering its brightness, the hourglass shape needs high power because of the nearby 9.5-magnitude star.

The major dark lane dividing the nebula in two, the one from which the Lagoon name is derived, can be seen in the smallest amateur telescopes. Under good skies more of the outer region becomes visible in the 8-inch, and it begins to take on the ragged-edge appearance seen in photographs. I've often thought the two bright stars superimposed on the nebula resemble eyes, the lagoon dark lane a mouth, and the star cluster on the other side, a beard on the chin of a face. The Hourglass region then appears like the brain, and the whole effect is that of a monster from some low-budget science fiction movie.

Photograph of M8. South is up. (Courtesy National Optical Astronomy Observatories.)

A VISUAL ATLAS OF DEEP-SKY OBJECTS

Drawing of M8.

Scale 1.2 arc-min/cm	Viewing Distance (cm)		air mass: 1.75, faintest star: 13.0 at zenith, 117×;
8-inch f/11.5 Cassegrain			no tracking
28mm Kellner (82×)	25×:115	200×:14	9/19/82 6:40–7:00 UT at Ewa Beach, Hawaii;
20mm Erfle (117×)	50×: 57	300×:10	R. Clark
	100×: 29	400×: 7	

M16 (NGC 6611), THE EAGLE NEBULA IN SERPENS

R.A. 18^h 18.8^m, Dec. $-13°$ $47'$ (2000.0)

Technical. M16 is another beautiful mixture of bright and dark nebulosity, not very far from M8 and M20. All three lie in the general direction of the center of our Galaxy and provide marvelous views to (Northern Hemisphere) summertime observers. M16, however, is fainter and probably farther away than M8 or M20. Its distance has been estimated as about 8000 light-years, and the mass of its gas as about 12 000 times that of the Sun.

The nebulosity has an embedded star cluster. Curiously, early observers noticed only the cluster and not the nebulosity. The cluster was discovered by P. L. de Cheseaux in 1746. Charles Messier in 1764 mentioned a weak light surrounding the stars, but he may have thought it came from unresolved stars. Later observers such as John Herschel and T. W. Webb mention only the cluster.

Like M8 and M20, M16 consists of hydrogen gas that is illuminated and blown about by newly formed stars near its center. The expanding gas is encountering dark, denser clouds, resulting in bright rims that show intricate and unusual detail on modern photographs. The cluster's age is estimated to be slightly less than 1 million years, with some stars only 50 000 years old.

Visual. M16 appears as a loose star cluster surrounded by a faint nebula. The nebulosity can be seen in small amateur telescopes, given dark skies, so it is a real mystery why early observers failed to report it. Did their optics scatter so much light that the glow of the stars hid it?

The magnitude of M16 is given as about 6.5, but this probably includes the integrated light of the cluster as well as the nebulosity, so the mean surface brightness of the nebulosity can't be stated. M16 is about 25 arcminutes in apparent diameter. The nebulosity is more difficult to detect than that of M20, but much easier than that surrounding Merope in the Pleiades. Professional photometry puts the surface brightness of the Merope Nebula at 21.6 magnitudes per square arc-second, but the calculated surface brightness for M20 is 24.4. However M20 appears brighter than the Merope Nebula. Photoelectric photometry needs to be done to resolve the discrepancy.

Through the 8-inch telescope under good skies, the general extent of the nebula was apparent but no small details such as dark patches could be seen. Even though the nebula was quite large, a magnification of 117× was best for detecting the nebulosity; it filled the field of view of the Erfle eyepiece, showing an apparent diameter of about 60°. A power of 188× was best for detecting faint stars in the cluster. In similar skies a 17-inch telescope showed some dark patches. Under excellent conditions some dark patches can be seen in 8- to 10-inch telescopes.

Photograph of M16. South is up. (Courtesy National Optical Astronomy Observatories.)

A VISUAL ATLAS OF DEEP-SKY OBJECTS

Drawing of M16.

Scale: 1.2 arc-min/cm	Viewing Distance (cm)		Drawing Method 2
8-inch f/11.5 Cassegrain			air mass: 1.23, faintest star: 14.5 at zenith, 188×;
20mm Erfle (117×)	25×:115	200×:14	no tracking
12.4mm Erfle (188×)	50×: 57	300×:10	8/12/83 7:40–8:00 UT at Waianae ranch, Hawaii;
	100×: 29	400×: 7	R. Clark

M17 (NGC 6618), THE OMEGA NEBULA AND OPEN CLUSTER IN SAGITTARIUS

R.A. 18^h 20.9^m, Dec. $-16°$ 11′ (2000.0)

Technical. M17 is a beautifully detailed nebula in the (Northern Hemisphere) summer Milky Way between M16 and M20. The west side resembles a capital Greek letter Omega. The brightest portion looks like the number 2, or perhaps the letter V, or perhaps a swan, from whence comes the other common name, Swan Nebula. M17 contains glowing gas and dark dust, some of which is cold, some warm. The warm dust is heated by stars embedded in the cloud to the west of the bright "2". The cold dust is to the east of the "2". There are at least 35 stars in and around the nebula, which is roughly 5 000 light-years away. Its estimated mass is about 4 000 suns, and its size, about 27 light-years across.

Visual. M17 shows a wealth of detail in telescopes large and small. Its total magnitude is about 6, and its faintest portions extend 45 by 35 arc-minutes. The mean surface brightness is thus 22.6 magnitudes per square arc-second, but the "2" is much brighter. Its base is about 13 arc-minutes long and its height 8 arc-minutes.

The "2" is detectable in telescopes as small as 2 inches, while larger telescopes begin to show additional detail. Through the 8-inch under moderate to good skies at medium powers, dark areas were seen crossing the base of the "2", and at the joint between the base and hook of the "2" a bright wedge was easily seen. Inside the hook the sky appears darker than elsewhere around the nebula. The detail in the brighter portions was best seen at 100× to 200×. This nebula provides a truly magnificent view under excellent skies through medium to large amateur telescopes.

Photograph of M17. South is up. (Courtesy National Optical Astronomy Observatories.)

Drawing of M17.

Scale: 1.1 arc-min/cm
8-inch f/11.5 Cassegrain
20mm Erfle (117×)
12.4mm Erfle (188×)

Viewing Distance (cm)

25×: 125   200×: 16
50×: 63   300×: 10
100×: 31   400×: 8

air mass: 1.31, faintest star: 13.8 at zenith, 188×;
no tracking
5/15/83 12:05–12:36 UT at Barbers Point, Hawaii;
R. Clark

M11 (NGC 6705), OPEN CLUSTER IN SCUTUM

R.A. $18^h 51.1^m$, Dec. $-06° 16'$ (2000.0)

Technical. M11 is a very tightly packed open cluster that appears within the Scutum star cloud. In reality, however, M11 is closer: about 5500 light-years away. Its density of stars is as great as some of the looser globular clusters. M11 is estimated be be about 500 million years old, however, much too young for it to be classed with the globulars.

The cluster's brightest part is about 10 arc-minutes, or about 15 light-years, across. Most of the stars are magnitudes 11 to 16. Our Sun at the distance of M11 would be magnitude 15.9, so the brightest stars in M11 have 100 times the Sun's luminosity. The cluster's total light is is estimated at 10 000 times that of the Sun, and its mass as nearly 3000 suns. M11 contains about 400 stars brighter than magnitude 14 and nearly 1000 brighter than 17.

Visual. M11 appears at magnitude 6, with a diameter of 12 arc-minutes and a mean surface brightness of about 20.0 magnitudes per square arc-second. It can be seen with the unaided eye under good to excellent skies. A 2-inch telescope will begin to show a few individual stars, and a 3-inch starts to resolve many. The cluster has a square shape, with an 8th-magnitude star near the southeast corner. In small telescopes this bright star suggests a comet nucleus and the cluster the surrounding coma. On the opposite side is a "bay" shaped like a heart.

Through the 8-inch M11 is spectacular at any power under skies from poor to excellent. The accompanying drawing, made under moderate skies, shows a square appearance at magnifications from 100× to 200×. The heart-shaped bay was also quite obvious. The "nebulosity" in the drawing represents unresolved background stars seen when averted vision was used – not an actual nebula.

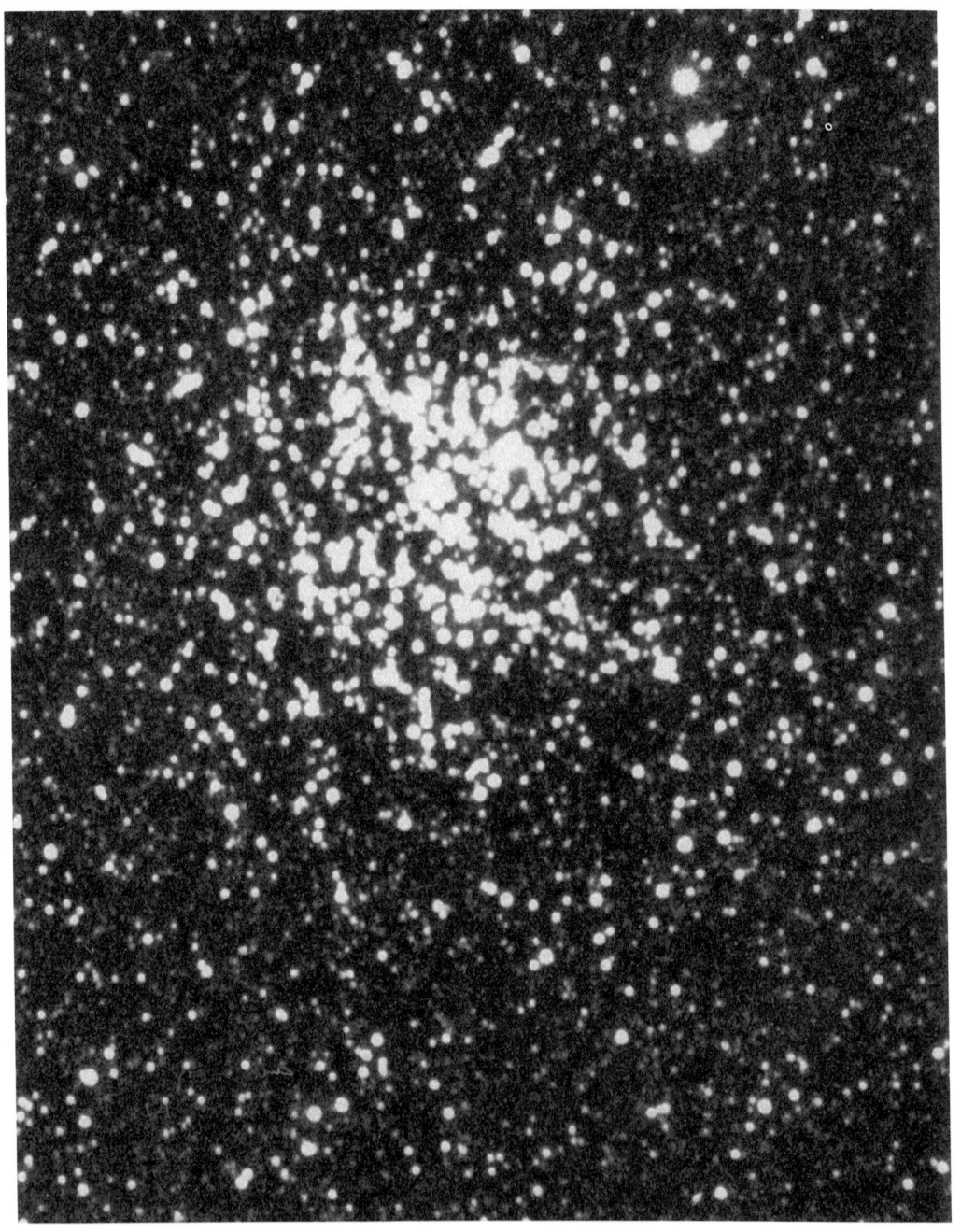

Photograph of M11. South is up. (Courtesy Ben Meyer.)

A VISUAL ATLAS OF DEEP-SKY OBJECTS

Drawing of M11.

Scale: 0.8 arc-min/cm
8-inch f/11.5 Cassegrain
20mm Erfle (117×)
12.5mm Orthoscopic
(187×)

Viewing Distance (cm)

25×: 172   200×: 21
50×: 86   300×: 14
100×: 43   400×: 11

air mass: 1.26, faintest star: 13.6 at zenith, 187×;
no tracking
9/19/82 7:05–7:20 UT at Ewa Beach, Hawaii;
R. Clark

M57 (NGC 6720), THE RING NEBULA IN LYRA

R.A. 18^h 53.6^m, Dec. $33° 02'$ (2000.0)

Technical. M57 is the classic example of a planetary nebula. It was the first one discovered: in 1779 by Antoine Darguier, who described it as "resembling a fading planet". There are probably 10 000 planetaries in our galaxy, though less than a thousand are catalogued. M57 consists of gas ejected from its central star less than 5500 years ago. It is now expanding at a rate of 19 kilometers per second and, at a distance of about 2000 light-years, is about 1/3 light-year in diameter.

The nebula's total mass is less than that of the Sun, but it emits about 50 times the Sun's light. Its composition has been found by detailed spectroscopic analysis. The percentages of atoms are:

Hydrogen	92.55	Sulfur	0.005
Helium	7.35	Argon	0.0007
Oxygen	0.054	Chlorine	0.00019
Nitrogen	0.027	Fluorine	0.00002
Neon	0.008		

Color photographs of M57 show a greenish center surrounded by yellow that turns to red on the outer edges. The different colors are due to the various atoms that are excited to emission. Near the center, light from oxygen and nitrogen dominates. Near the edge, the ultraviolet radiation from the central star is too weak to excite these elements, so emission from hydrogen dominates.

The central star, shining at magnitude 14.8, is either a white dwarf or evolving toward white dwarf status. It is extremely hot, with an estimated surface temperature of 100 000 kelvin, and is several thousand times the density of our Sun.

Visual. M57 is magnitude 9 and has an apparent size of 1.3 by 1.0 arc-minutes. Its mean surface brightness is very high at 17.9 magnitudes per square arc-second. The nebula inside the ring is much dimmer but still can be seen easily in medium-size telescopes. This central glow consists of several striations parallel to the major axis. Resolving the striations requires a large amateur telescope and excellent skies.

Through the 8-inch, the ends of the ellipse were fainter than the other edges. The "center of the doughnut" was markedly brighter than the surrounding sky, but no fine detail could be seen under good skies. The best view was at powers near 200×, while 334× gave about the same detail.

The central star could not be seen. It is suspected of being variable because it is sometimes easy in a 12-inch telescope, at other times difficult in a 40-inch.

The ring shape can be detected in very small amateur telescopes. I conducted an experiment to find the minimum aperture needed to do so. On the same night as the drawing was made, I used the 8-inch telescope at 188× with various masks over the front to simulate smaller apertures. Since sky conditions, observer and magnification were all constant, the only variable was aperture. The ring shape was easily detected when the aperture was just two inches. At the next lowest aperture, 1 inch, M57 could not be seen at all. The minimum aperture required under good skies is probably near 1.5 inches.

A VISUAL ATLAS OF DEEP-SKY OBJECTS

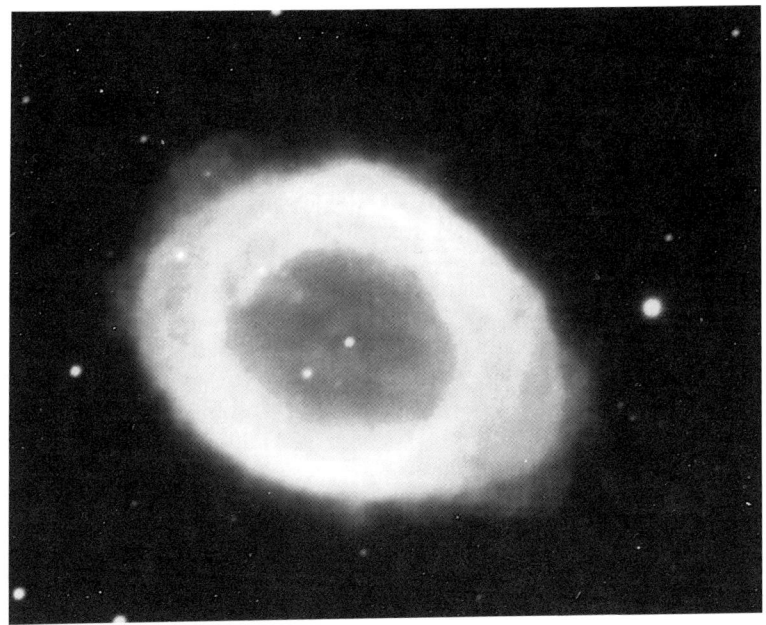

Photograph of M57. South is up. (Courtesy Laird A. Thompson, Canada–France–Hawaii Telescope Corporation.)

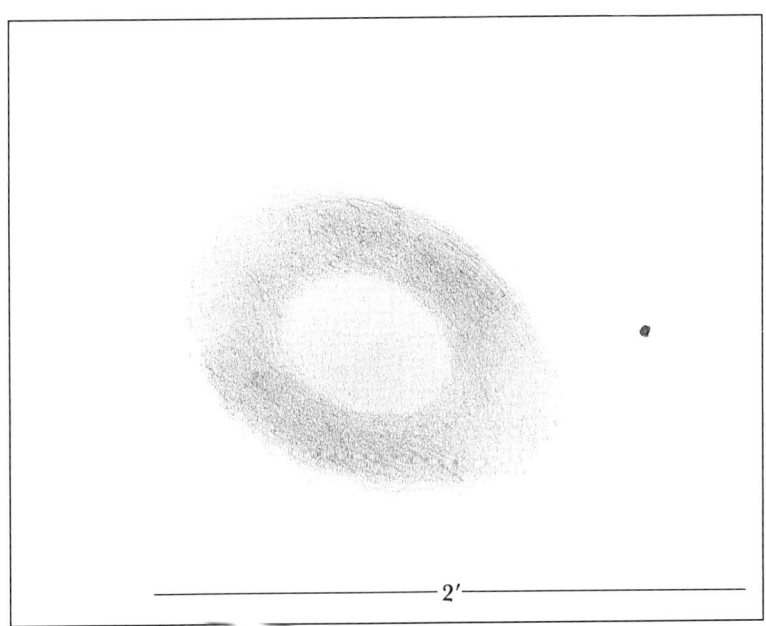

Drawing of M57.

Scale: 0.25 arc-min/cm
8-inch f/11.5 Cassegrain
12.4mm Erfle (188×, best view)
7mm Erfle (334×)

Viewing Distance (cm)

50×:275   300×:46
100×:138 400×:34
200×: 69 600×:23

air mass: 1.02, faintest star: 14.2 at zenith, 188×; no tracking
6/17/83 11:35–11:50 UT at Barbers Point, Hawaii; R. Clark

M27 (NGC 6853), THE DUMBBELL NEBULA IN VULPECULA

R.A. 19^h 59.6^m, Dec. 22° 43' (2000.0)

Technical. M27 was discovered in 1764 by Charles Messier. This is one of the largest and brightest planetary nebulae in the sky. Astronomers have been able to measure an annual growth in its apparent diameter of 0.068 arc-second per year. Spectroscopic measurements show a physical expansion rate of about 27 kilometers per second. If we assume that both velocities are measuring the same expansion, the nebula's distance can be found: about 275 light-years. This is about three times closer than distances obtained a few years ago. This also means that M27 has an age close to the mean for planetaries, about 20 000 years. The actual size would be a little under one light-year.

However, this way of finding a planetary's size and distance should be treated with caution. The outer edge of the nebula may not be the actual outer edge of the gas but just the zone where it ceases to glow. Therefore the visible expansion of the outer edge may have little or nothing to do with the outflow speed of the gas.

The central star is very hot, with a calculated surface temperature of 85 000 K. Anything this hot radiates mostly in the ultraviolet, and it is this radiation that excites the gas in the nebula to glow. The gases involved and the colors produced are similar to those described for M57 on page 208.

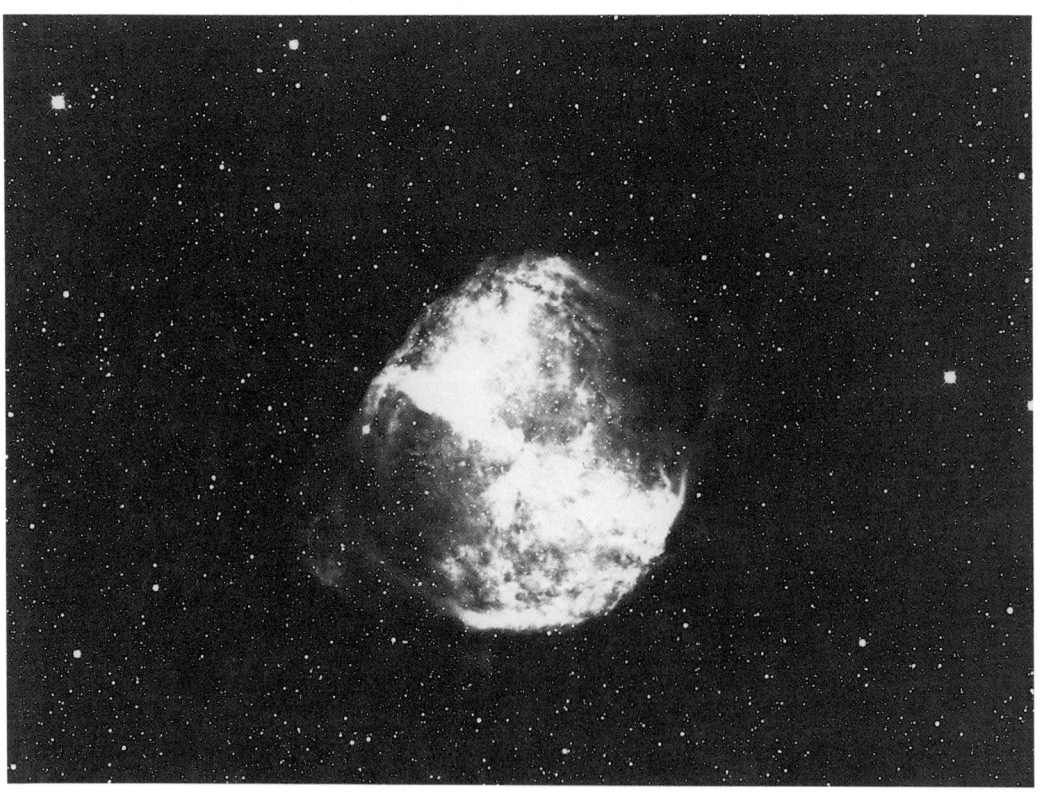

Photograph of M27. South is up. (Courtesy National Optical Astronomy Observatories.)

Visual. M27 is very bright and easily visible in 7-power binoculars as a slightly out-of-focus star. The nebula has a total magnitude of 8 and an apparent size of 8 by 6 arc-minutes. Its mean surface brightness is 20.8 magnitudes per square arc-second. The brightest portion forms a rectangle or hourglass shape 6 by 4 arc-minutes. It is this hourglass shape that gave the Dumbbell its name in the 19th century. In small telescopes the bright portion looks more like a rectangle. The faint outer parts require a medium size telescope and at least moderate skies, or a small telescope and excellent skies.

Through the 8-inch under moderate skies, the 13th-magnitude central star was seen only with averted vision at magnifications near 200×. Because the nebula has such a high surface brightness, better skies won't really improve the central star's visibility. An 8-inch is probably the minimum size telescope in which it can be detected.

The hourglass shape was easy near 200x. The faint extensions could be seen at all magnifications used, and a few other stars could be seen superimposed on the nebula at high powers. Under excellent skies, structure corresponding to the sharp borders at the ends of the hourglass could be glimpsed occasionally. Larger telescopes show this detail under good skies. Many observers have reported a distinct greenish color to M27 even in small telescopes, though I have observed only a dull gray–white through the 8-inch.

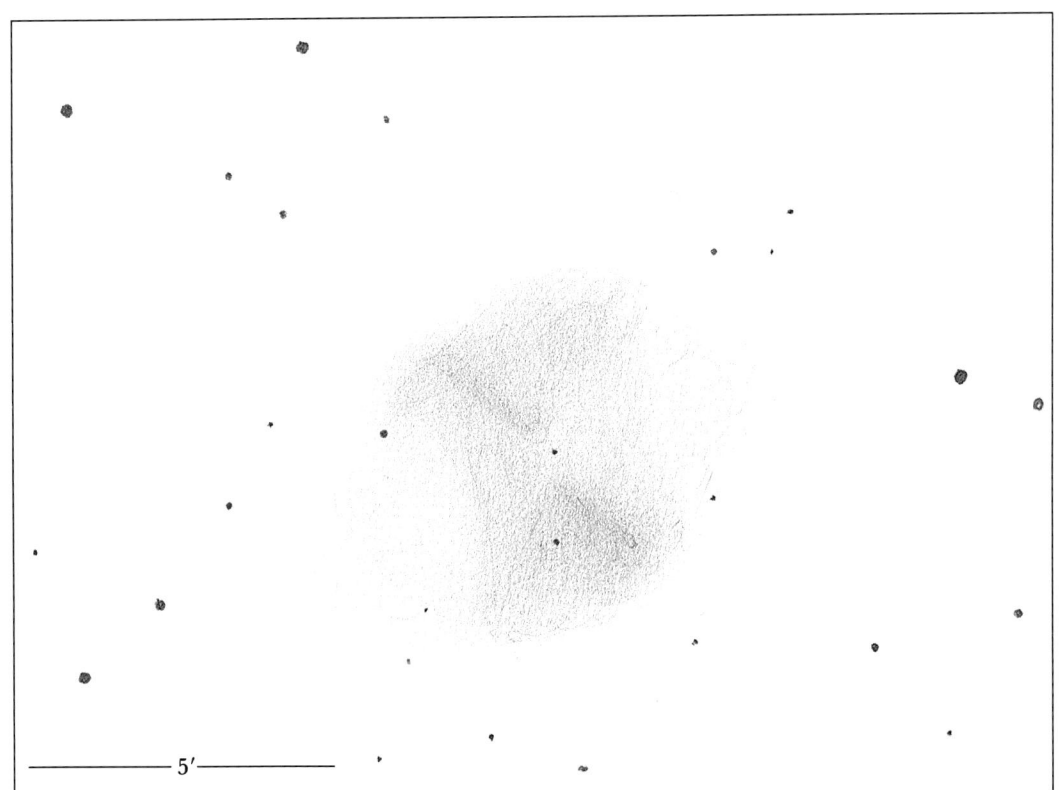

Drawing of M27.

Scale: 1.2 arc-min/cm	Viewing Distance (cm)	air mass: 1.03, faintest star: 13.0 at zenith, 188×; no tracking
8-inch f/11.5 Cassegrain		
28mm Erfle (82×)	25×:115 200×:14	8/22/82 9:00–9:40 UT at Ewa Beach, Hawaii;
20mm Erfle (117×)	50×: 57 300×:10	R. Clark
100×: 29 400×: 7		
12.5mm Ortho. (187×, best view)		

NGC 6888, THE CRESCENT NEBULA IN CYGNUS

R.A. 20^h 12.5^m, Dec. 38° 25' (2000.0)

Technical. NGC 6888 is a diffuse shell of gas being blown from an unusual object: a Wolf-Rayet star, a rare class of star (spectral type W) that is extremely hot, massive, and hydrogen-poor. The nebula is rather large, with a true size of 16 by 24 light-years, and contains about 2 solar masses of gas. Its origin is somewhat unclear. It may all have been ejected from the central star, or most of it may already have existed in the neighborhood and is just being pushed away from the star by a powerful "stellar wind."

The 7th-magnitude central star is known as HD 192163. It is one of only about 150 Wolf-Rayet stars known in our Galaxy. One reason they are unusual is that the absorption lines in their spectra are very broad instead of narrow. The strongest spectral features of this star are due to nitrogen, and no features to indicate the presence of carbon. A second type of Wolf-Rayet star has the spectral signature of carbon but no nitrogen.

HD 192163 is about seven times larger than the Sun and seven times hotter: 40 000 kelvin. Wolf-Rayet stars constantly shed mass as a thick stellar wind flowing outward at great speed. The wind's velocity is similar to that of an expanding nova: about 3000 kilometers per second. If this were to keep up, the typical Wolf-Rayet star would lose all of its mass in just a few million years.

Photograph of NGC 6888. South is up. (Courtesy Martin Germano.)

Visual. NGC 6888, often called the Crescent Nebula, is harder to see than the Veil Nebula 10° to the southeast. No magnitude or surface brightness estimates are available for NGC 6888. The star HD 192163 appears off-center in the egg-shaped loop. Much of the loop is very faint and probably cannot be detected visually. The brightest portion is a crescent on the north side. Its shape is rather reminiscent of the much larger Veil.

The Crescent is easily seen through moderate-size telescopes under good skies. It should be visible through small telescopes in excellent conditions. A magnification of 117× on the 8-inch in good skies gave the best view. The crescent was dimmest near the middle of the arc. A number of stars appeared superimposed upon it.

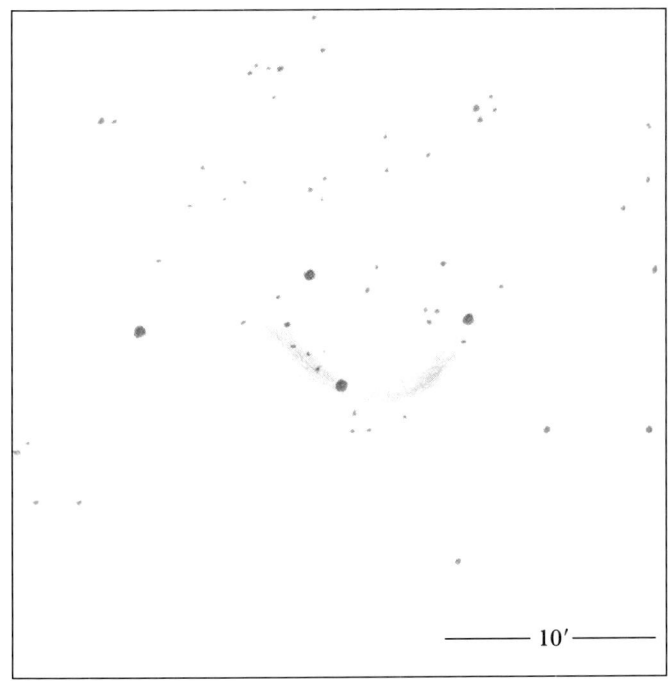

Drawing of NGC 6888.

Scale: 3.5 arc-min/cm	Viewing Distance (cm)	
8-inch f/11.5 Cassegrain		
20mm Erfle (117×	25×:39	200×:5
best view)	50×:20	300×:3
12.4mm Erfle (188×)	100×:10	400×:2

air mass 1.24, faintest star 14.0 at zenith, 188×; no tracking
9/11/83 9:35–10:20 UT at Barbers Point, Hawaii; R. Clark

NGC 6946, GALAXY IN CEPHEUS

R.A. 20^h 35.0^m, Dec. 60° 08' (2000.0)

Technical. NGC 6946 is a beautiful multi-arm spiral galaxy only 11° from the plane of our Galaxy at a distance of 10 to 20 million light-years. Because it is so close to the Milky Way on our sky, it is much dimmed by dust in our Galaxy. This absorption has hampered studies of the galaxy. NGC 6946 has a total luminosity estimated at about 100 million suns. It is also a radio source. There are at least four major spiral arms, as well as several smaller ones that branch off from them. On high-resolution observatory photographs, the many bright and dark patches show marvelous detail.

Visual. NGC 6946 has a total magnitude of 11.1 and an apparent diameter of 8 arc-minutes. The mean surface brightness is very low at 24.2 magnitudes per square arc-second. Because of this low surface brightness, many amateurs have bypassed the galaxy. Some observing books state that only the brighter nuclear region can be detected at all in amateur telescopes.

Thus the following observation came as a surprise to me. Through the 8-inch telescope under good skies, NGC 6946 did appear fainter than either side of the Veil Nebula (NGC 6960 and 6992), but brighter than the Merope Nebula in the Pleiades. And spiral arms could be seen! This view, depicted in the drawing, was about the same at 117× and 188×. It was the first time I had seen the galaxy at all. The large angular size of NGC 6946 is what enables the arms to be seen. Only moderate magnification (about 100×) brings them near the optimum magnified visual angle. They are about 1 arc-minute wide, so 100× enlarges them to almost 2° as seen by the eye – near the optimum.

Through the three-inch finder at 31× under excellent skies, NGC 6946 was visible as a faint fuzzy object. And the spiral arms, though not resolved, formed a fainter glow around the nucleus. This galaxy is another lesson in observing: just because the surface brightness is low, do not restrict your observations to low magnifications.

Photograph of NGC 6946. South is up. (Courtesy National Optical Astronomy Observatories.)

A VISUAL ATLAS OF DEEP-SKY OBJECTS

Drawing of NGC 6946.

Scale: 0.75 arc-min/cm   Viewing Distance (cm)
8-inch f/11.5 Cassegrain
20mm Erfle (117×)   25×:183   200×:23
12.4mm Erfle (188×)   50×: 92   300×:15
                         100×: 46   400×:11

air mass: 1.30, faintest star: 14.5 at zenith, 188×;
no tracking
8/12/83 10:20–10:35 UT at Waianae ranch,
Hawaii; R. Clark

NGC 6960 AND NGC 6992-5 THE VEIL NEBULA IN CYGNUS

NGC 6960:   *R.A. $20^h 45.7^m$, Dec. $30° 43'$*
NGC 6992–5: R.A. $20^h 56.4^m$, Dec. $31° 42'$
(2000.0)

Technical. The Veil Nebula is a spectacular supernova remnant showing beautiful, intricate detail. Observers have given it many names in attempts to convey its delicate appearance: the Cirrus, Bridal Veil, Filamentary, Network, and Loop Nebula. Each is appropriate.

The nebula, discovered by William Herschel in 1784, is fully 2.7° in diameter. Modern investigations have shown it to be expanding by 0.06 arc-second per year. Spectroscopic analysis shows its present physical expansion rate to be 45 kilometers per second. These two values allow us to compute a distance of 1 300 light-years.

The original expansion rate after the supernova explosion must have been well over 1000 kilometers per second. The expansion has slowed because the gas has been plowing into interstellar matter. Using a reasonable estimate of the deceleration rate, the time since the supernova can be calculated as about 30 000 years. No remnant star has been identified, and none may exist. The energy to light the Veil probably comes from its ongoing collision with interstellar matter, a process known as shock excitation.

Visual. The Veil Nebula is a prized object among amateur astronomers and is sure to be a target for medium to large telescopes at any (Northern Hemisphere) summer star party. NGC 6960 and 6992–5 are arcs each with a total magnitude of about 8. NGC 6960 has a length of about 70 arc-minutes, an average width of 6 arc-minutes, and a mean surface brightness of 23.2 magnitudes per square arc-second. NGC 6992–5 has a size of 78 by 8 arc-minutes and a mean surface brightness of 23.6 magnitudes per square arc-second. Both appear brighter than the Merope Nebula in the Pleiades or the outer loop of the Orion Nebula, M42. They are fainter than the Trifid Nebula, M20.

The Veil is divided into three parts. The brightest is NGC 6992–5, which is centered 2.7° northeast of the 4th-magnitude star 52 Cygni. The next brightest part is NGC 6960, which passes right next to 52 Cygni. The third and faintest is an unnamed triangular patch between the brighter two. There are other, smaller and fainter fragments of the nebula that have not been reported visually by amateurs in recent years.

How hard it is to detect the Veil is somewhat controversial. Some observing manuals say that NGC 6960 and NGC 6992–5 require a 6- or 8-inch telescope at low powers under an excellent sky, and that the triangular patch cannot be seen visually in any amateur telescope. Others say the brighter parts are visible in 7 × 50 binoculars. The reason for this disparity is unclear. I suspect it lies in particular observers' experiences trying to *find* the objects. Under only moderate skies, the nebula may indeed be difficult to detect in an 8-inch. But slight improvements in sky quality show the Veil's beauty in smaller telescopes. The drawings of August 12, 1983, illustrate the view through the 8-inch telescope under good skies. Only one magnification, 117×, was tried because clouds interrupted the observing session. Even so considerable detail could be seen.

Of NGC 6960, the part north of 52 Cygni was easy. The edges were sharp, and the width slowly decreased to a fine point at the northernmost extreme. The bend in the nebula 12 arc-minutes north of 52 Cygni was easily detected, as was a small fork that extended an arc-minute or so to the northeast. The part south of the star (actually curving to the southeast) could not be detected. That portion becomes very wide and dim. The glare from 52 Cygni makes the observation difficult by adding stray light to the field.

NGC 6992–5, the brighter arc, showed considerable detail through the 8-inch under good skies. The brightest portion, near the northern extreme, is NGC 6992. The southern end becomes broad and looked forked in the 8-inch. That end is NGC 6995. The surface brightness varied along the length of the arc, an indication that if the sky were slightly better, the individual filaments that make up the nebula could be resolved.

Under good skies, none of the Veil was visible in the 2.4-inch, 7.9× finder. But under skies showing stars only about 0.3 magnitude fainter, the Veil showed much better. The change resulting from this slight increase in sky quality was amazing. Now NGC 6992-5 could be seen through the 2.4-inch finder, and in the 3-inch finder at 31×, all three portions of the Veil were visible! This observation, on August 26, 1984, was the first time I had seen the triangular patch.

Under these improved conditions, the 8-inch showed the faint part of NGC 6960 south of 52 Cygni easily at 117×. At 188×, the narrow tip at the northern end had sharper edges than when seen the year before. NGC 6992–5 contained a wealth of detail. At the northern end of this arc, in NGC 6992, individual filaments appeared. Throughout the Veil, the filaments range from only 1 to 5 arc-seconds in width. Resolving them demands excellent skies and at least an 8-inch telescope.

For the drawing made that night, the best view was at 188×. At that power only portions of NGC 6992–5 could be seen at one time. I had the impression that the filaments coincided with star chains. In fact, about two-thirds of the way toward the southern end I drew a nearly circular loop. This is not nebulosity but many faint stars that gave that impression. Farther south, where the nebula's width increases, several additional arcs could be seen.

Some amateurs using large telescopes have reported seeing colors of individual filaments. The Veil is truly a beautiful structure that will reveal its intricate splendor to medium size instruments under good to excellent skies.

Photograph of NGC 6960. South is up. (Courtesy Mount Wilson and Las Campanas Observatories, Carnegie Institution of Washington.)

A VISUAL ATLAS OF DEEP-SKY OBJECTS

Drawing of NGC 6960.

Scale: 3.0 arc-min/cm
8-inch f/11.5 Cassegrain
20mm Erfle (117×)

Viewing Distance (cm)

10×: 115 100×: 11
25×: 46 200×: 6
50×: 23 300×: 4

air mass: 1.08, faintest star: 14.5 at zenith, 188×;
no tracking
8/12/83 8:20–8:35 UT at Waianae ranch, Hawaii;
R. Clark

Photograph of NGC 6992–5. South is up. (Courtesy Mount Wilson and Las Campanas Observatories, Carnegie Institution of Washington.)

A VISUAL ATLAS OF DEEP-SKY OBJECTS

Drawing of NGC 6992-5.

Scale: 3.4 arc-min/cm
8-inch f/11.5 Cassegrain
20mm Erfle (117×)

Viewing Distance (cm)

10×: 101	100×: 10
25×: 40	200×: 5
50×: 20	300×: 3

Drawing Method 2
air mass: 1.12, faintest star: 14.5 at zenith, 188×; no tracking
8/12/83 8:05–8:20 UT at Waianae ranch, Hawaii; R. Clark

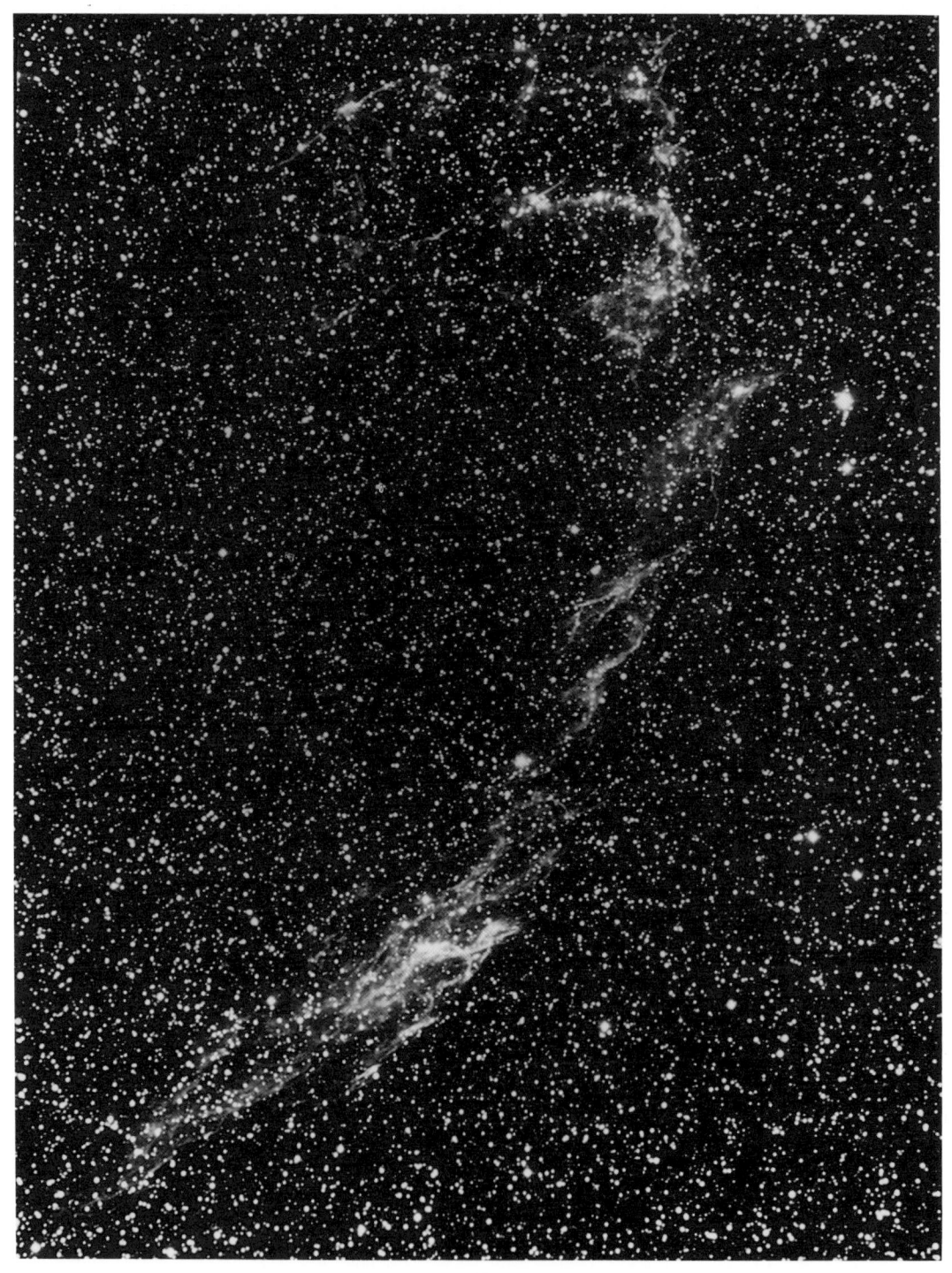

Photograph of NGC 6992–5. South is up. (Courtesy Mount Wilson and Las Campanas Observatories, Carnegie Institution of Washington.)

A VISUAL ATLAS OF DEEP-SKY OBJECTS

Drawing of NGC 6992–5. Note the greater detail that became apparent with the modest improvement in sky quality over the observation of 8/12/83.

Scale: 3.4 arc-min/cm	Viewing Distance (cm)		Drawing Method 2
8-inch f/11.5 Cassegrain			air mass: 1.05, faintest star: 14.7 at zenith, 188×;
20mm Erfle (117×)	10×:101	100×:10	no tracking
12.4mm Erfle (88×)	25×: 40	200×: 5	8/26/84 6:45–7:20 UT at Jones Hill Quadrangle,
7mm Erfle (334×)	50×: 20	300×: 3	Colorado; R. Clark

NGC 7000, THE NORTH AMERICA NEBULA IN CYGNUS.

R.A. 20^h 58.8^m, Dec. $44°$ $20'$ (2000.0)

Technical. NGC 7000 is a huge and beautiful mixture of emission, reflection and dark nebulae. It was first noted by William Herschel in 1786. The whole complex is estimated to be about 2300 light-years away and about 60 light-years across. For many years NGC 7000 was thought to be illuminated primarily by 1st-magnitude Deneb, but that type A2 star produces too little ultraviolet light. Astronomers now believe that a faint star largely hidden by intervening dust provides the main illumination. This star is near the "Atlantic coast".

Another part of the nebular complex, lying west of NGC 7000, is known as the Pelican Nebula: IC 5067, 5068, and 5070. The Pelican and North America are actually one object; a dark cloud in the foreground merely seems to split them.

Visual. The North America Nebula has an angular width of about 100 arc-minutes. Recent estimates give NGC 7000 a total magnitude of 5 and a mean surface brightness of 23.6 magnitudes per square arc-second. Its size is about 100 arc-minutes in diameter. It's often stated that NGC 7000 can be seen with the unaided eye under excellent skies. This may be because its shape is outlined by foreground dark clouds that block our view of both stars and nebulosity in the background. The many faint stars seen in the nebula itself contribute to the apparent contrast of "North America" with the surrounding sky, and their light may be what's seen with the unaided eye. The faint stars may also have biased the magnitude estimates.

The nebula itself is easily visible with 7 × 50 binoculars under good skies, but the best views are under excellent skies with a telescope at about 30×. The drawing shows the view through the 8-inch's finder: a mere 2.4-inch refractor working at 7.9×. The nebula appeared as a soft glow intermingled with many faint stars. The "North America" outline was clearly visible. Through the 8-inch Cassegrain itself, the nebula could not be detected at 82×, the lowest power tried. Under less than ideal conditions, lower power wide-field instruments are the only ones that show this unusual object. Under excellent skies, higher powers may be tried in an effort to see some of the detail within the nebula, but again, low powers yielding a field of view of at least 2° give the best overall view.

In August 1984 I used the 3-inch finder under excellent skies of the Colorado Rocky Mountains. The North America Nebula was a beautiful sight at 31×. It stood out like a puff of cotton among the stars. The Pelican was also faintly visible. Under such excellent conditions, detail could be detected through the 8-inch in the "Mexico–Central America" region at 82×. That region has the highest surface brightness and contrast with the surrounding sky.

VISUAL ASTRONOMY OF THE DEEP SKY

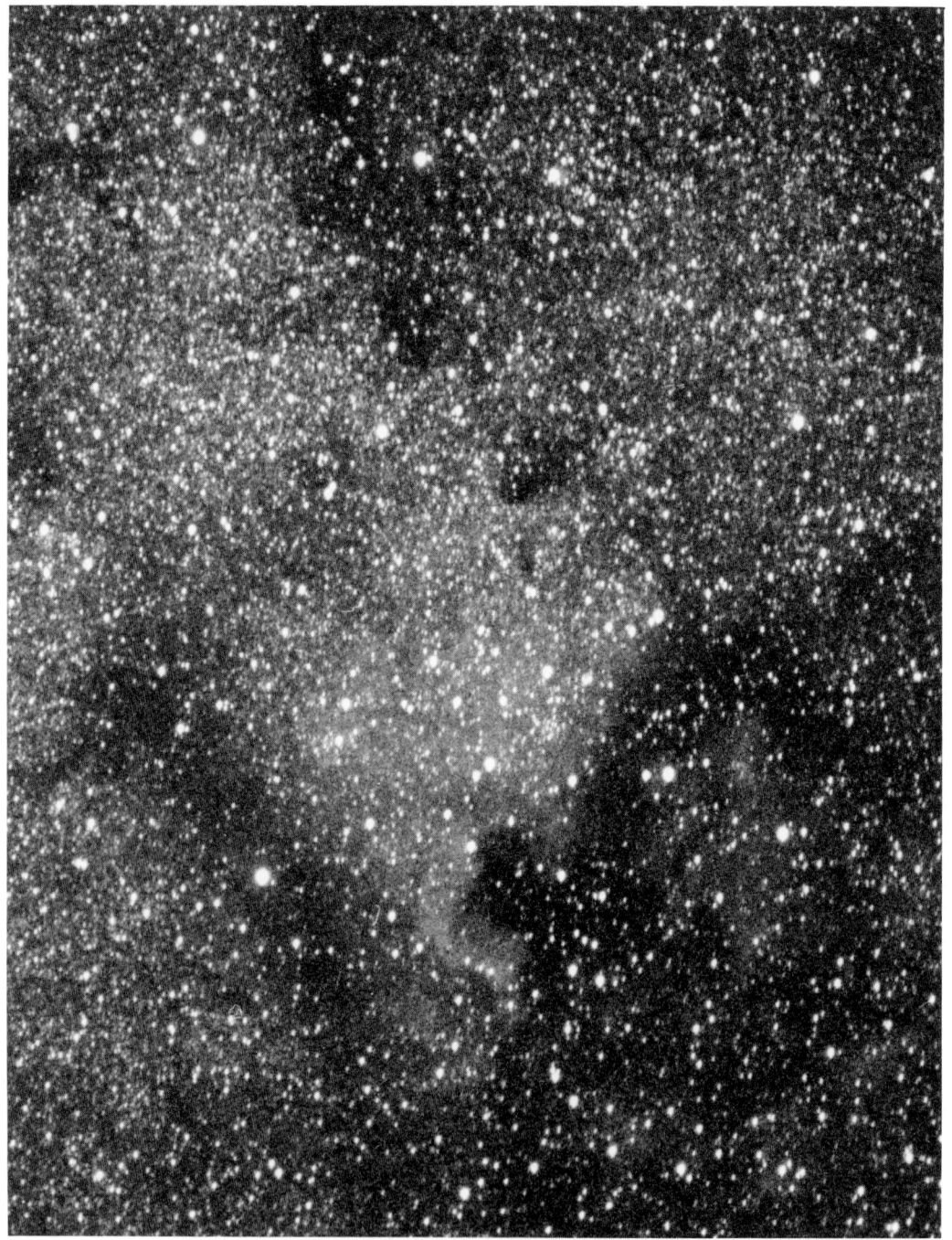

Photograph of NGC 7000. North is up. (Courtesy Ron Pearson.)

Drawing of NGC 7000.

Scale: 25. arc-min/cm
60mm f/5 refractor
38mm Erfle (7.9×)

Viewing Distance (cm)

1×:138	10×:14
5×: 28	15×: 9
8×: 17	20×: 7

air mass: 1.11, faintest star: 11 at zenith, 8×; no tracking
10/8/83 7:10–7:40 UT at Waianae ranch, Hawaii; R. Clark

M15 (NGC 7078), GLOBULAR CLUSTER IN PEGASUS

R.A. 21^h 30.0^m, Dec. $12°$ $10'$ (2000.0)

Technical. M15 is an unusual globular cluster in that it is the only one known to contain a planetary nebula. The cluster was discovered by J. D. Maraldi in 1746, but it was not until 1927 that the small planetary (K 648) was found, by F.G. Pease on photographs made with the 100-inch Mount Wilson reflector. The nebula seems definitely to be a true member of M15. The reason only one planetary nebula is known among the millions of stars in globular clusters is that planetary nebulae are extremely short-lived compared with the age of the ancient stars that populate the globulars.

M15 also contains an X-ray source. Some astronomers have suggested that this source, at the core of the cluster, is a black hole. Its mass is thought to be about 800 suns.

The cluster is 30 000 to 40 000 light-years away, and its total luminosity is about 200 000 times that of the Sun. It is about 130 light-years in diameter.

Photograph of M15. South is up. (Courtesy National Optical Astronomy Observatories.)

Visual. M15 is one of the brightest globulars in the northern sky. With a total magnitude of 6.5, it can be seen by the unaided eye under excellent conditions. Its size of about 10 arc-minutes yields a mean surface brightness of 20.1 magnitudes per square arc-second – though of course it is much brighter near the center.

M15 is very rich, with its stars tightly packed. So it takes a medium-size telescope to begin resolving the stars. Through the 8-inch in good skies, the best view was at 188×. At this power the cluster was resolved and the stars appeared like many pinpoints. In the drawing the brighter stars are positioned correctly; the many faint ones are placed only to show the overall impression. The center of the cluster is very bright, with the intensity decreasing rapidly outward. The small planetary nebula is magnitude 13.8 and only one arc-second across; I can find no reports of it ever having been seen visually.

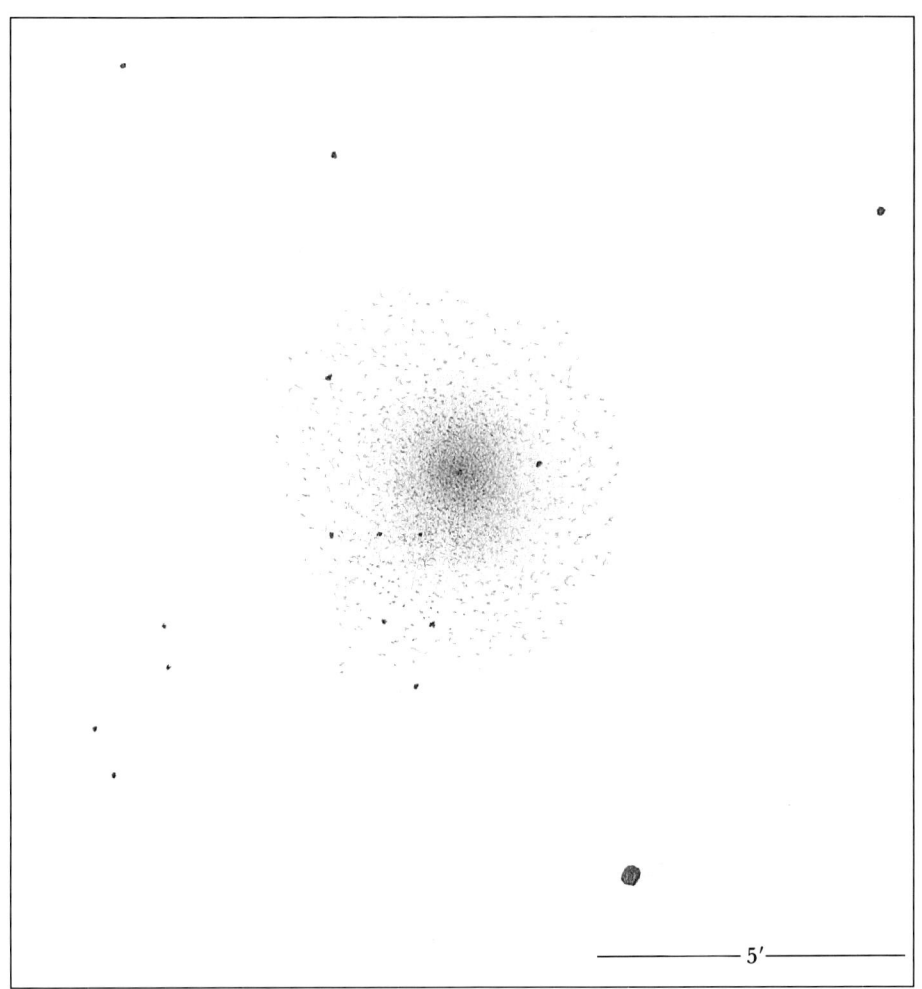

Drawing of M15.

Scale: 1.2 arc-min/cm
8-inch f/11.5 Cassegrain
20mm Erfle (117×)
12.4mm Erfle (188×, best view)

Viewing Distance (cm)

25×:115   200×:14
50×: 57   300×:10
100×: 29 400×: 7

air mass: 1.07, faintest star: 14.5 at zenith, 188×; no tracking
8/12/83 9:10–9:30 UT at Waianae ranch, Hawaii; R. Clark

M2 (NGC 7089), GLOBULAR STAR CLUSTER IN AQUARIUS

R.A. 21^h 33.5^m, Dec. $-00°$ $50'$ (2000.0)

Technical. M2 is a bright globular cluster discovered by J. D. Maraldi in 1746. Its distance is about 50 000 light-years, its diameter 150 light-years. M2 is very compact; the central arc-minute emits about 30 percent of the light. The cluster contains at least 100 000 stars larger and more luminous than our Sun – which at the distance of M2 would appear as faint as magnitude 20.7. The cluster contains many thousands of red and yellow giant stars brighter than 15th magnitude.

M2 contains very few variable stars compared with other globulars. Only 17 variables were known as of 1947, whereas other globulars often contain 200 or more. One variable star was discovered by an amateur in 1897. A. Chevremont found an RV Tauri type variable fluctuating between magnitude 12.5 and 14.0, with a period of 67.09 days. This variable is said to significantly alter the appearance of the cluster when bright. It is on the east edge, slightly north of center. Careful monitoring with medium or large telescopes for a two-month period should reveal its changes.

Photograph of M2. South is up. (Courtesy Evered Kreimer, The Messier Album.)

Visual. M2 can be seen with the unaided eye under excellent skies. It has a total magnitude of 6.0 and a diameter of 7 arc-minutes, which gives a very high mean surface brightness, 18.9 magnitudes per square arc-second – and the center is about four magnitudes brighter still. Through the 8-inch telescope in moderate skies, M2 was mostly resolved at 187×. Note the cluster's smaller apparent size in the drawing compared with the photograph. Only the central 3.5 arc-minutes could be seen, but since this portion of the cluster mimics the overall appearance, the view resembles the photo.

Amateur astronomer and author John Mallas reported resolving the few brightest members in his 4-inch refractor, as well as seeing a dark lane crossing the cluster's northeast corner. But I have failed in many attempts to detect the dark lane in the 8-inch. Mallas suggested the lane could be seen in the photograph by Evered Kreimer in their co-authored book *The Messier Album*. However, that "lane" is near the edge of the cluster in the photo and is most probably beyond the limits of visibility in the 8-inch as well as the 4-inch. Mallas had wondered if the "lane" might be an illusion.

Drawing of M2.

Scale: 1.2 arc-min/cm	Viewing Distance (cm)	air mass: 1.19, faintest star: 13.6 at zenith, 187×;
8-inch f/11.5 Cassegrain		no tracking
12.5mm Orthoscopic	25×:115 200×:14	9/19/82 9:45–10:03 UT at Ewa Beach, Hawaii;
(187×)	50×: 57 300×:10	R. Clark
	100×: 29 400×: 7	

NGC 7293, THE HELIX NEBULA IN AQUARIUS

R.A. 22^h 29.7^m, Dec. $-20°$ 51' (2000.0)

Technical. NGC 7293 is a large, dim, beautiful and complex planetary nebula. It is the biggest planetary in the sky, with an angular size half that of the Moon. Because it appears so large, it is considered the nearest planetary. Estimates of its distance vary considerably, but 450 light-years seems a reasonable compromise. At that distance the nebula would have a diameter of 1.7 light-years. As with all planetaries, there is a hot, blue central star that puffed off much of its mass to form the nebula, very recently as astronomical timescales go. The remaining core of the star resembles a white dwarf, but is so hot that it produces copious ultraviolet light – enough to make the nebula fluoresce. In this case the star's surface temperature is about 100 000 Kelvin.

Because NGC 7293 is so large, deep photographs show a wealth of detail. The expanding gas has the appearance of two overlapping circles or a helix, thus it is commonly called the Helix Nebula. On the inner side of the helix are many nebulous streaks that point to the central star. One explanation for these almost comet-like filaments is that a stellar wind is sweeping past small clumps of gas in the nebula and blowing tails from them.

This planetary, like many others, has a very non-spherial structure. Such shapes suggest that a planetary's form often depends on such things as the central star's rotation rate, whether or not it is binary (many central stars are indeed double), whether a disk of gas orbits it closely, and possibly whether it has a strong magnetic field.

Visual. NGC 7293 has a deceptively bright total magnitude, 6.5. But its angular size of 12 by 16 arc-minutes means this light is spread over a huge area. The average surface brightness is quite low compared with other planetaries, 20.8 magnitudes per square arc-second. Nevertheless, the difficulty of seeing the Helix is often exaggerated. Its surface brightness is greater than that of many other deep-sky objects discussed in this book.

Under excellent skies, the Helix can easily be seen through binoculars as a faint patch. With larger telescopes, there has been some controversy over whether the dark center is visible. Since the surface brightness is low, many observing guides recommend very low powers, and this is probably why observers fail to detect the dark center.

Through the 8-inch under good skies at a magnification of 117×, the Helix appeared as a ring of uneven brightness. The north and south portions of the ring were strongest. The west side was very faint, sometimes giving the nebula the appearance of a horseshoe on its side. The 13th-magnitude central star was easy.

Under only moderate skies, the ring shape is still easy in the 8-inch at 117×, but the ring's unevenness cannot be detected. Under excellent skies, the 3-inch finder at 31× showed the Helix as a faint patch with a dark center, but the central star was invisible.

Photograph of NGC 7293. South is up. (Courtesy National Optical Astronomy Observatories.)

A VISUAL ATLAS OF DEEP-SKY OBJECTS

Drawing of NGC 7293.

Scale: 1.2 arc-min/cm	Viewing Distance (cm)
8-inch f/11.5 Cassegrain	
20mm Erfle (117×)	25×:115 200×:14
	50×: 57 300×:10
	100×: 29 400×: 7

Drawing Method 2
air mass: 1.58, faintest star: 14.5 at zenith, 188×;
no tracking
8/12/83 9:40–9:56 UT at Waianae ranch, Hawaii;
R. Clark

NGC 7331, SPIRAL GALAXY IN PEGASUS
R.A. 22^h 37.1^m, Dec. $34°$ $26'$ (2000.0)

Technical. NGC 7331 is a spiral galaxy tilted about 15° to 20° from edge-on. It is similar in structure to M31, but at about 50 million light-years it is some 20 times more distant. The mass of NGC 7331 is estimated as about 140 billion times the mass of our Sun, and its luminosity as some 50 billion times the Sun's light.

Visual. NGC 7331 has a total magnitude of 10.4 and an angular size of 10 by 2.4 arcminutes, for a mean surface brightness of 22.5 magnitudes per square arc-second. Under good skies the galaxy is visible in small amateur telescopes as a fuzzy patch. Through the 8-inch under good skies, it appears as an oval extending north–south with a small, bright nucleus. The west side is closest to us, as the photograph shows; dark dust patches here are silhouetted against the galaxy's bright inner part. This side had a sharp boundary as seen through the 8-inch at magnifications of 117× to 188×. No other detail was detected. But large amateur telescopes and good skies may bring out some detail, such as dark patches and some of the faint surrounding galaxies.

A half degree to the south-southwest are five galaxies in a tight group known as Stephan's Quintet. These galaxies and their magnitudes are: NGC 7317 (14.5), NGC 7318A (13.8), NGC 7318B (13.5), 7319 (13.5) and NGC 7320 (13.1). None could be found with the 8-inch during the observing session for NGC 7331. However, in the clear skies of the Colorado Rockies, all but NGC 7317 were detected in the 8-inch at magnifications of 117× and 188×. There the limiting magnitude was 14.6.

Photograph of NGC 7331. South is up. (Courtesy Palomar Observatory.)

A VISUAL ATLAS OF DEEP-SKY OBJECTS

Drawing of NGC 7331.

Scale: 0.6 arc-min/cm
8-inch f/11.5 Cassegrain
20mm Erfle (117×)
12.4mm Erfle (188×)

Viewing Distance (cm)

25×:229   200×:29
50×:115   300×:19
100×: 57 400×:14

Air mass: 1.08, faintest star: 14.0 at zenith, 188×; no tracking
9/11/83 11:00–11:20 UT at Barbers Point, Hawaii; R. Clark

NGC 7662, PLANETARY NEBULA IN ANDROMEDA

R.A. 23^h 25.9^m, Dec. $42°$ $33'$ (2000.0)

Technical. NGC 7662 is a small planetary nebula somewhere between 2000 and 6000 light-years away. (The distances of all planetary nebulae are poorly known.) Its diameter is between 0.3 and 0.8 light-years, depending on the distance adopted. The central star has a surface temperature around 75 000 Kelvin and has been suspected of variability, though no systematic photometric study has been completed. The nebula consists of an inner, broken ellipse surrounded by a fainter elliptical disk. The dark area in the center is perhaps similar in brightness to the outer ellipse.

Visual. NGC 7662 has a total magnitude of 8.5 and an angular size of 32 by 28 arc-seconds, for a very high mean surface brightness: 15.6 magnitudes per square arc-second. The planetary is visible in 7 × 50 binoculars and small telescopes as a star, but will begin to show its disk at powers near 50×. Through the 8-inch telescope under moderate skies, the darker center was suspected at 334× and seen fairly well at 592×.

This object provides a good example of how the upper limit to useful magnification depends on image brightness. In the dim light levels of deep-sky astronomy, the eye has very poor resolution, so small detail may have to be magnified very highly to be detectable. The 8-inch had no clock drive for tracking, so examination of this object was rather difficult at very high powers. Telescopes of similar size with clock drives could use higher powers still. The lesson is that if you are observing an object and feel that higher magnification may bring out more detail, go ahead and ignore the "accepted" magnification limit of 60× per inch of objective diameter (480× on an 8-inch telescope). See Chapter 3 for more on this subject.

A VISUAL ATLAS OF DEEP-SKY OBJECTS

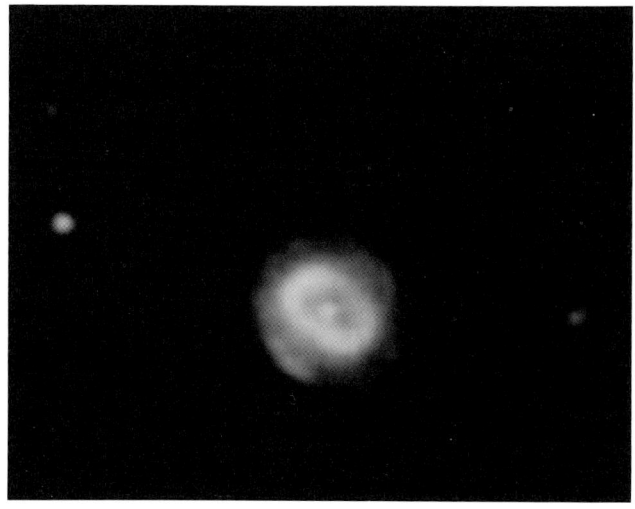

Photograph of NGC 7662. South is up. (Courtesy Jack B. Marling.)

Drawing of NGC 7662.

Scale: 0.25 arc-min/cm
8-inch f/11.5 Cassegrain
12.4mm Erfle (188×)
7mm Erfle (334×)
12.4mm Erfle + 3.15× barlow (592×)

Viewing Distance (cm)
50×:275 300×:46
100×:138 400×:34
200×: 69 600×:23

air mass: 1.41, faintest star: 13.8 at zenith, 188×; no tracking
9/5/83 14:00–14:20 UT at Hawaii Kai, Hawaii; R. Clark

CONCLUSIONS

A tremendous amount of detail is visible in deep-sky objects through amateur telescopes if plenty of time is taken to observe an object and several different magnifications are used. The greatest variable for seeing such detail is not the size of the telescope but the condition of the sky. For deep-sky work, a 4-inch telescope in the country will beat a 16-inch in the city.

Appendix A: Suggested reading

RECOMMENDED BOOKS

No single book can cover all of amateur astronomy. This appendix describes some sources of material not in this book that amateurs need to know. Of course, there are many more good publications than can be listed here.

The two main amateur magazines are *Sky & Telescope* (Sky Publishing Corp., P.O. Box 9111, Belmont, MA 02178-9111, USA) and *Astronomy* (AstroMedia, a division of Kalmbach Publishing Co., 1027 N. 7th St., Milwaukee, WI 53233, USA). Both are monthly. They have articles on all aspects of observing, photography, and current astronomical research, as well as book reviews and descriptions of upcoming sky events. They also contain advertisements for the latest books, star atlases, and other literature. Any active amateur should subscribe to one of these magazines – or both, at least until you decide which better suits your needs. *Astronomy* is aimed more at beginners and the general public; *Sky & Telescope* is more substantial. It contains articles of interest to the serious amateur but is still excellent for beginners.

Another magazine of special interest to the visual observer is *Deep Sky* (AstroMedia, a division of Kalmbach Publishing Co., 1027 N. 7th St. Milwaukee, WI 53233, USA). It is produced by many of the same people who produce *Astronomy*. As the title suggests, it is limited to visual and photographic observations of deep-sky objects. Published quarterly, it contains articles on observing that are too long and detailed for *Astronomy* or *Sky & Telescope*. I have found *Deep Sky* interesting because the many observers who write letters and articles for it are constantly pointing out new objects to look for, many of which have not previously been considered within the amateur's realm.

The Astronomical Society of the Pacific (390 Ashton Avenue, San Francisco, CA 94112, USA) is an international, non-profit scientific and educational organization. It sells books, introductory pamphlets, a teachers' newsletter, slides, posters, computer software, bumperstickers, and other materials. It holds yearly meetings in the United States of America. Its bimonthly magazine *Mercury* features non-technical articles on new developments in astronomy. The *Publications of the ASP* is a technical journal of research reports from professional, and sometimes amateur, observatories; it may be of interest to advanced amateurs.

BEGINNING: LEARNING THE SKY

There are many books, pamphlets and planispheres for learning the constellations, so only a few are mentioned. A good bookstore should have several to choose from.

Whitney's Star Finder by Charles A. Whitney (Alfred A. Knopf, Inc., New York, 102 pages, 1981). This book teaches about finding planets, stars and constellations, as well as the basics of eclipses, phases of the Moon, and more. The star finder (planisphere) is too small, less than seven inches in diameter, and it may be difficult to read with a flashlight while trying to identify constellations.

The Stars by H.A. Rey (Houghton Mifflin Co., Boston, 160 pages, 1952, revised 1980). This is probably the most "user-friendly" guide to learning the stars and constellations. It also covers the basics of star and planet motions, Moon phases, and so on. It's easily understood by teenagers. Possible drawbacks are its heavy use of English constellation names ("Serpent Holder" for Ophiuchus) and Rey's charming but sometimes too elaborate stick-figure constellation patterns. It has all-sky maps individually drafted for each month to reduce distortion.

The Night Sky by David Chandler (Sky Publishing Corp., Cambridge, Mass., 1977) This 10-inch-diameter planisphere reduces the distortion of star patterns that affects most planispheres. It has two sides: one minimizes distortion north of the celestial equator, the other south of it.

Astro-dome: A Three Dimensional Map of the Night Sky by Klaus Hunig (Sunstone Publications, Cooperstown, NY, 1983). This is an innovative cutout book from which a paper sky-dome is constructed. The stars are printed with phosphorescent paint to glow in the dark. The dome shape keeps the constellations from having to be distorted to fit onto a flat piece of paper. It is quite large: 20 inches in diameter, giving the appearance of a miniature planetarium. The cutout and construction work is a full evening's task, and the paper dome is too flimsy to stand heavy use or breezy nights.

STAR ATLASES

A star atlas is a larger, more serious set of charts than the simple maps for finding constellations. Just as a mariner may run aground without excellent sea charts, you need accurate, detailed star charts to navigate the skies. The better the charts, and the fainter the stars they show, the more precisely you will be able to locate a difficult object's position with a telescope.

The most widely used general-purpose atlas is *Sky Atlas 2000.0* by Wil Tirion (Sky Publishing Corp., Cambridge, Mass., 1981). As the name implies, its coordinate grid is drawn for equinox 2000.0. Over 43 000 stars to as faint as magnitude 8 are plotted on 26 charts. Only the deluxe edition is recommended: the charts are wire bound inside a cover and are more durable than the slightly smaller, unbound charts of the black-and-white editions. In the deluxe edition each chart is 12.5 by 17.5 inches, the 2500 deep-sky objects are color coded, and there is a plastic coordinate-grid overlay for reading or plotting accurate positions.

Surpassing even this work is Tirion's latest opus (with B.Rappaport and G. Lovi), *Uranometria 2000.0* (Willmann-Bell, Inc., Richmond, Va., 1987 and 1988). This atlas plots 332 000 stars to about magnitude 9.5, as well as some 10 300 deep-sky objects including the entire NGC. It consists of 473 black-and-white charts, each about 6 by 8 inches, bound in two hardback volumes covering the northern and southern sky, respectively. The cartography follows the style of *Sky Atlas 2000.0*, making the step up from it easy. The sheer number of charts makes this no atlas for the novice; you have to know the sky fairly well not to get lost among them.

To improve their usefulness Appendix E, a catalog of deep-sky objects, includes a list of which chart to use to find a particular object. To use *Uranometria 2000.0* you will usually need a more general atlas like *Sky Atlas 2000.0* simply to locate where the *Uranometria* field is in the sky. Trying to find which chart to use if you have only an object's coordinates can take quite a long time in the dark. The time spent can be worth it because once you have the right field area, the details on *Uranometria 2000.0* will virtually assure you that you are pointing your telescope at the right place.

The most detailed (and expensive) atlases are photographic: high-contrast copies of actual photographs of the sky. These are for the very advanced amateur, since they generally omit star names, deep-sky object symbols, constellation borders, and all the other clarifications and conveniences at the mapmaker's disposal. They show vast numbers of stars, but deep-sky objects may or may not have been recorded by the camera.

The lowest-price photographic atlas is *Photographic Star Atlas 1950.0* (also called *The Falkau Atlas*) by Hans Vehrenberg (Treugesell-Verlag, Dusseldorf, 1972). It covers the sky in 428 charts down to about magnitude 13. Plastic overlays can be used to determine a star's coordinates. No objects are labelled. Thus the amateur must very carefully plot any faint object to be sought. But once it's plotted, the many stars can greatly assist in locating it. The *Photographic Star Atlas* costs about three times as much as the deluxe *Sky Atlas 2000.0*.

Also published by Hans Vehrenberg is the even larger and more detailed *Atlas Stellarum 1950.0* (Treugesell-Verlag, Dusseldorf, 1970).

It is twice the scale of the *Photographic Star Atlas* and reaches stars at least a magnitude fainter. Its star images are smaller, rounder and cleaner. The 486 charts are each 13 by 13 inches, but even so you may need a magnifier for the swarms of the very smallest stars. It too has no labels.

HANDBOOKS AND OBSERVING GUIDES

The Messier Album by John H. Mallas and Evered Kreimer (Sky Publishing Corp., Cambridge, Mass., 216 pages, 1978, revised 1980). This is an observing guide to the 110 Messier objects. Evered Kreimer photographed each one with his 12.5-inch reflector and John Mallas made drawings of most with a 4-inch refractor. It's interesting to compare Mallas' observations with those in this book. Unfortunately he did not include the scale of his drawings, or even a few reference stars, so comparisons with photographs or other drawings are often difficult.

Burnham's Celestial Handbook by Robert Burnham, Jr. (Dover Publications, Inc., New York, 2139 pages in three volumes, 1978). This is a magnificent collection of facts and observational notes on thousands of celestial objects. It's an excellent companion to this book and to any star atlas. There are over 600 black-and-white photographs but no drawings. The descriptions of many objects include pages of history and folklore, as well as scientific facts and theories gleaned from the literature up to the 1970s. The technical descriptions in the book you are holding include many newer scientific explanations, often based on observations that were impossible when Burnham wrote. Any "latest scientific explanations" tend to become dated quickly, but even so, the rich contents of *Burnham's Celestial Handbook* guarantee that it will always remain a classic.

Amateur Astronomer's Handbook by J.B. Sidgwick, revised by James Muirden (Enslow Publishers, Hillside, New Jersey, 586 pages, 4th edition, 1980). This handbook from England covers telescope construction, operation, drives, eyepieces, spectroscopes, micrometers, and many advanced topics, all in great depth. Some sections are still outdated, despite the recent revisions. For instance, the chapter on eyepieces (or oculars as the British call them) dwells on many that are no longer in common use and cannot be purchased anywhere in the United States of America. But in general, this is still a valuable and thorough treatment of astronomical instruments and their use.

Chapter 3 of the fourth edition has several mistakes concerning magnification and apparent image brightness. None of the concepts presented in this book were discussed by Sidgwick. For example, in Chapter 24 (page 427 of the fourth edition), Sidgwick says that reducing the brightness (implied by a magnification increase) will worsen the eye's contrast-detection threshold. Remember that in Chapter 2 of this book, we saw that the contrast threshold depends on the surface brightness of the background. The threshold will *improve* for most faint objects as power is raised, because they are small and usually below the optimum magnified visual angle. Sidgwick wrongly says that lower powers will make low-contrast detail visible. With these cautions in mind (which cover only a few pages of the book), the *Amateur Astronomer's Handbook* is an excellent work.

Observational Astronomy for Amateurs by J.B. Sidgwick, revised by James Muirden (Enslow Publishers, Hillside, New Jersey, 348 pages, 4th edition, 1982). This companion book to the one above treats practical aspects of observing. Though it too is growing dated, it is still a good, detailed resource for observing the Sun, Moon, planets (one chapter for each planet), asteroids, zodiacal light, aurorae, meteors, comets, and variable and binary stars. It contains only one-half page on nebulae and clusters, doesn't mention galaxies (as if they were just a kind of nebula), and part of this page is on photography. For the subjects the book does cover well, it's one of the best in its class. This lopsidedness is part of the reason I wrote the present book: the typical observing handbook did not properly treat deep-sky objects.

Atlas of Deep Sky Splendors by Hans Vehrenberg (Sky Publishing Corp., Cambridge, Mass.; Cambridge University Press, Cambridge, England; and Treugesell-Verlag, Düsseldorf; 242 pages, 4th edition, 1983). This beautiful collection of wide-field, mostly black-and-white photographs shows over 400 galaxies, star clusters, and nebulae. Most of the photos are a standard 3.5 degrees on a side, a scale that makes them super finder charts but renders most objects too small to show much detail. The book was designed for the purpose of finding faint objects, and at this it succeeds superbly.

Webb Society Deep-Sky Observer's Handbooks, compiled by the Webb Society, Kenneth Glyn Jones, editor (Enslow Publishers, Hillside, NJ, and Lutterworth Press, London, 1979-1987). Published in seven volumes:

1 Double Stars
2 Planetary and Gaseous Nebulae
3 Open and Globular Clusters
4 Galaxies
5 Clusters of Galaxies
6 Anonymous Galaxies
7 The Southern Sky

This series is a fine addition to the serious amateur observer's reference collection. Each volume has descriptions and drawings of many objects, including many that are rarely described. Volume 5, Clusters of Galaxies, covers a topic not found elsewhere in amateur literature. Unfortunately, each book is quite expensive for a paperback, the text is reproduced from typewriter manuscript, and the drawings tend to be reproduced poorly and too small. Often it is hard to correlate a sketch with a photograph.

Edmund Scientific (101 East Gloucester Pike, Barrington, NJ 08007, USA) published a series of short paperback books on astronomy in the mid 1960s, all written by Sam Brown. The following is a list:

Homebuilt Reflector Telescopes. 36 pages. Catalog no. 9066
Photography With Your Telescope. 36 pages. Catalog no. 9078
How to Use Your Telescope. 36 pages. Catalog no. 9055
Time in Astronomy 36 pages. Catalog no. 9065
Mounting Your Telescope. 36 pages. Catalog no. 9082
Telescope Optics. 32 pages. Catalog no. 9074

The above books were compiled into one work with only a few omissions:

All About Telescopes. 192 pages. Catalog no. 9094

These books describe nearly all aspects of the use of a telescope, finding objects in the sky, the use of setting circles, photography, and even making your own telescope. The text is clear and many information-packed figures appear throughout. Students in grade school or junior high can easily understand most of the concepts. These books are highly recommended for the beginning amateur astronomer. The price is also less than most books on the market today, costing less than one low-grade eyepiece.

The books are, however, dated. The title "All About Telescopes" implies material on many types, but Schmidt-Cassegrains, popular today, are not mentioned. Also the sections on photography do not discuss modern films. In spite of this they are a good source of basic material for the beginner.

Appendix B
Star clusters for finding your limiting magnitude

On the following pages are maps of open star clusters whose individual stars have had their magnitudes determined by Hoag *et al.* (1961). For each cluster, two charts are presented: one showing stars labelled with their visual magnitudes; the second, on the facing page, with stars only. Magnitudes are given to the nearest tenth, with the decimal point omitted so it won't be confused with a faint star. Thus, magnitude 12.2 is written 122.

Most magnitude labels are placed to the upper right of each star, the same distance from the center of the star dot regardless of the dot size. In some cases, crowding required the label be above or to the left of the star. Occasionally, the label had to be put below the star. By remembering these rules, and by refering to the key for star sizes with each chart, every magnitude label can be matched to the correct star – even when the label is closest to a different star.

Chapter 4 discusses in more detail how to use the charts to determine your telescope's limiting magnitude on a given night. In particular, avoid using faint stars close to other stars, especially bright ones. Magnification strongly affects the visibility of faint stars; each chart lists viewing distances from the page that correspond to various magnifications in a telescope (see Chapter 5 and equation 5.2).

North is up and astronomical east is to the left on each chart. The scales on the sides are approximate offsets in arc-minutes from a star that was selected to be the chart's center. These offsets are not exactly right ascension and declination. True lines of right ascension on these charts would converge slightly toward the poles, and most true declination lines would be slightly curved. But these effects are small, and for ordinary visual purposes, the "X" direction can be considered right ascension and the "Y" direction declination.

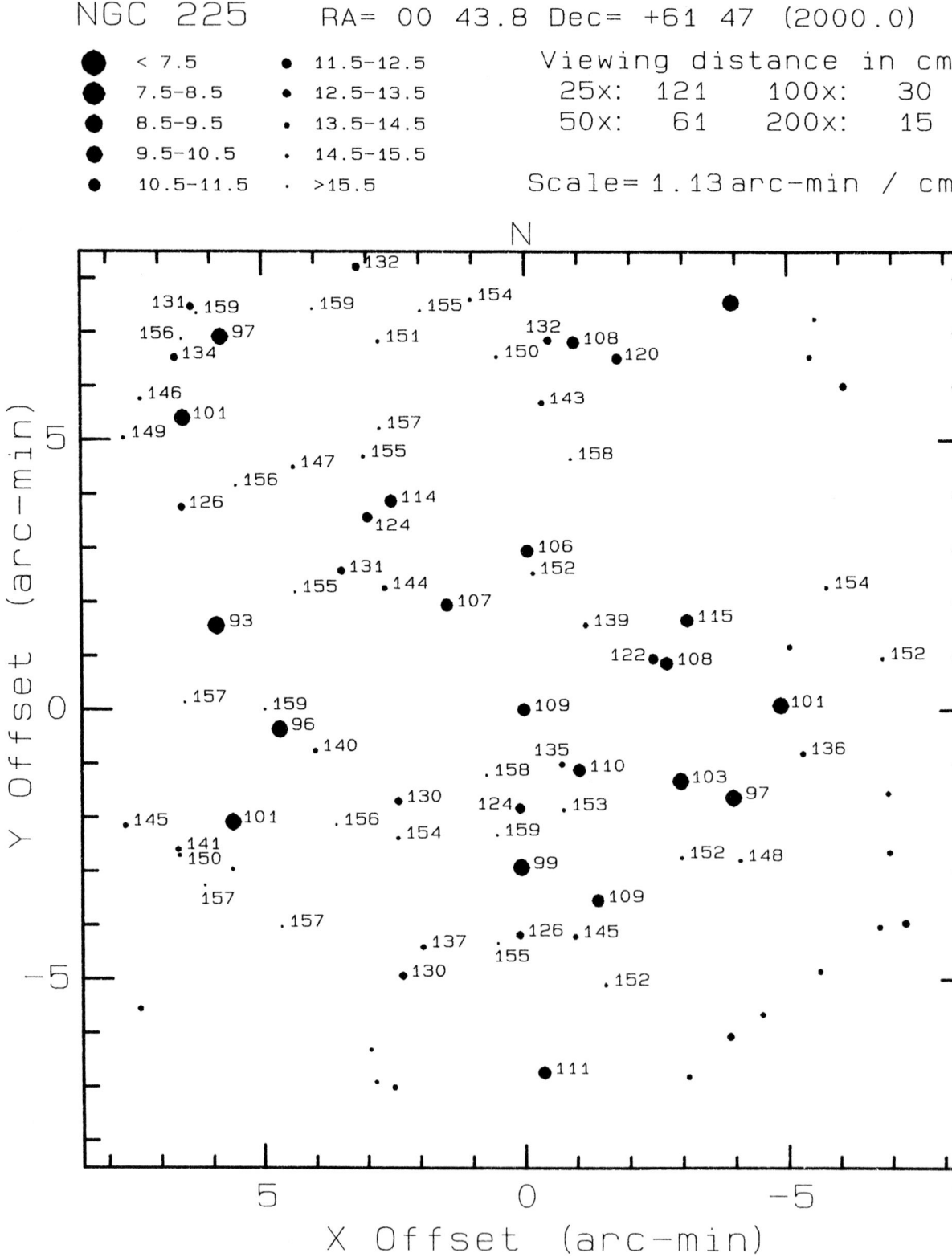

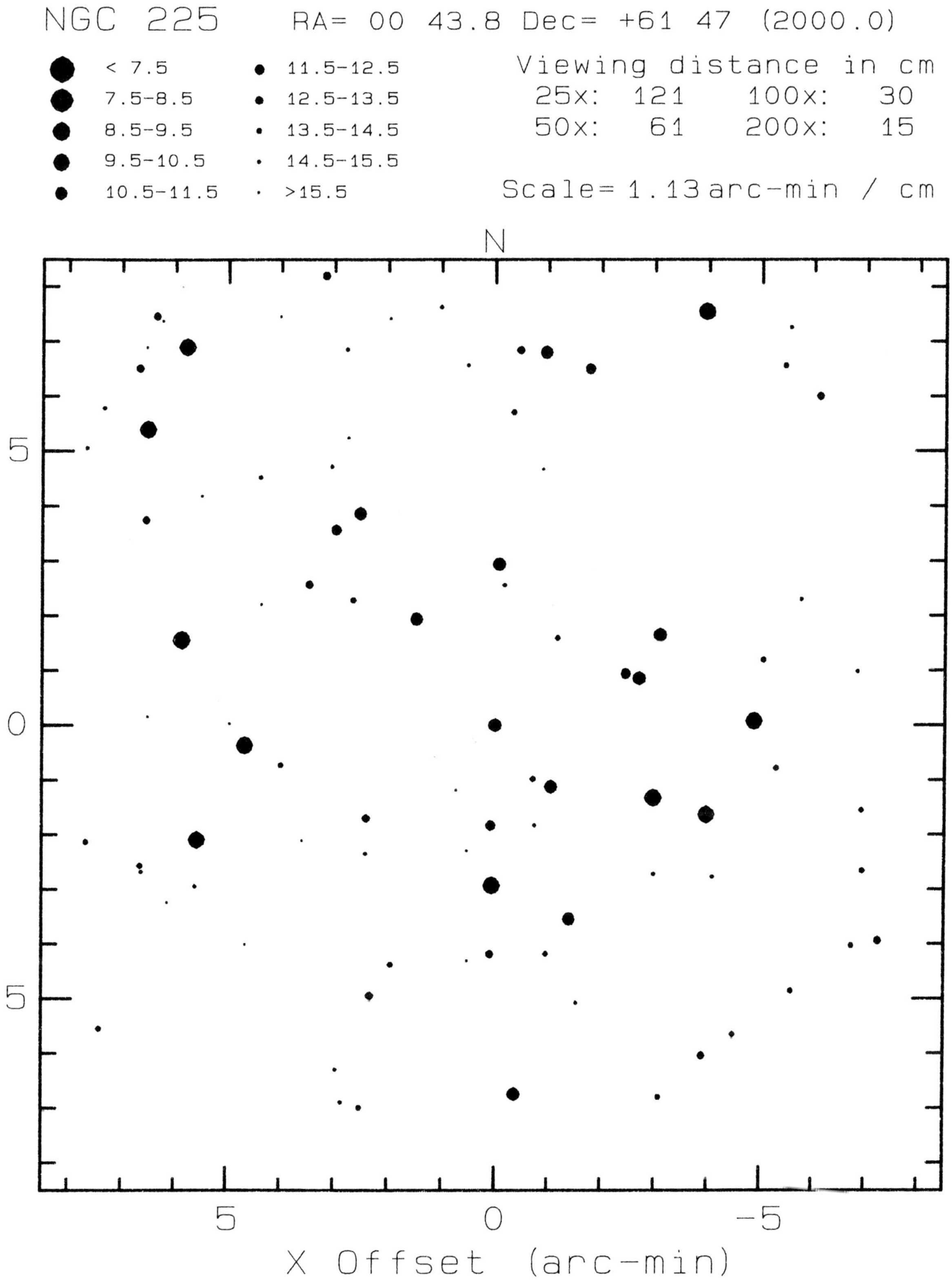

APPENDIX B: STAR CLUSTERS FOR FINDING YOUR LIMITING MAGNITUDE

NGC 225 RA= 00 43.8 Dec= +61 47 (2000.0)

Symbol	Magnitude	Symbol	Magnitude
●	< 7.5	•	11.5–12.5
●	7.5–8.5	·	12.5–13.5
●	8.5–9.5	·	13.5–14.5
●	9.5–10.5	·	14.5–15.5
●	10.5–11.5	·	>15.5

Viewing distance in cm
25x: 121   100x: 30
50x: 61   200x: 15

Scale= 1.13 arc-min / cm

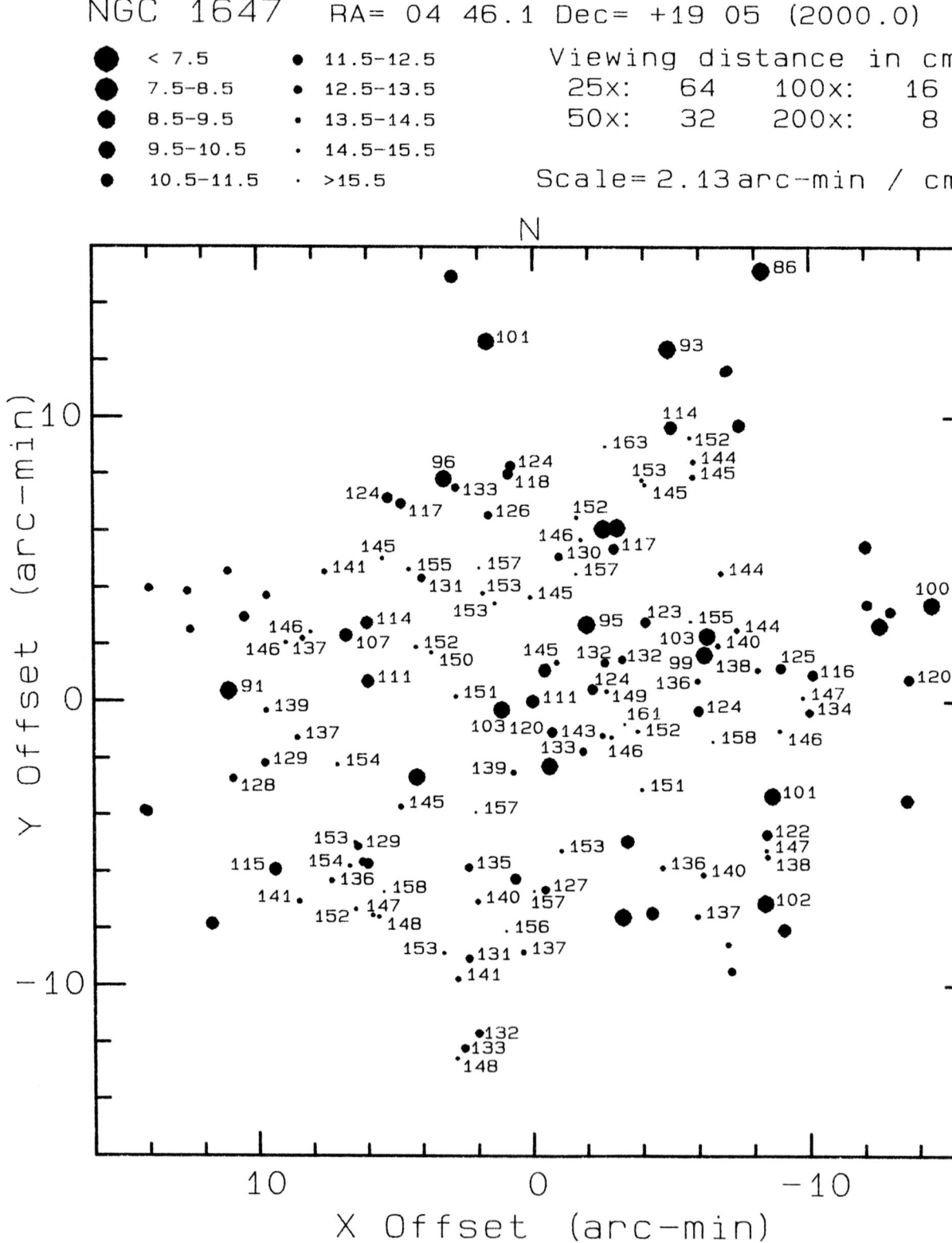

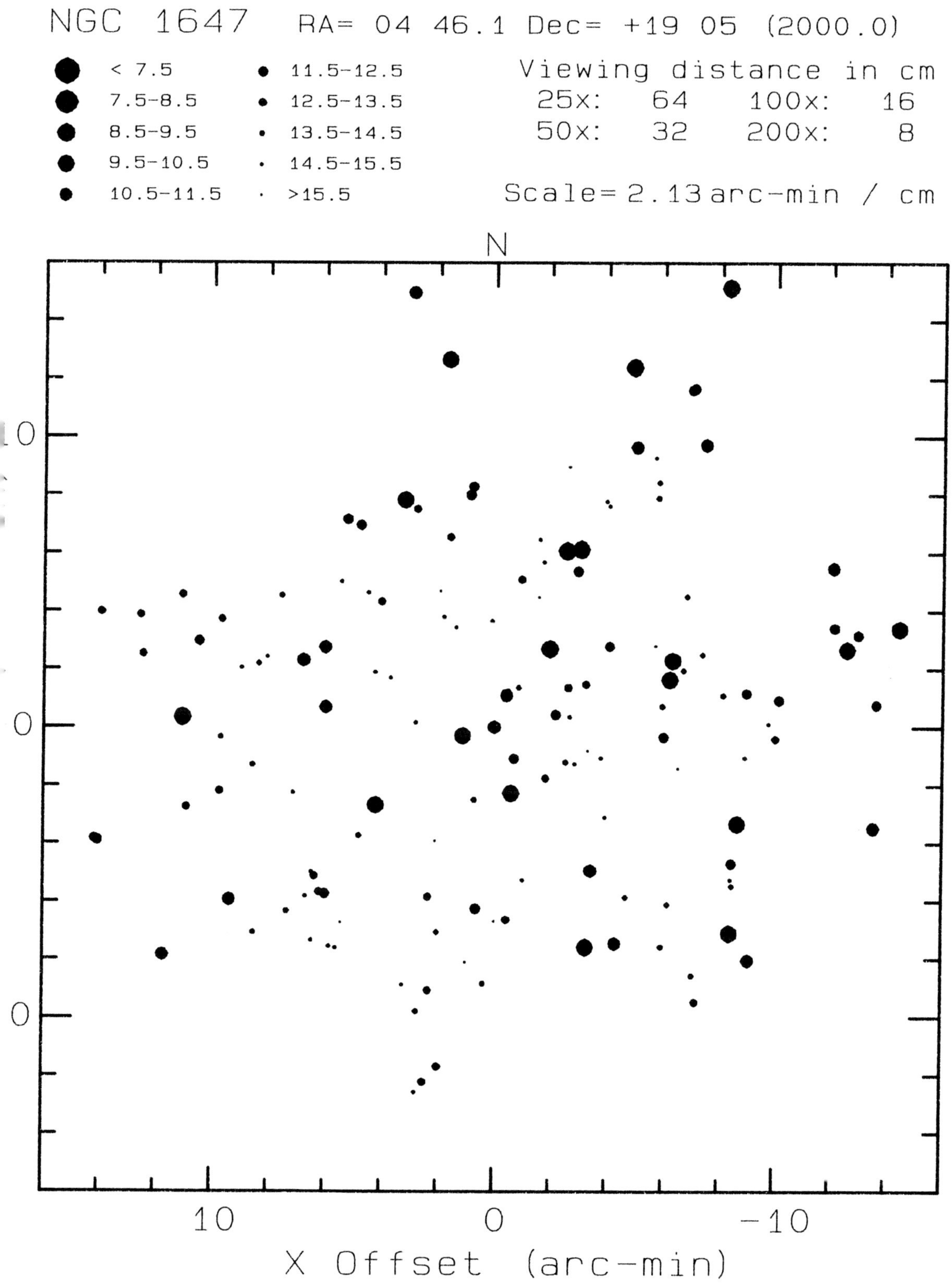

APPENDIX B: STAR CLUSTERS FOR FINDING YOUR LIMITING MAGNITUDE

NGC 1647   RA= 04 46.1 Dec= +19 05 (2000.0)

- < 7.5
- 7.5-8.5
- 8.5-9.5
- 9.5-10.5
- 10.5-11.5
- 11.5-12.5
- 12.5-13.5
- 13.5-14.5
- 14.5-15.5
- >15.5

Viewing distance in cm
25x: 64   100x: 16
50x: 32   200x: 8

Scale= 2.13 arc-min / cm

X Offset (arc-min)

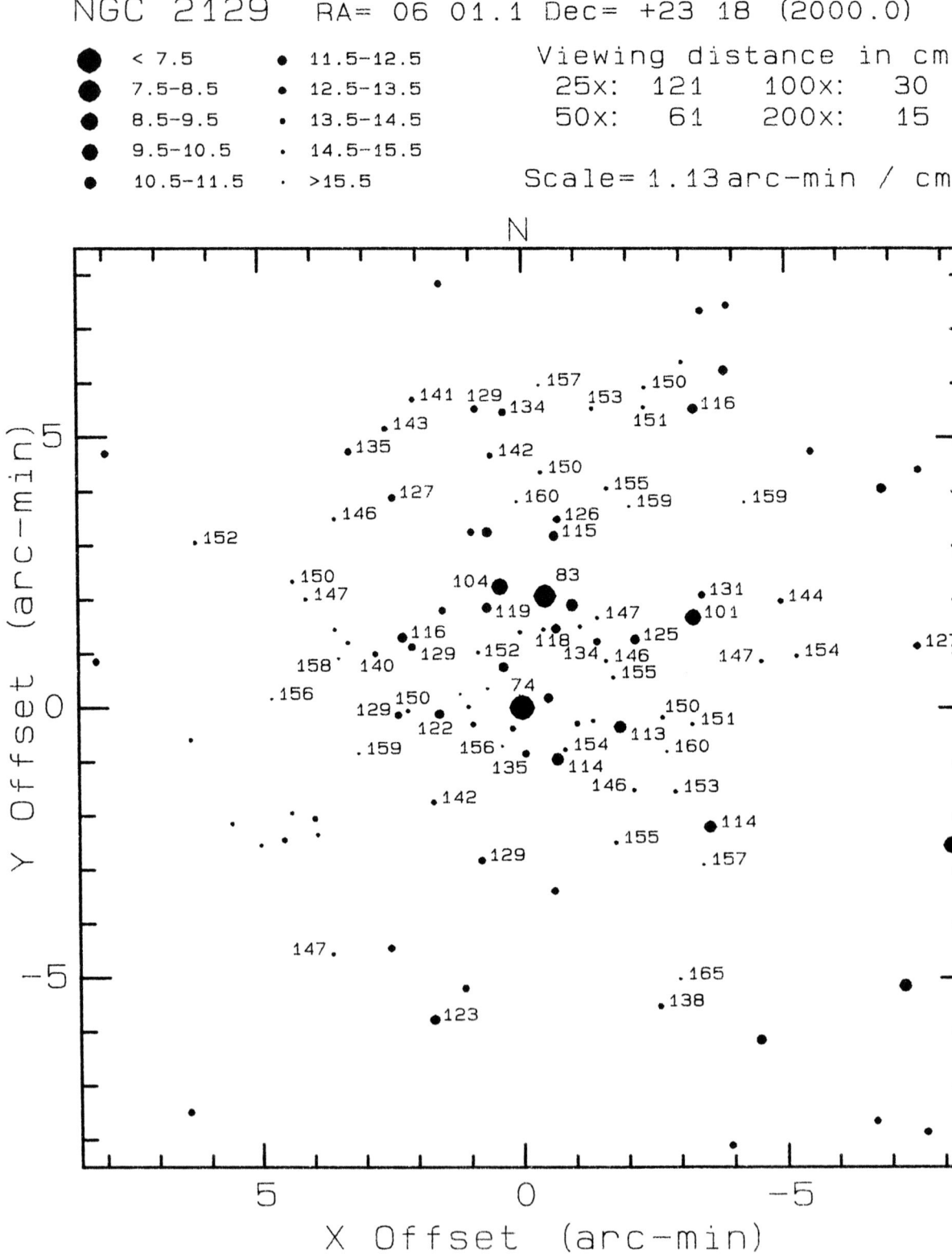

APPENDIX B: STAR CLUSTERS FOR FINDING YOUR LIMITING MAGNITUDE

NGC 2129   RA= 06 01.1 Dec= +23 18 (2000.0)

Symbol	Magnitude	Symbol	Magnitude
●	< 7.5	•	11.5–12.5
●	7.5–8.5	•	12.5–13.5
●	8.5–9.5	·	13.5–14.5
●	9.5–10.5	·	14.5–15.5
●	10.5–11.5	·	>15.5

Viewing distance in cm
25x: 121   100x: 30
50x: 61   200x: 15

Scale = 1.13 arc-min / cm

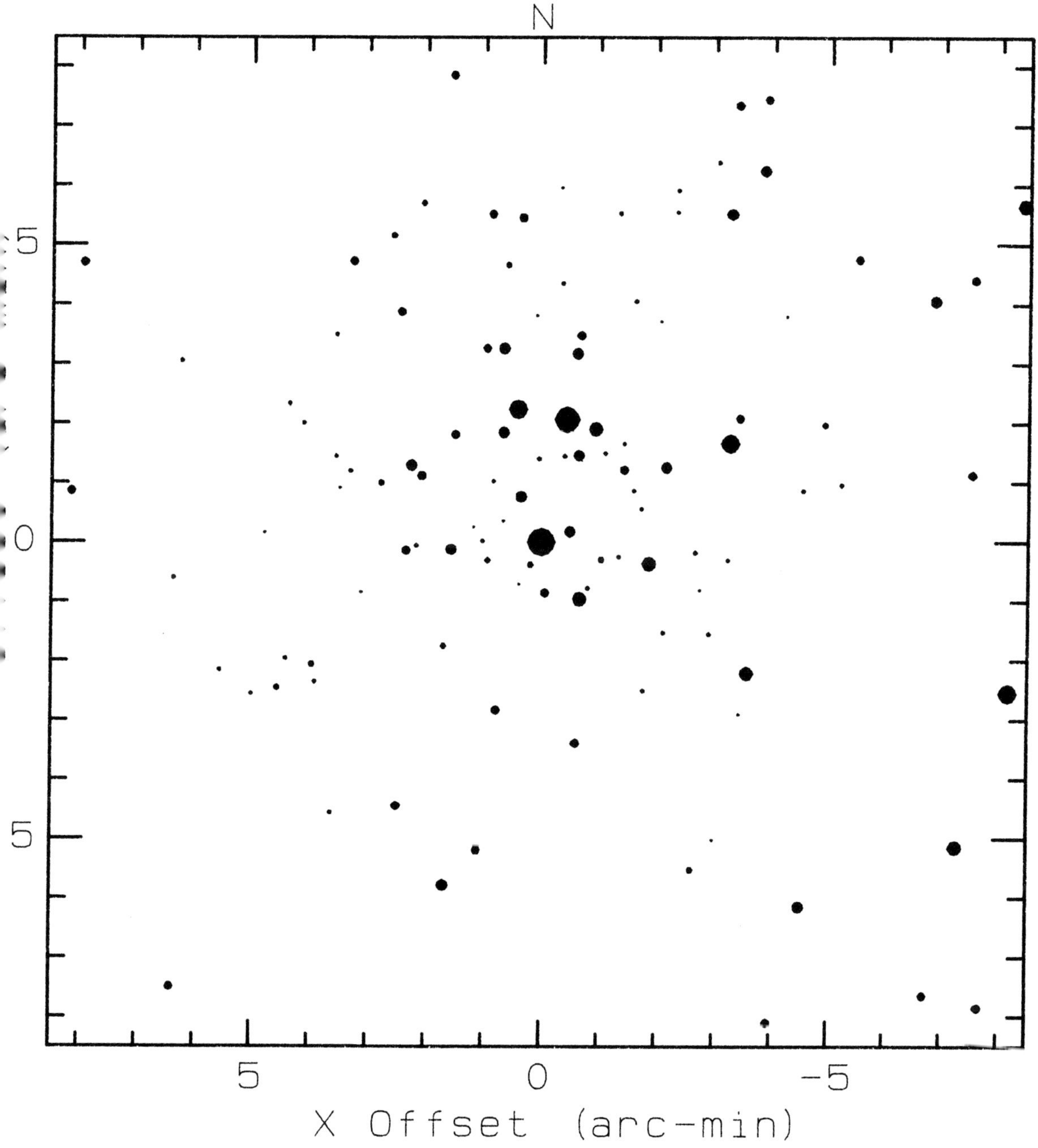

X Offset (arc-min)

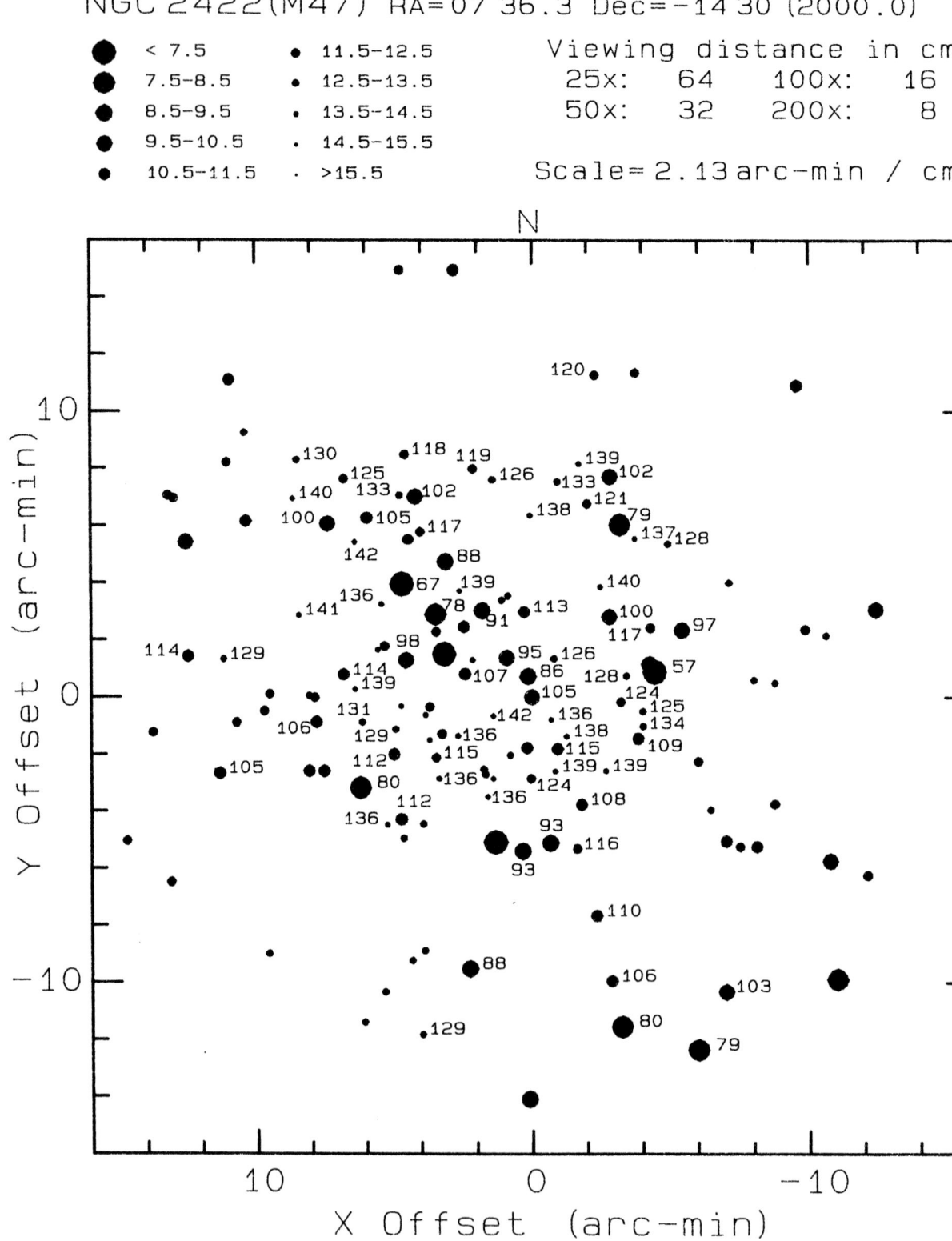

APPENDIX B: STAR CLUSTERS FOR FINDING YOUR LIMITING MAGNITUDE

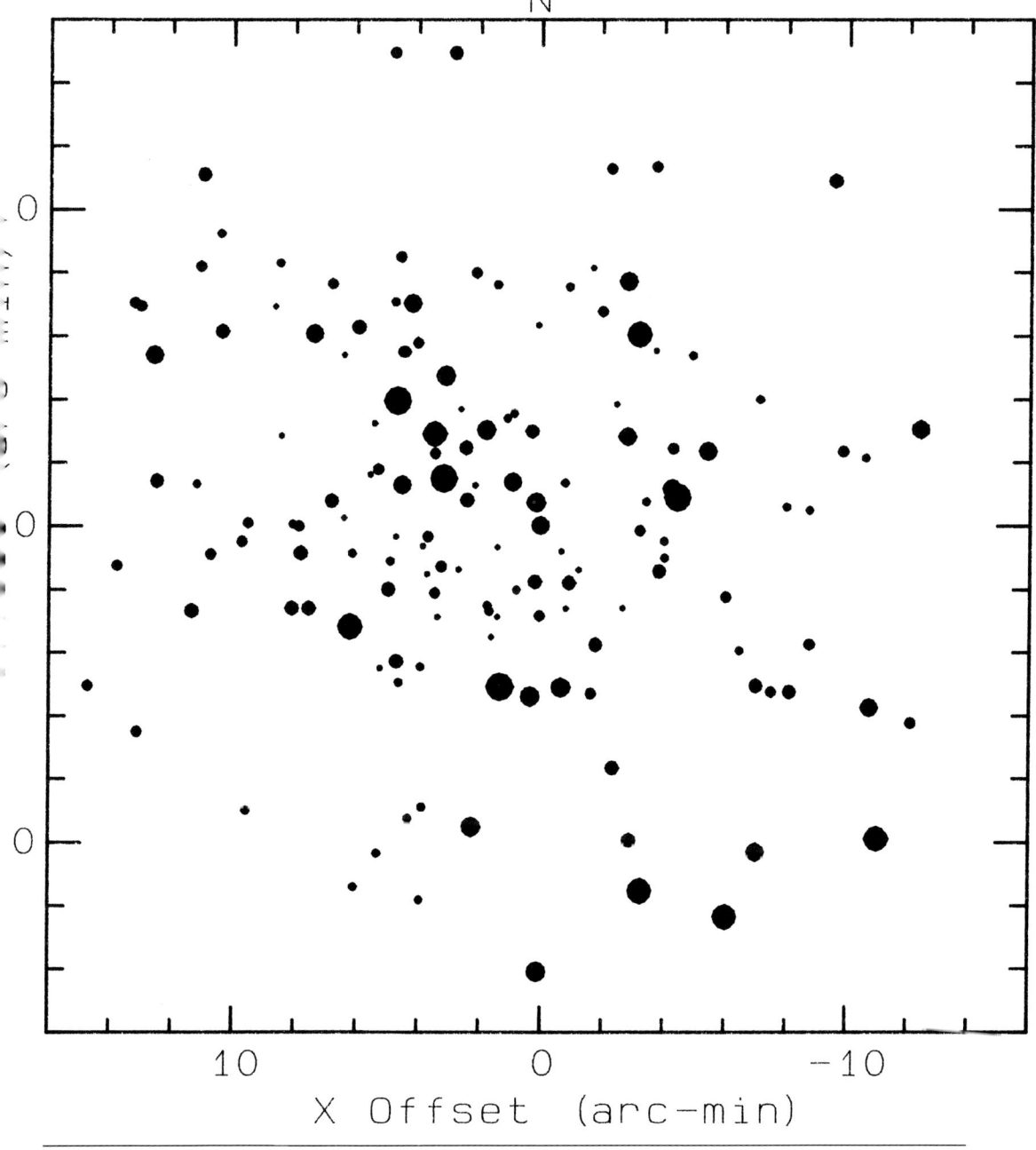

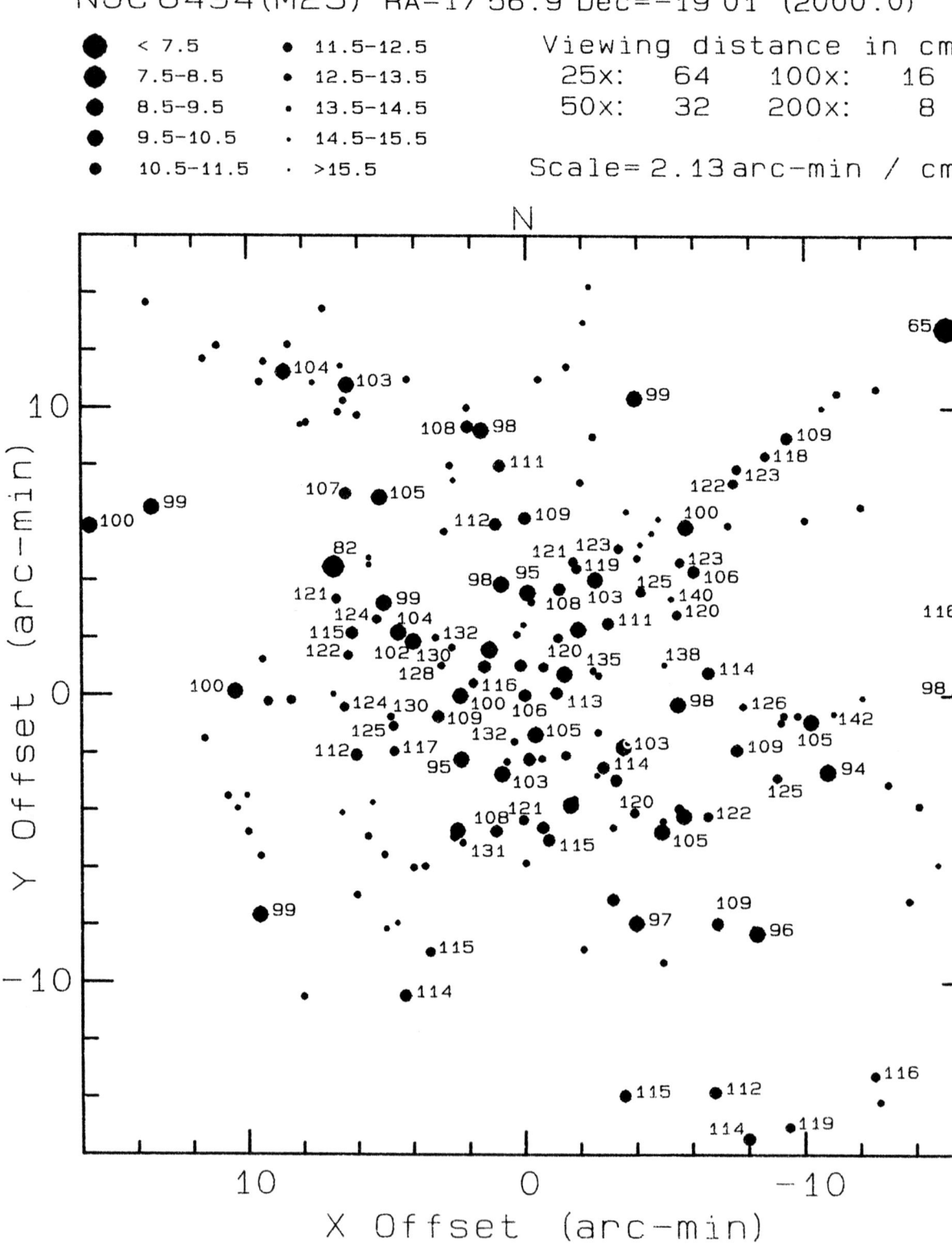

APPENDIX B: STAR CLUSTERS FOR FINDING YOUR LIMITING MAGNITUDE

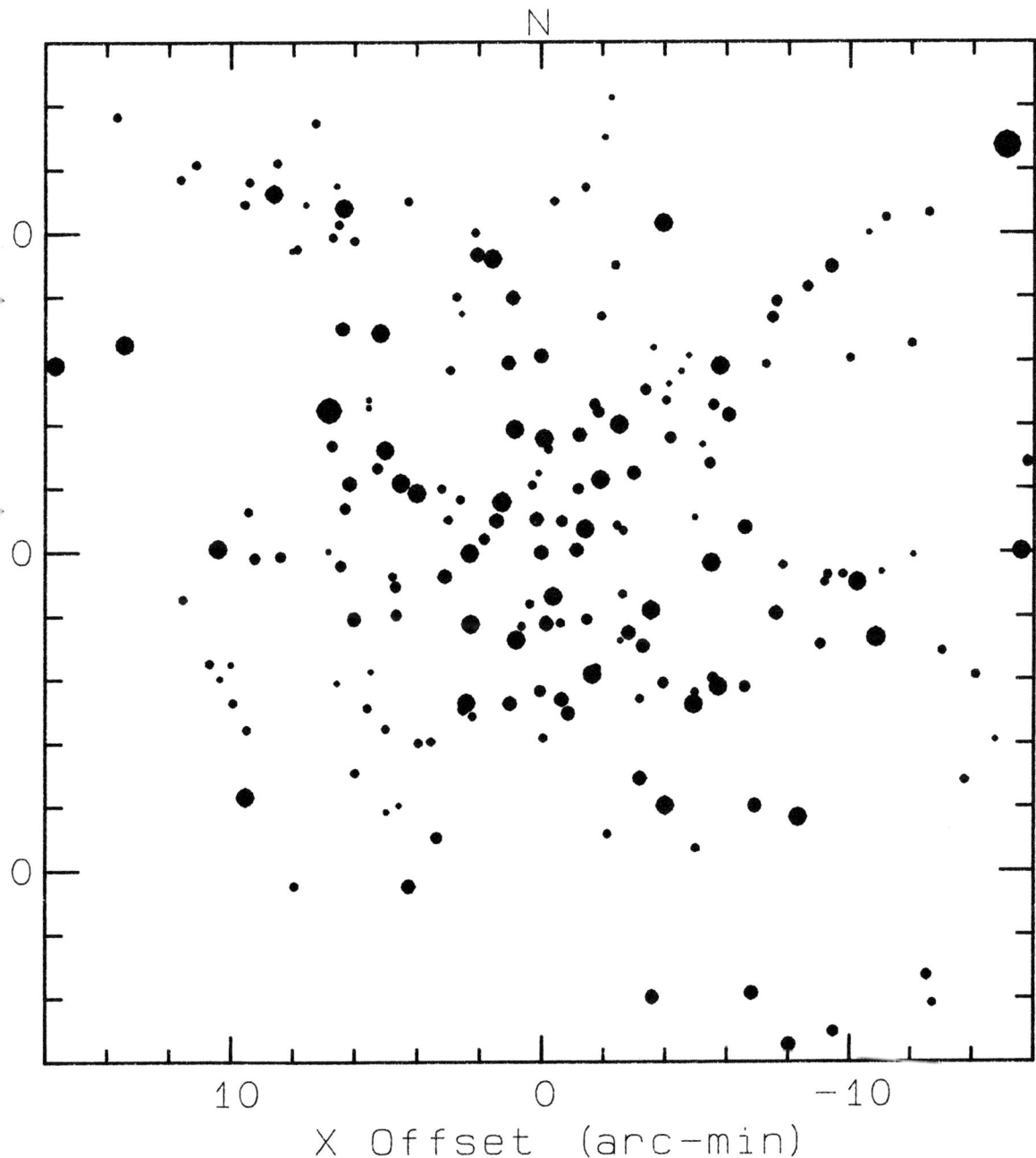

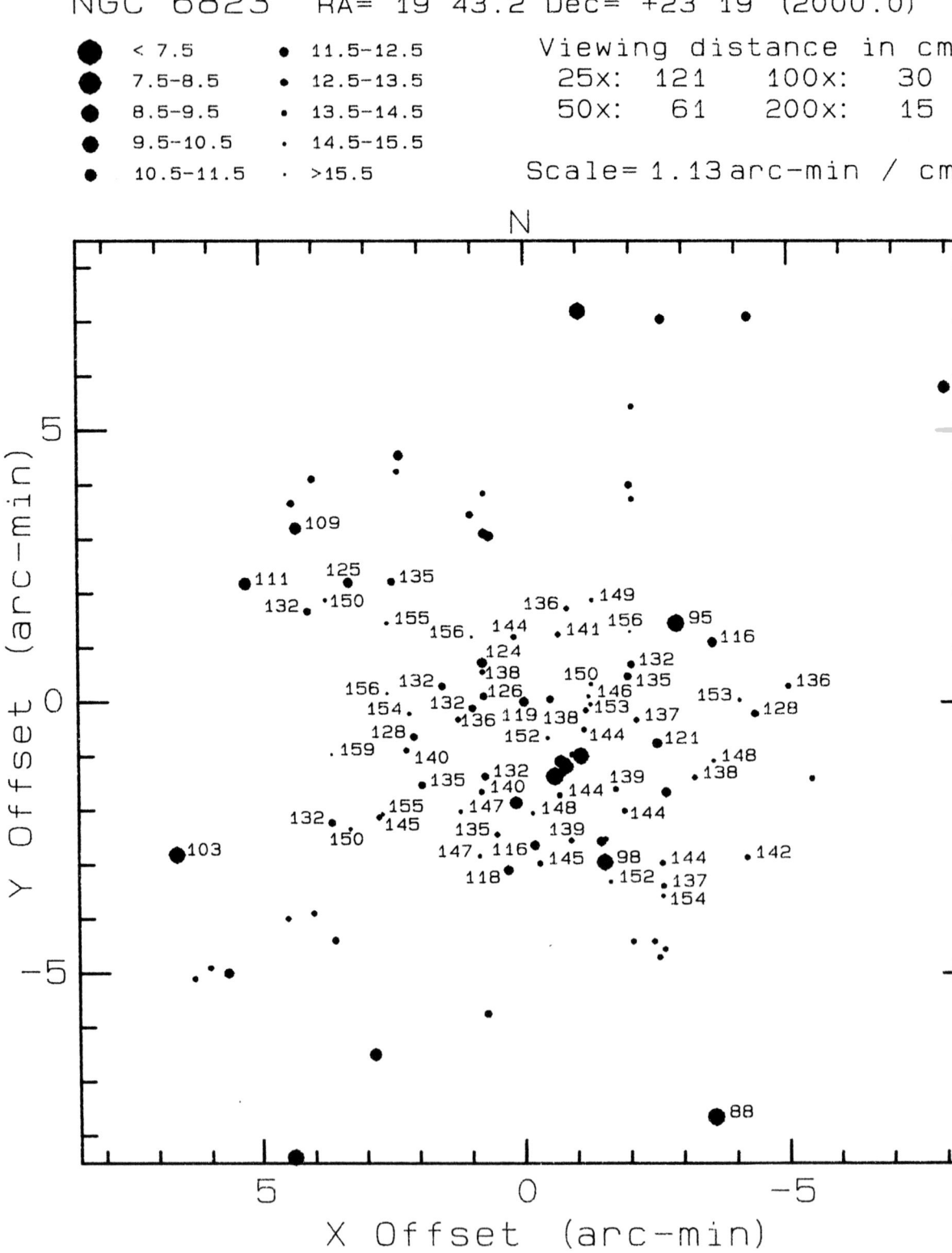

APPENDIX B: STAR CLUSTERS FOR FINDING YOUR LIMITING MAGNITUDE

NGC 6823   RA= 19 43.2 Dec= +23 19 (2000.0)

●	< 7.5	•	11.5-12.5	Viewing distance in cm
●	7.5-8.5	•	12.5-13.5	25x: 121 100x: 30
●	8.5-9.5	•	13.5-14.5	50x: 61 200x: 15
●	9.5-10.5	·	14.5-15.5	
•	10.5-11.5	.	>15.5	Scale= 1.13 arc-min / cm

X Offset (arc-min)

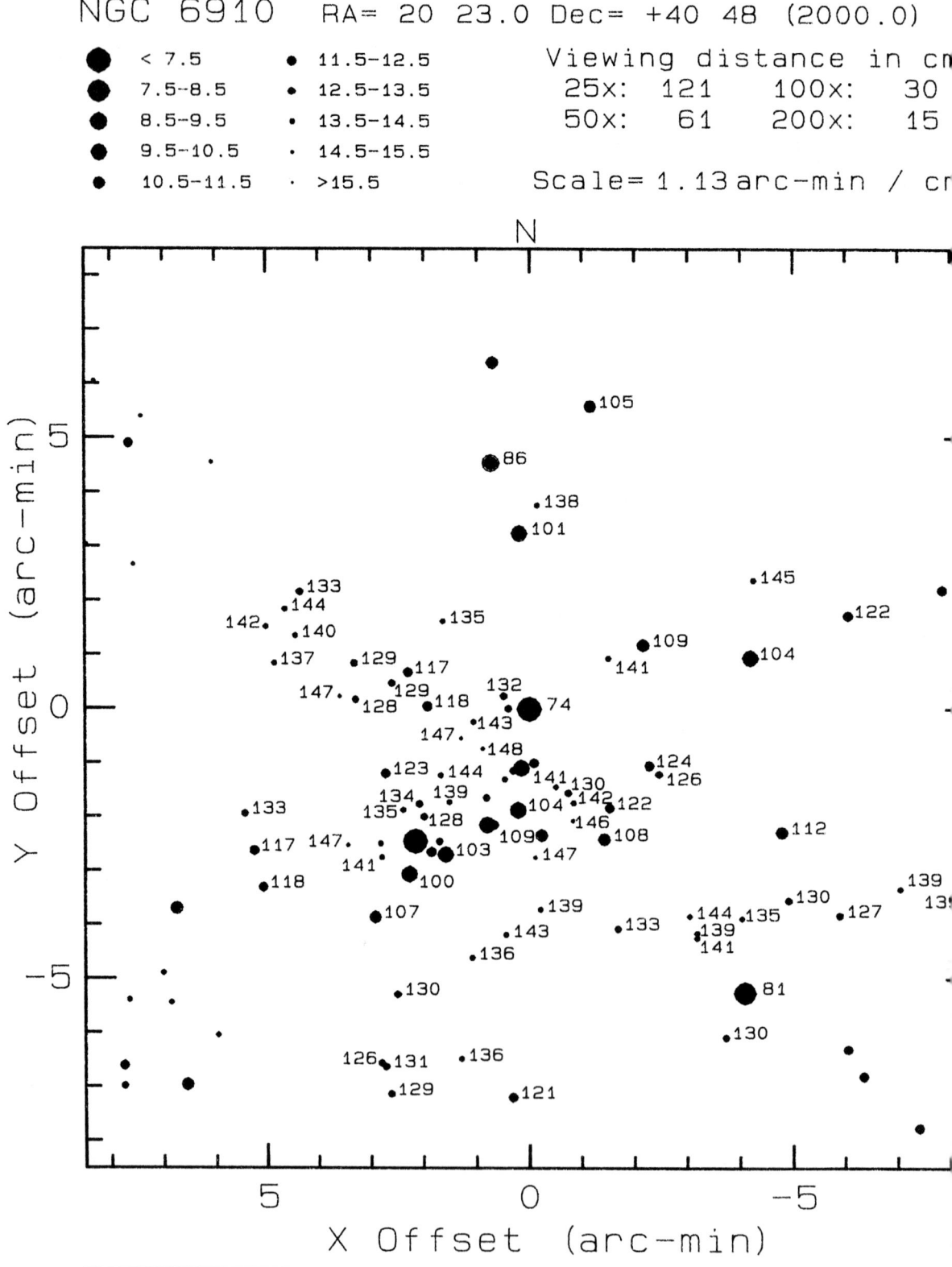

APPENDIX B: STAR CLUSTERS FOR FINDING YOUR LIMITING MAGNITUDE

NGC 6910   RA= 20 23.0 Dec= +40 48 (2000.0)

	< 7.5		11.5-12.5
	7.5-8.5		12.5-13.5
	8.5-9.5		13.5-14.5
	9.5-10.5		14.5-15.5
	10.5-11.5		>15.5

Viewing distance in cm
25x: 121   100x: 30
50x: 61   200x: 15

Scale = 1.13 arc-min / cm

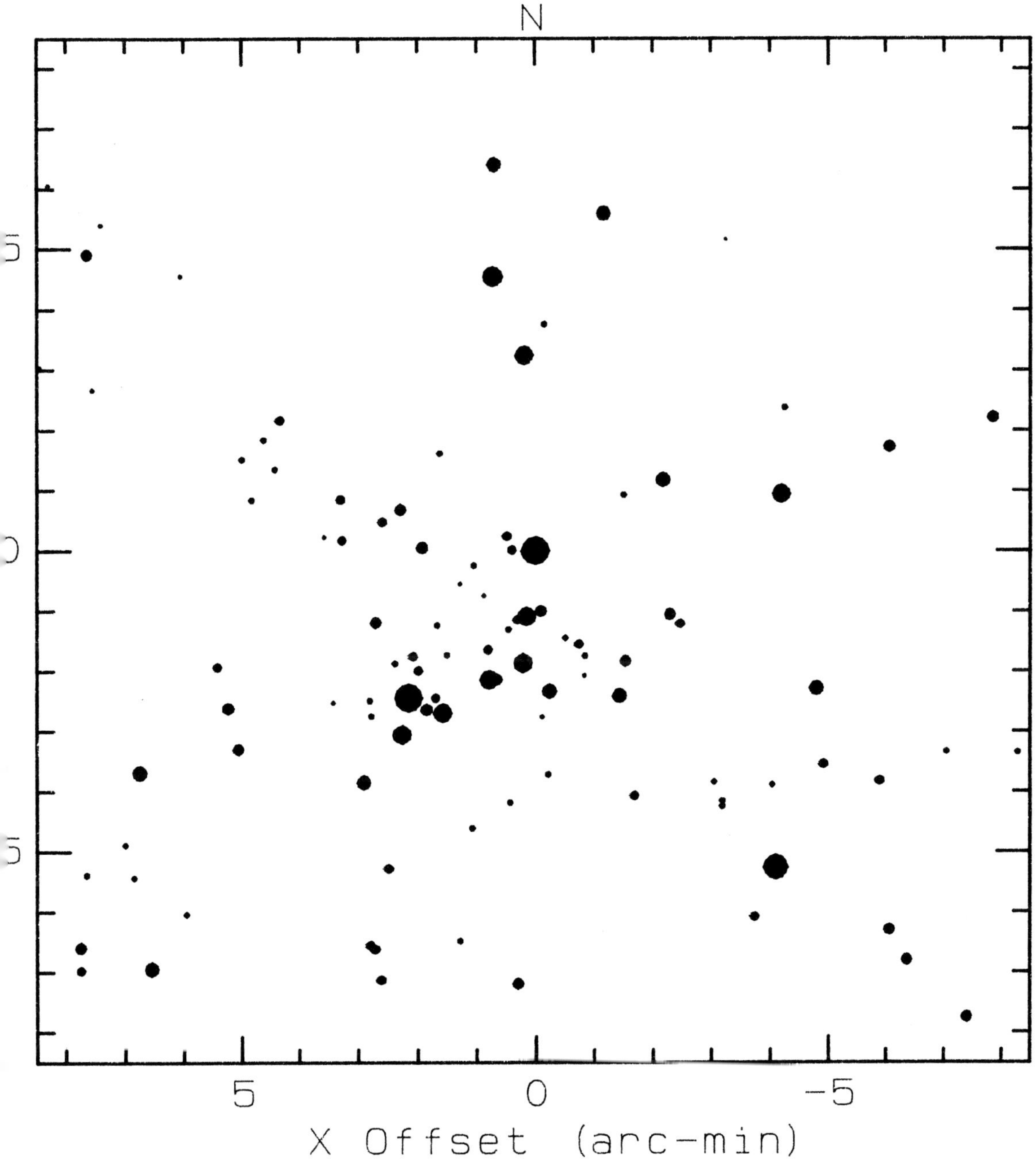

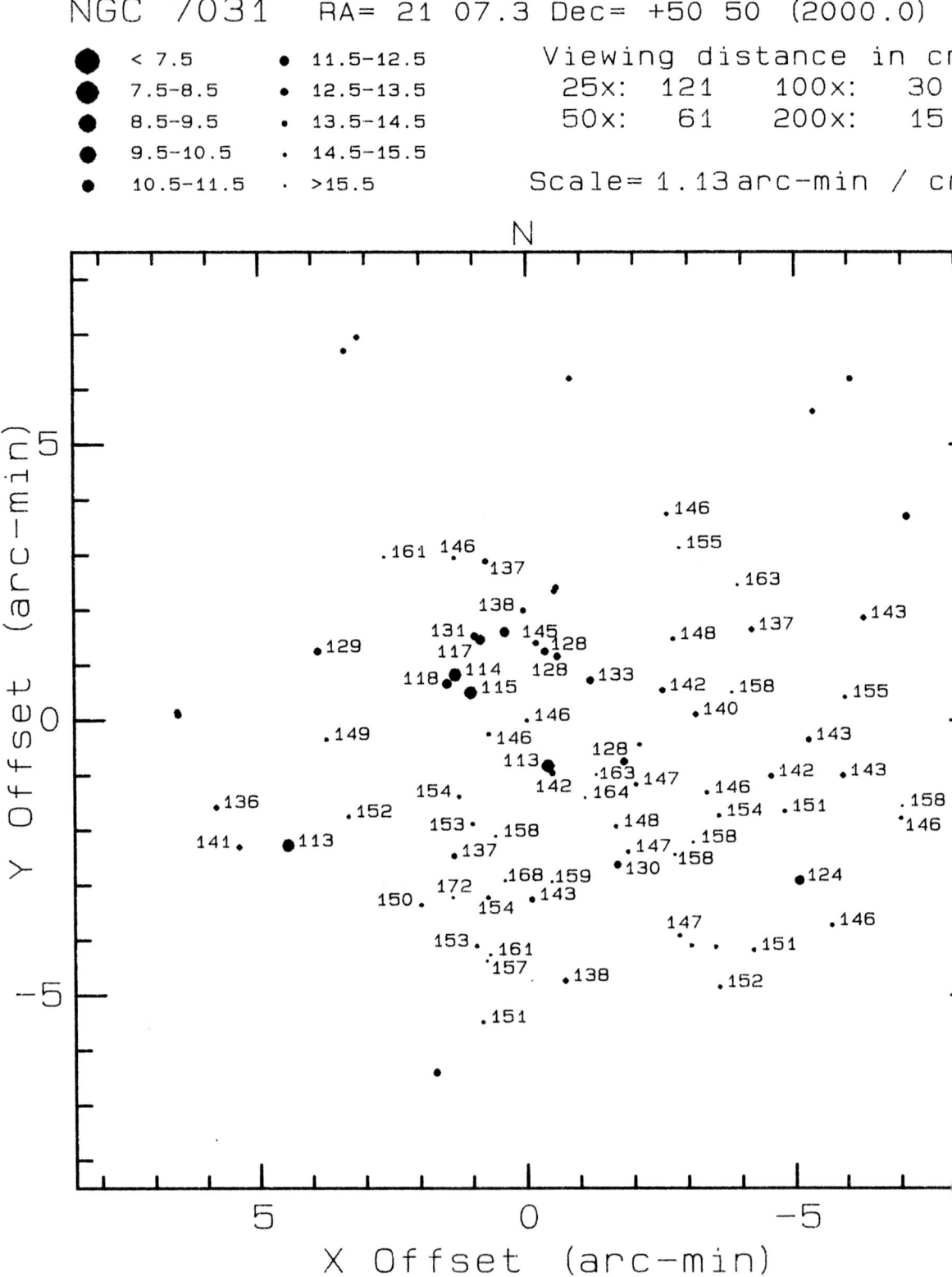

APPENDIX B: STAR CLUSTERS FOR FINDING YOUR LIMITING MAGNITUDE

NGC 7031   RA= 21 07.3 Dec= +50 50 (2000.0)

- ● < 7.5
- ● 7.5-8.5
- ● 8.5-9.5
- ● 9.5-10.5
- ● 10.5-11.5
- • 11.5-12.5
- • 12.5-13.5
- • 13.5-14.5
- • 14.5-15.5
- • >15.5

Viewing distance in cm
25x: 121   100x: 30
50x: 61   200x: 15

Scale= 1.13 arc-min / cm

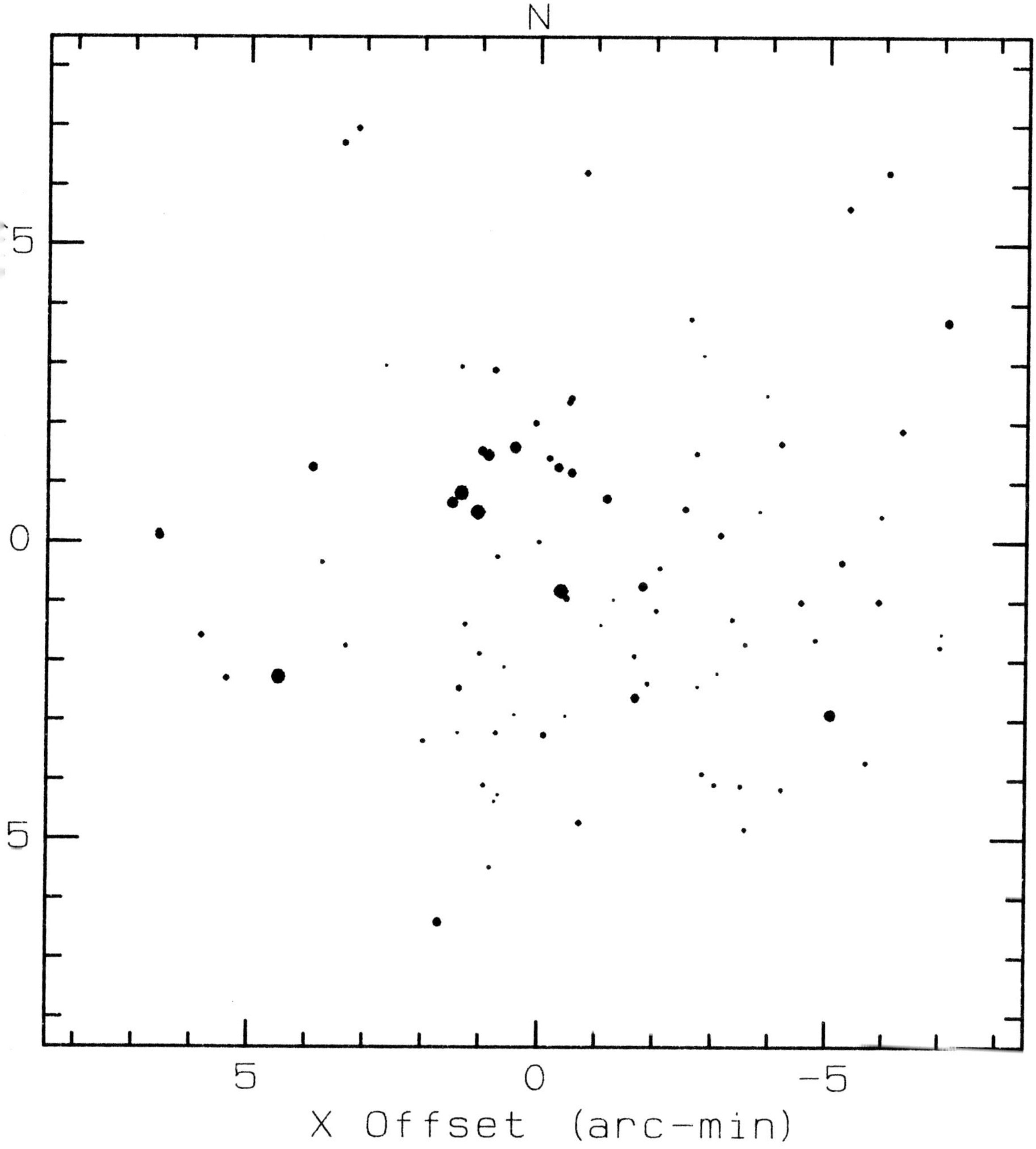

X Offset (arc-min)

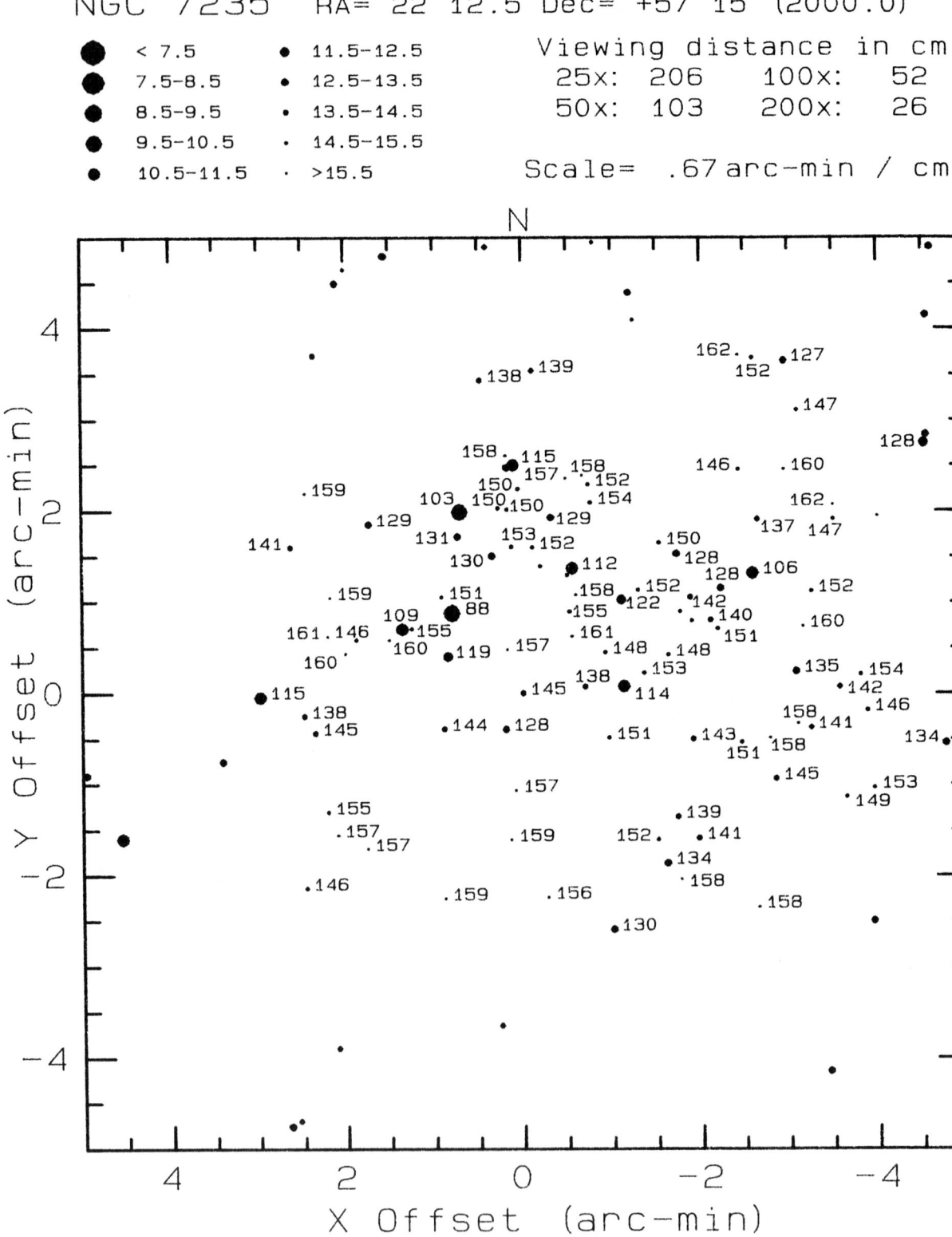

APPENDIX B: STAR CLUSTERS FOR FINDING YOUR LIMITING MAGNITUDE

NGC 7235   RA= 22 12.5 Dec= +57 15 (2000.0)

- ● < 7.5
- ● 7.5-8.5
- ● 8.5-9.5
- ● 9.5-10.5
- ● 10.5-11.5
- ● 11.5-12.5
- ● 12.5-13.5
- • 13.5-14.5
- • 14.5-15.5
- • >15.5

Viewing distance in cm
25x: 206   100x: 52
50x: 103   200x: 26

Scale= .67 arc-min / cm

X Offset (arc-min)

Appendix C
Air mass, atmospheric extinction, and other calculations

The Earth's atmosphere absorbs and scatters some of the light from every celestial object. The longer the light's path is through the atmosphere – and the more dust or haze is in the air – the greater is this *atmospheric extinction*.

The shortest path that light can follow from space to ground is straight down; that is, when an object is at the observer's zenith. Whenever it is some angle from the zenith, its light traverses more air and the dimming is greater. Near the horizon extinction becomes very great. This is why the Sun is so much dimmer when setting than when overhead.

Another effect of air and its contaminants is to absorb or scatter blue light more than red. This is why the Sun appears red when near the horizon, while the light of a clear sky (which is scattered sunlight) is blue.

If we call the angle of a celestial object from the zenith z_a (assuming no atmospheric refraction), and define the atmospheric thickness directly overhead for any observer to be 1, then at any other zenith angle the relative thickness of the atmosphere – commonly called the *air mass* – is greater than 1. Its value, T_{atm} is given approximately by

$$T_{atm} = 1/\cos(z_a). \quad \text{(equation C.1)}$$

Air mass is shown in Figure C.1.

The actual amount of air overhead at "air mass 1" depends on the observer's height above sea level and on the barometric pressure. A high mountaintop site will have less absorption and scattering than one at sea level under the same atmospheric conditions. A parameter known as k specifies the amount of light (in magnitudes) that is absorbed for an object at the zenith. A typical value for k in clear air at visual wavelengths is only about 0.15 to 0.25 magnitudes. In other words, you would gain only a very little light by doing astronomy above the atmosphere.

If the atmosphere absorbs k stellar magnitudes at the zenith, then at an angle of 60° from the zenith, where $\cos(z_a)$ is 0.5 and the air mass is 2.0, $2k$ magnitudes are absorbed. In equation form, the absorption of light in magnitudes, A_m, is

$$A_m = kT_{atm} \quad \text{(equation C.2)}$$

CALCULATING ZENITH ANGLE AND AIR MASS

An object's zenith angle is a fairly involved function of its right ascension and declination, the observer's latitude and longitude, and the local sidereal time. But this computation is rendered easy with a home computer or programmable calculator. Here's how.

The first step is to find the sidereal time. Some observers keep a sidereal-time clock. Alternatively, sidereal time can be computed to great accuracy as follows:

First, convert your standard time to Universal Time (UT) by adding or subtracting the number of hours difference between your time zone and that of Greenwich, England (0° longitude). *Sky & Telescope* describes how to convert to UT in each issue; see Appendix A.

Next, the mean sidereal time at Greenwich (known as the GMST) in seconds is computed for 0 hours UT on the date desired, by using a standard formula:

$$\text{GMST0} = 24110.54841 + 8640184.812866K + 0.093104K^2 - 0.0000062K^3,$$
$$\text{(equation C.3)}$$

where K is the number of Julian centuries elapsed from 12:00 UT January 1, 2000. To compute K, find the number of days from this date and time and divide by 36525.

APPENDIX C: AIR MASS, ATMOSPHERIC EXTINCTION

An easy way to find the number of days between any two dates is to convert the calendar month, day and year to the Julian Day number. The following is a simple procedure for converting a date based on the Gregorian calendar (the one currently in use) to the Julian Day:

$tempa = -\text{INT}(7*(\text{INT}((\text{MONTH}+9)/12)+\text{YEAR})/4)$

$tempb = \text{INT}(\text{YEAR} + \text{SGN}(\text{MONTH}-9)\,\text{INT}(\text{ABS}(\text{MONTH}-9)/7))$

$tempc = -\text{INT}((\text{INT}(tempb/100)+1)3/4)$

$tempd = tempa + \text{INT}(275\,\text{MONTH}/9) + \text{DAY} + tempc$

$\text{JD} = tempd + 1721028.5 + 367\,\text{YEAR}$

(equation C.4)

where MONTH, DAY and YEAR are integers. The SGN function gives the sign (+ if greater than or equal to 0, − if less than 0). The INT function converts to an integer (for example, $5/2 = 2$ and $-10/6 = -2$) the same way as all microcomputer Basic programming languages. The equation is valid at 0 hours UT on any given date. See Sinnott (1984) for a more general approach and programs written in Basic.

For example, on January 1, 2000 at 0 hours UT the JD will be 2451544.5, so at 12 hours UT it will be 2451545.0. On January 1, 1987 at 0 hours UT the JD was 2446796.5, so the number of days between the two is −4748.5 days.

On January 1, 1987 at 0 hours UT the value of K was −0.130006844627. Using equation C.3, the GMST0 on that date was 24027.3855 sidereal seconds. The number of seconds must be converted to a value between 0 and 86400 (the number of seconds in a day) by adding or subtracting multiples of 86400. In this example it already is, so we find that the GMST at 0 hours UT January 1, 1987 equals $6^h\,40^m\,27.3855^s$, the exact value given in The *Astronomical Almanac* for 1987. This is far more precision than virtually anyone needs.

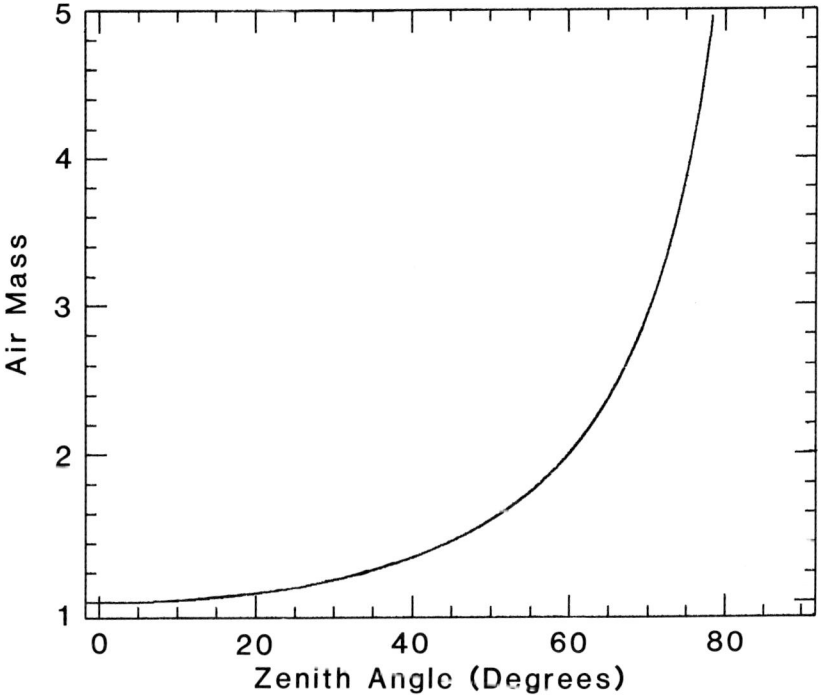

Figure C.1. The relative air mass as a function of zenith angle. Air mass increases rapidly farther from the zenith than about 60°, with a correspondingly rapid loss of an object's brightness. Airglow also increases with zenith angle, degrading the viewing conditions further. At this scale, there is no significant difference between equations C.1 and C.14.

Your own mean sidereal time, ST, at any time on the date in question is then computed from

ST = GMST0 − long/15
  + 1.0027379093 UT,   (equation C.5)

where "long" is your west longitude in degrees, and GMST0, UT, and ST are expressed in hours.

The next step is to find the particular object's hour angle, HA. The HA is the difference between the object's right ascension, RA, and the right ascension of the meridian, which equals ST. The hour angle is zero when the object is on the meridian, negative beforehand, and positive afterward. HA is found from the simple equation

HA = ST − RA.   (equation C.6)

Note that HA must be presented in a form that falls between −12 hours and +12 hours. If it does not fall in this range, add or subtract 24 hours until it does.

Now the quantity $\cos(z_a)$ can be computed:

$\cos(z_a) = \sin(lat)\sin(Dec)$
  $+ \cos(lat)\cos(Dec)\cos(HA)$
   (equation C.7)

where "lat" is the geographic latitude of the observing site and "Dec" is the object's declination.

Finally, the zenith angle, z_a, can be found by taking the arc-cosine of $\cos(z_a)$. Or the air mass can be found without going through this step by plugging directly into equation C.1:

$T_{atm} = 1/\cos(z_a)$.   (equation C.8)

OTHER USEFUL COMPUTATIONS

Having gone to the work of finding so many valuable parameters, we might as well put them to further use. Here are some examples.

Rise and set times. The time when an object rises and sets can be found by first finding its hour angle at rising or setting, HA′, as follows:

$\cos(HA') = \dfrac{\cos(z_h) - \sin(lat)\sin(Dec)}{\cos(lat)\cos(Dec)}$,
   (equation C.9)

where z_h can be the zenith angle of any "horizon" you wish, be it a mountaintop, treetop, roof, etc.

One brief aside: At sea level the zenith angle of the true horizon (such as a sea horizon) is 90°. But as seen from higher elevations on Earth, the true horizon "dips" downward owing to the Earth's curvature, and z_h increases:

$z_h = 90° + 0.0162(h)^{0.5}$,   (equation C.10)

where h is the observer's elevation in feet.

The time an object sets, t_{set}, in UT hours, is

t_{set} = RA + HA′ − GMST0 + long/15
   (equation C.11)

The time it rises, in UT hours, is

t_{rise} = RA − HA′ − GMST0 + long/15
   (equation C.12)

Atmospheric refraction. The atmosphere not only dims starlight but bends it. This bending, or atmospheric refraction, makes objects appear slightly higher than they really are. Thus the "apparent zenith angle", where we see an object, is slightly less than the "true zenith angle," where it actually is and where we would see it if the Earth had no air.

So far, we have been calculating entirely with true zenith angles. The apparent zenith angle, z, can be found more accurately by a simple equation using the true zenith angle z_a expressed in decimal degrees, the air temperature, T_c, in degrees Celsius, and the pressure P in millibars:

$z = z_a - 0.05717[P/(273+T_c)](z_a$
  $- \arcsin[0.99860\sin(0.99676\, z_a)])$
  $- 0.00058\, z_a$.   (equation C.13)

At sea level, P is about 1013 millibars.

Technically, the above equation is used for computing the true zenith angle from the apparent zenith angle (reversing z and z_a). However, the correction is so small that the calculation works either backward or forward. The accuracy is about 0.1 arc-minute for zenith angles between 0 and 75°, 1 arc-minute between 75° and 90°. Near the horizon the refraction amounts to about 0.5°.

Atmospheric refraction is slightly different at different wavelengths of light (giving rise to *atmospheric dispersion*, which makes bright stars and planets show false color fringes near the horizon). It also varies somewhat with

APPENDIX C: AIR MASS, ATMOSPHERIC EXTINCTION

Table C.1. *Example computation of apparent positions*

Object: Veil Nebula (NGC 6992–5)
Position: RA = $20^h 56^m 24^s$ Dec = $31° 42'$
Date of observation: 8/14/1987 Julian day at 0^h UT = 2447021.5
Observation site: Longitude $105° 0'$ Latitude $39° 48'$
Elevation = 6,000 feet Pressure = 1,000 millibar Temperature = 5 °C
Zenith angle to horizon = 91.3°
Sidereal time at Greenwich at 0^h UT: $21^h 27^m 32.343^s$
Rise time (UT): $1^h 44^m 5^s$
Set time (UT): $14^h 41^m 48^s$

Time (UT)			Time (ST)			Air mass	Altitude (degrees)	Azimuth (degrees)	Hour angle		
h	m	s	h	m	s				h	m	s
4	0	0	18	28	11.77	1.165	60.38	93.45	−2	28	12
6	0	0	20	28	31.48	1.015	81.39	142.98	−0	27	53
8	0	0	22	28	51.20	1.066	70.91	253.78	1	32	27
10	0	0	0	29	10.91	1.371	48.00	276.48	3	32	47
11	0	0	1	29	20.76	1.722	36.66	284.32	4	32	57
12	0	0	2	29	30.62	2.403	25.69	291.92	5	33	07
13	0	0	3	29	40.48	5.78	15.31	299.82	6	33	16
14	0	0	4	29	50.33	26.0	5.78	308.43	7	33	26

temperature, barometric pressure, and height above sea level.

Altitude and Azimith. An object's altazimuth coordinates are often useful to know. The altitude of an object above the horizon is simply

altitude = $z_h - z$ (equation C.14)

Its azimuth is given by:

tempa = −cos(Dec)sin(HA)
tempb = cos(lat)sin(Dec)
         − sin(lat)cos(Dec)cos(HA)
tempc = arctan(tempa/tempb),

azimuth = 360° + tempc if tempb > 0 and tempa < 0,
         = tempc if tempb > 0 and tempa > 0,
         = 180° + tempc if tempb < 0,
         = undefined if tempb = 0
         (object at zenith)   (equation C.15)

The azimuth of the north point on the horizon is 0°, the east point 90°, south 180°, and west 270°.

More accurate air mass. A more exact equation for the air mass can be expressed as

$T_{atm} = x - 0.0012(x^3 - x)$,   (equation C.16)

where $x = 1 / \cos(z)$ (equation C.17)

and z is less than about 75°.

These equations and a small computer or programmable calculator will provide most of the observational parameters an observer might want – though of course none are necessary for successful observing projects.

To aid you in checking your own programs a sample computation is given in Table C.1. The results for this case show that at a site near Denver, Colorado in mid-August, observations of the Veil nebula would be best around 6 to 7 hours UT. That is when it is near the meridian at that date and observation site.

Finally, a BASIC computer program is given in Table C.2. The program called ASTROPOS will do the computations discussed in this appendix. It was developed on a machine that allows variable names up to 40 characters long. It will have to be rewritten for BASICS that accept only 2-character variables.

Table C.2. *BASIC Program ASTROPOS*

```
1 REM Definitions:
2 REM      LONG#= longitude in decimal degrees
3 REM      LONGR# = long in radians = LONG#*57.29577951
4 REM      LONGD = degree portion of longitude, west= +, east= -
5 REM      LONGM = minutes portion of longitude
6 REM      LONGS = seconds portion of longitude
7 REM
8 REM
9 REM      LAT# = latitude in decimal degrees
10 REM     LATR#, LATD, LATM, LATS= same as in longitude, but corresponding
11 REM           to latitude of the observer on the earth
12 REM
13 REM     S0# = sidereal time at Greenwich at 0 hours UT on date of observation
14 REM     ST# = Sidereal time, decimal days
15 REM     UT# = Universal time in decimal days
16 REM     HA# = hour angle in radians
17 REM     RA# = object Right Ascension in radians
18 REM     DEC# = object Declination in radians
19 REM     CZA# = cosine of the zenith angle, uncorrected for refraction
20 REM     ZH# = zenith angle to the horizon
21 REM     TSET# = set time of an object in decimal days
22 REM     TRISE# = rise time of an object in decimal days
23 REM     Zcorr# = zenith angle of an object, corrected for refraction
24 REM     ZA# = zenith angle of an object, not corrected for refraction
25 REM     ALT# = altitude of an object
26 REM     AZIM# = azimuth of an object (0=north, 90=east, 180=south, 270=west)
27 REM     AIRMAS = air mass of an object, relative to overhead (where =1.0)
28 REM     H = height (elevation) of observer in feet
29 REM ---------------------------------------------------------------
30 REM
31 REM ********** the value of PI:
32     PI# = 3.1415926536#
33 REM
34 REM ********** the value of one half PI:
35     PIHALF# = PI# /2!
36 REM
37 REM ********** number of degrees per radian:
38     ONERAD# = 57.29577951#
39 REM
99 REM *** Begin Program ***
100 GOSUB 1500 REM input long lat elevation pressure tempc
104 REM setup day
106 INPUT "date of observation in MM,DD,YYYY"; MM,DD,YY
107 Y=YY: M=MM: D=DD
108 GOSUB 2100    REM Cal to JD and compS0
109 JJJ# = J1#
110 REM setup RA
115 GOSUB 1400   REM inp RA DEC
120 REM setup name
```

APPENDIX C: AIR MASS, ATMOSPHERIC EXTINCTION

Table C.2 (cont.)

```
125 INPUT "name of object"; OBJNAM$
130 REM setup times
135 INPUT "begin time in H,M,S"; BH$,BM,BS
140 INPUT "end time in H,M,S"; EH$,EM,ES
145 INPUT "time increment in H,M,S"; A$,A2,A3
150 GOSUB 1600  REM DMS to Decimal
155 TINC# = A#/24#
160 A$=BH$: A2=BM: A3=BS
165 GOSUB 1600  REM DMS to decimal
170 BTIM# = A#/24#
175 A$=EH$: A2=EM: A3=ES
180 GOSUB 1600  REM DMS to decimal
185 ETIM# = A#/24#
190 CLS
193 PRINT "          ****************** ASTROPOS **********************"
195 PRINT "Object:    ";OBJNAM$
200 DDEG#= RA#*ONERAD#/15#: GOSUB 1650
201 DEG1=DEG: MIN1=MIN: SEC1=SEC
202 DDEG#= DEC#*ONERAD#: GOSUB 1650
203 DEG2=ABS(DEG): MIN2=MIN: SEC2=SEC
205 PRINT USING "Position: RA= ##_h ##_m ##.###_s";DEG1;MIN1;SEC1;
206 IF XSIGN=1 THEN XCHAR$="+" ELSE XCHAR$="-"
207 PRINT USING "     DEC= ##";DEG2;: PRINT CHR$(248);
208 PRINT USING " ##_' ##.##";MIN2;SEC2;: PRINT """"
210 PRINT " ";
215 PRINT USING "Date: ##_/##_/####   ";MM;DD;YY;
216 PRINT USING "   Julian Day at 0 hours ########.##";JJJ#
220 REM
225 PRINT "Observation site:";
230 IF SLONG=1 THEN XCHAR$=" " ELSE XCHAR$="-"
235 PRINT " Longitude: ";XCHAR$;
236 PRINT USING "###";LONGD;: PRINT CHR$(248);
237 PRINT USING " ##_' ##.##";LONGM;LONGS;: PRINT CHR$(34);
240 IF SLAT=1 THEN XCHAR$=" " ELSE XCHAR$="-"
241 PRINT "  Latitude: ";XCHAR$;
242 PRINT USING "###";LATD;: PRINT CHR$(248);
243 PRINT USING " ##_' ##.##";LATM;LATS;: PRINT CHR$(34)
245 PRINT USING "   Elevation: ######. feet";H;
246 PRINT USING "   Pressure: ####. mbar";PRESSURE;
247 PRINT USING "   Temperature: #### C";TEMPC
250 REM
254    GOSUB 1270 REM ZanglHoriz
255    GOSUB 2300 REM cosineHA (of horizon)
260    GOSUB 2400 REM settime
265    GOSUB 2450 REM risetime
267 REM
270 PRINT USING "zenith angle to horizon: ####.### degrees";ZH#*ONERAD#
271 DDAY# = S0#: GOSUB 2030 REM ddaytoHMS
272 PRINT "Sidereal time at 0h UT at Greenwich: ";
273 PRINT USING "##_h ##_m ##.###_s";HOURS;MIN;SEC
275 DDAY# = TRISE#: GOSUB 2030 REM ddaystoHMS
280 PRINT "Rise time (UT):";HOURS;"h ";MIN;"m ";SEC;"s"
285 DDAY# = TSET#: GOSUB 2030 REM ddaystoHMS
```

Table C.2 (*cont.*)

```
290 PRINT "Set time (UT): ";HOURS;"h ";MIN;"m ";SEC;"s"
325 PRINT " Time (UT)         (ST)   Air Mass   Altitude   Azimuth   Hour Angle"
330 PRINT "-----------------------------------------------------------------"
400 REM ** do computations **
405 REM
409 TIM# = BTIM# - TINC#
410 ILAST = INT((ETIM#-BTIM#)/TINC#+1)
415 FOR I = 1 TO ILAST
420    TIM# = TIM# + TINC#
430    UT# = TIM#
440    GOSUB 1350 REM UTtoST
450    DDAY# = UT#
460    GOSUB 2030   REM DDAYtoHMS
465    UTH=HOURS: UTM=MIN: UTS=SEC
470    DDAY# = ST#
475    GOSUB 2030   REM DDAYtoHMS
480    STH=HOURS: STM=MIN: STS=SEC
485 REM
490 REM compute ha, alt, azim, airmas
495    GOSUB 1700   REM compAltAzAirm
500 REM
550 REM
555    DDAY# = HA# * ONERAD#/360#
560    GOSUB 2030   REM DDAYtoHMS
570    HAH=HOURS: HAM=MIN: HAS=SEC
575    IF XSIGN=1 THEN XCHAR$=" " ELSE XCHAR$="-"
585 REM
590    PRINT USING "## ";UTH;UTM;: PRINT USING "##.## ";UTS;
592    PRINT USING "## ";STH;STM;: PRINT USING "##.## ";STS;
594    PRINT USING " ###.###   ";  AIRMAS;ALT#;AZIM#;
595    PRINT " ";: PRINT XCHAR$;
596    PRINT USING "## ## ##"; HAH;HAM;HAS
600 NEXT I
610 GOTO 135
999 END
1200 REM ********** compute HA
1201 REM subroutine to find HA
1202 REM
1205    HA# = ST#*2#*PI# - RA#
1207    IF HA# > PI# THEN HA# = HA# -2 * PI# : GOTO 1207
1210 RETURN
1250 REM ********** Zenithangle
1251 REM subroutine to compute cosine of the zenith angle, CZA
1252 REM
1255    CZA# = SIN(LATR#)*SIN(DEC#) + COS(LATR#)*COS(DEC#)*COS(HA#)
1257    X#=1#-CZA#*CZA#  :IF X#<0! THEN X#=1E-37
1258    ZA# = PIHALF# - ATN(CZA#/SQR(X#))
1260 RETURN
1270 REM ********** Zangl Horiz
1271 REM subroutine to compute the zenith angle of the horizon ZH
1272 REM
1275    ZH# = (90# + .0162*SQR(H))/ONERAD#
1280 RETURN
```

APPENDIX C: AIR MASS, ATMOSPHERIC EXTINCTION

Table C.2 (cont.)

```
1300 REM ********** input UT
1301 REM subroutine to request UT input
1302 REM
1305      INPUT "UT in hours, min, sec" A$,A2,A3
1310      IF A2>59 OR A3>=60 THEN PRINT "out of range, reenter" : GOTO 1305
1311      GOSUB 1600 REM convert input to decimal
1315      UT# = A#/24#
1320 RETURN
1350 REM ********** UTtoST
1351 REM subroutine to convert decimal UT to decimal ST
1355 REM
1360      ST# = S0# - LONG#/360# +1.002737915# *UT#
1365 RETURN
1370 REM ********** STtoUT
1371 REM subroutine to convert decimal ST to UT
1372 REM
1375      UT# = (ST# - S0# + LONG#/360#)/1.002737915#
1380 RETURN
1400 REM ********** input RA, DEC
1401 REM subroutine to request input of right ascension and declination
1402 REM
1405      INPUT "Right Ascension in hours, min, sec"; A$,A2,A3
1410      IF A2>59 OR A3>=60 THEN PRINT "out of range, reenter" : GOTO 1405
1415      GOSUB 1600     REM DMS to Decimal
1420      RA# = A#*15#/ONERAD#
1425 REM
1430      INPUT "declination in deg, min, sec"; A$,A2,A3
1435      IF A2>59 OR A3>60 THEN PRINT "out of range, reenter" : GOTO 1430
1436      GOSUB 1600     REM DMS to Decimal
1440      DEC# = A#/ONERAD#
1445 RETURN
1500 REM **********  input long, lat
1501 REM subroutine to request input of longitude and latitude
1502 REM
1505      INPUT "longitude in deg, min, sec"; A$,A2,A3
1506      IF A2>59 OR A3>=60 THEN PRINT "out of range, reenter" : GOTO 1505
1510      GOSUB 1600       REM convert input to decimal radians
1515      LONG#=A#: LONGD=A1: LONGM=A2: LONGS=A3: SLONG=XSIGN
1520      LONGR# = LONG#/ONERAD#
1525 REM
1530      INPUT "latitude in deg, min, sec"; A$,A2,A3
1531      IF A2>59 OR A3>=60 THEN PRINT "out of range, reenter" : GOTO 1530
1535      GOSUB 1600       REM DMS to decimal
1540      LAT#=A#: LATD=A1: LATM=A2: LATS=A3: SLAT=XSIGN
1545      LATR# = LAT# /ONERAD#
1546 REM
1547      INPUT "elevation of site in feet";H
1548      INPUT "pressure at site in millibars (1 atm = 1013 mbar) ";PRESSURE
1549      INPUT "temperature in degrees C";TEMPC
1550 RETURN
1600 REM ********** DMS to Decimal
1601 REM subroutine to convert sexagesimal to decimal
1602 REM
```

Table C.2 (cont.)

```
1605       XSIGN = 1: A1 = ABS(VAL(A$))
1610       IF LEFT$(A$,1)="-" THEN XSIGN= -1
1615       A# = XSIGN*(A1+A2/60#+A3/3600#)
1620 RETURN
1650 REM ********** Ddeg to DMS
1651 REM subroutine to convert decimal degrees (+ or -) to Deg, Min, Sec.
1652 REM
1655       XSIGN =SGN(DDEG#): X#=ABS(DDEG#): IF XSIGN=0 THEN XSIGN=1
1660       DEG = INT(X#)*XSIGN: X#=(X#-INT(X#))*60#
1665       MIN = INT(X#): X#=(X#-INT(X#))*60#
1670       SEC = X#
1675       IF SEC > 59.99 THEN SEC = SEC-60: MIN = MIN + XSIGN
1680       IF MIN > 59.99 THEN MIN = MIN=60: DEG = DEG + XSIGN
1685 RETURN
1700 REM ********** compute Alt, Az, Airm
1705 REM subroutine to compute altitude, azimuth, airmass, hour angle
1710 REM
1715 REM input: ST, RA, DEC, LATR, S0
1720 REM output is corrected for refraction
1725 REM
1730       GOSUB 1200    REM compHA
1735       GOSUB 1250    REM Zenithangle
1736       GOSUB 2500    REM refcorr
1740       GOSUB 2550    REM Altitude
1745       GOSUB 2600    REM Azimuth
1750       GOSUB 2650    REM Airmass
1755 RETURN
1799 REM ********** Cal to JD
1800 REM SUBROUTINE to convert Calendar day to Julian Day
1805     REM adapted from Sky and Telescope May 1984, page 454-455.
1810     REM input: year=y month=m day=d
1812     REM g=flag for gregorian calendar (begin 1752 in england, colonies)
1813     REM       (begin 1873 in japan, 1927 in turkey)
1814     REM g=1 gregorian calendar, =0 for Julian calendar
1815       G=1
1816       IF Y<1572 THEN G=0: REM set time for change to Gregorian calendar
1820       D1= INT(D): F=D-D1-.5
1825       J=-INT(7*(INT((M+9)/12)+Y)/4)
1830       IF G=0 THEN GOTO 1850
1835           S=SGN(M-9): A=ABS(M-9)
1840           J1=INT(Y+S*INT(A/7))
1845           J1=-INT((INT(J1/100)+1)*3/4)
1850       J=J+INT(275*M/9)+D1+G*J1
1855       J=J+1721027!+2*G+367*Y
1860       IF F>=0 THEN RETURN
1865       F=F+1: J=J-1
1870 RETURN
1899 REM ********** JD to Cal
1900 REM SUBROUTINE to convert Julian Day to Calendar Day.
1901 REM
1902 REM Adapted from Sky and Telescope May, 1984 page 454-455.
1905 REM input: J= Julian Day (integer part), F= fractional part of day
1906 REM output: Y= year, M= month, D=day (including fractional part)
```

APPENDIX C: AIR MASS, ATMOSPHERIC EXTINCTION

Table C.2 (cont.)

```
1910      G=1:FTMP=F:JTMP=J
1915      IF J<2360964! THEN G=0
1920      FTMP=FTMP+.5
1925      IF FTMP<1 THEN GOTO 1935
1930          FTMP=FTMP-1: JTMP=JTMP+1
1935      IF G=1 THEN GOTO 1945
1940          A=JTMP
1941      GOTO 1955
1945          A1=INT((JTMP/36524.25)-51.12264):
1950          A=JTMP+1+A1-INT(A1/4)
1955      B=A+1524
1960      C=INT((B/365.25)-.3343)
1965      D=INT(365.25*C)
1970      E=INT((B-D)/30.61)
1975      D=B-D-INT(30.61*E) +FTMP
1980      M=E-1: Y=C-4716
1985      IF E>13.5 THEN M=M-12
1990      IF M<2.5 THEN Y=Y+1
1995 RETURN
1999 REM ********** HMS to Dday
2000 REM SUBROUTINE TO CONVERT TIME TO DECIMAL DAYS
2010      DDAY# = HOURS/24# + MIN/1440# +SEC/86400#
2020 RETURN
2021 REM
2029 REM ********** Dday to DMS
2030 REM subroutine to convert decimal days to hours, min, seconds.
2035      XSIGN=SGN(DDAY#): X#=ABS(DDAY#): IF XSIGN=0 THEN XSIGN=1
2040      DHOURS# = (X# - INT(X#))*24#
2050      DMIN# = (DHOURS# - INT(DHOURS#))*60#
2060      HOURS = DHOURS# - DMIN#/60#
2070      SEC = (DMIN# - INT(DMIN#))*60#
2080      MIN = DMIN# - SEC/60
2082      IF SEC > 59.995 THEN SEC = SEC-60: MIN = MIN + XSIGN
2084      IF MIN > 59.995 THEN MIN = MIN-60: HOURS = HOURS + XSIGN
2090 RETURN
2100 REM ********** compS0
2110 REM subroutine to compute sidereal time at Greenwich at 0 hours UT
2120      REM       on the date DD MM YY (day, month, year)
2130      REM J= Julian Centuries elapsed from DD MM YY
2140      Y= YY
2150      M= MM
2160      D= DD
2170      GOSUB 1800   REM Cal to JD
2173      JJ# = J
2176      JF# = F
2180      J1# = JJ# + JF#
2185      REM Julian day 2451545.0 is January 1, 2000 at 12 hours UT
2190      J0# = (J1# - 2451545#)/36525#
2200      S0# = 24110.54841#+8640184.812866#*J0#+.093104#*J0#*J0#
2201      S0# = S0# - .0000062#*J0#*J0#*J0#
2205      S0# = S0# /86400#
2210      S0# = S0# - INT(S0#)
2220 RETURN
```

Table C.2 (cont.)

```
2300 REM ********** cosineHA
2301 REM subroutine to compute the cosine of the zenith angle when the
2302 REM              object is at the horizon = COSHA
2305      COSHA# =(COS(ZH#) - SIN(LATR#)*SIN(DEC#))/(COS(LATR#)*COS(DEC#))
2307      X#= SQR(1#-COSHA#*COSHA#)
2308      IF X#=0! THEN X#=1E+37
2310      HAHORZ# = PIHALF# - ATN(COSHA#/X#)
2315 RETURN
2400 REM ********** settime
2401 REM subroutine to compute the set time of an object
2402 REM              note ra, hahoriz, longr are in radians, s0 in decimal days
2405 REM
2410      TSET# = (RA# + HAHORZ# + LONGR#)/(2#*PI#) - S0#
2415 RETURN
2450 REM ********** risetime
2451 REM subroutine to compute rise time of an object
2455      TRISE# = (RA# - HAHORZ# + LONGR#)/(2#*PI#) - S0#
2460 RETURN
2500 REM ********** refcorr
2501 REM subroutine to compute refraction correction to zenith angle.
2502 REM
2503      REM the following computes the arcsin of (0.9986sin(0.9967za))
2505      X# = .9986047 * SIN(.9967614 * ZA#)
2510      X# = ATN(X#/SQR(1#-X#*X#))
2512      XTMP# = -1!* PRESSURE * .0571716 /(273! * TEMPC)
2515      XCORR# = XTMP#*(ZA#- X#)*ONERAD#-.000579084#*ZA#*ONERAD#
2517      ZCORR# = ZA# + XCORR#/ONERAD#
2520 RETURN
2550 REM ********** Altitude
2551 REM subroutine to compute altitude in decimal degrees
2552 REM
2555      ALT# = (ZH# - ZCORR#)*ONERAD#
2560 RETURN
2600 REM ********** Azimuth
2601 REM subroutine to compute azimuth in decimal degrees
2602 REM
2603      X1# = -1#*COS(DEC#)*SIN(HA#)
2605      X2# = COS(LATR#)*SIN(DEC#) - SIN(LATR#)*COS(DEC#)*COS(HA#)
2606      IF X2# = 0! THEN AZIM# = 0# : RETURN
2607      X# = ATN(X1#/X2#)*ONERAD#
2608      IF X2# > 0# AND X1# < 0# THEN AZIM# = 360# + X#
2609      IF X2# > 0# AND X1# > 0# THEN AZIM# = X#
2610      IF X2# < 0#               THEN AZIM# = 180# + X#
2620 RETURN
2650 REM ********** Airmass
2651 REM subroutine to compute air mass of an object
2652 REM
2655      X = 1! / COS(ZCORR#)
2660      IF ZCORR#<1.3089 THEN AIRMAS = X - .0012*(X*X*X - X)
2661      XTMP = ZCORR# - 1.3089
2662 REM the following is a VERY ROUGH APPROXIMATION when object < 75 deg ZCORR
2665      IF ZCORR# >= 1.3089 THEN AIRMAS = 3.7978 + XTMP * 120
2667      IF ZCORR# > ZH#      THEN AIRMAS=999!
2670 RETURN
```

Appendix D
Symbols and their definitions

a	The angular size of the telescope's diffraction pattern in arc-seconds. (Chapter 3)
a_e	The maximum apparent field of view seen in an eyepiece. (Chapter 3)
a_p	The apparent diameter of an object as viewed by the human eye. (Chapter 3)
a_s	The angular diameter of the diffraction disk in the telescope, defined as the diameter where the light falls to zero in the first dark ring of the diffraction pattern. (Chapter 3)
a_t	The true angular diameter of an object in the sky. (Chapters 3, 5)
altitude	The angular height of an object above the local true horizon. (Appendix C)
azimuth	The position angle of an object counting around the horizon; 0° when due north, 90° when due east, 180° when due south, and 270° when due west. (Appendix C)
A_m	The amount of light absorbed by the atmosphere, in stellar magnitudes. (Appendix C)
B	The surface brightness of an object. (Chapters 2, 6; Appendices E, F)
B_0	The surface brightness of the background around an object. (Chapters 2, 6; Appendices E, F)
C	The contrast between an object and its background. (Chapters 2, 6; Appendices E, F)
$C1$	Logarithm of the contrast between the main surface brightness of an object and sky background of 24.25 magnitudes per square arc-second. (Appendices E, F)
D	The diameter of a telescope objective or primary mirror. (Chapters 3, 4; Appendix F)
D_e	The diameter of the fully dilated human eye, usually taken to be about 7.5 millimeters. (Chapter 4)
Dec	The declination of an object. (Chapter 3; Appendix C)
ΔDec	The difference between two declinations. (Chapter 3)
e_p	The diameter of a telescope's exit pupil. (Chapter 3)
f	The focal length of an eyepiece. (Chapter 3)
f_m	The maximum usable focal length of an eyepiece. Larger than this, the apparent field would be restricted by the eyepiece tube assembly. (Chapter 3)
F	The effective focal length of a telescope. In simple reflectors or refractors, this is the focal length of the lens or primary mirror. In compound systems, it is the focal length of the primary times the magnification of one or more secondaries. (Chapter 3)
GMST0	The mean sidereal time at Greenwich, England, calculated for 0 hours UT on a given date. (Appendix C)
h	The observer's elevation above sea level in feet. (Appendix C)
HA	The hour angle of an object from the local meridian. (Appendix C)
HA'	The hour angle of an object on the horizon. (Appendix C)
JD	Julian Day. (Appendix C)
k	The atmospheric absorption per air mass in stellar magnitudes. (Appendix C)
K	The number of Julian centuries from 12:00 UT January 1, 2000. A

	Julian century is exactly 36 525 days long. (Appendix C)	P	The atmospheric pressure in millibars. (Appendix C)
l_e	The linear size of an eyepiece field of view on a drawing or photograph for a given image scale and viewing distance. (Chapter 5)	p_s	The image scale of a photograph or drawing, sometimes called plate scale. (Chapters 5, 7)
		RA	Right ascension
l_s	A linear distance on a drawing or photograph, which corresponds to a certain angle on the sky. (Chapter 5)	ΔRA	The difference between two right ascensions. (Chapter 3)
		s	The linear size of an image in the telescope focal plane. (Chapter 3)
lat	The observer's latitude on the Earth. (Appendix C)	S.B.	Surface brightness in magnitudes per square arc-second. (Appendices E, F)
long	The observer's longitude on the Earth. (Appendix C)	ST	The observer's local mean sidereal time. (Appendix C)
m	The magnification of a telescope. (Chapters 3, 4; Appendix F)	t	The telescope transmission factor (assumed to be 0.7 in this book) (Chapter 4)
m_m	The minimum usable magnification of a telescope. (Chapter 4)		
M	The surface brightness of an object in stellar magnitudes per square arc-second. (Chapter 6; Appendices E, F)	t_{rise}	The rising time in UT of an object. (Appendix C)
		t_{set}	The setting time in UT of an object. (Appendix C)
M_b	The reduction factor in an object's surface brightness owing to transmission losses and magnification, expressed in stellar magnitudes per square arc-second. (Chapter 4; Appendix F)	T_{atm}	The relative air mass factor. Defined to be always 1.0 overhead for any observer under any conditions. (Appendix C)
		T_c	Temperature in degrees Celsius. (Appendix C)
M_e	The limiting magnitude of the naked eye. (Chapter 4)	UT	Universal Time; the local mean time at 0° longitude. Formerly called Greenwich Mean Time (GMT). What astronomers often mean by "Universal Time" is actually *Coordinated Universal Time (UTC)*, which is broadcast by radio time services such as WWV and is the basis for setting clocks worldwide. UTC is kept within 0.9 second of true UT by the occasional introduction of leap seconds to compensate for the changing rate of the Earth's rotation. (Appendix C)
M_m	The reduction in an object's surface brightness in a telescope at the minimum usable magnification. (Chapter 4)		
M_o	The surface brightness of the background around an object, in stellar magnitudes per square arc-second. (Chapter 6; Appendices E, F)		
M_t	The visual limiting magnitude of a telescope. (Chapter 4)		
MDM	minimum optimum detection magnification. (Appendix E)		
ODM	optimum detection magnification. The best power for detecting a particular faint object or detail in a given telescope under given sky conditions. (Appendix F)	v	The viewing distance to a drawing or photograph at which the image appears the same angular size as in a telescope at a given magnification. (Chapters 5, 7)
		z	Apparent zenith angle; the zenith angle observed through the Earth's atmosphere. See z_a. (Appendix C)
OMVA	optimum magnified visual angle. The apparent size to which a faint object or detail should be magnified in a given telescope under given sky conditions to make it most easily detectable. (Chapter 2; Appendix F)	z_a	True zenith angle; the angle from the zenith to an object, uncorrected for atmospheric refraction. (Appendix C)

APPENDIX D: SYMBOLS AND THEIR DEFINITIONS

z_h Normally, the zenith angle to the true horizon. This angle is 90° at sea level, but increases slightly at higher elevations. (Appendix C)

Appendix E
A catalog of deep-sky objects

This appendix lists the 611 deep-sky objects that, in the author's opinion, are the most interesting for amateur astronomers. The list represents a search of the literature for objects that are both interesting to observe and also have photographs widely available. I culled the information from many sources: *Burnham's Celestial Handbook*, magazines such as *Sky & Telescope, Astronomy*, and *Deep Sky; Sky Catalog 2000.0 (Volume 2)*, and professional journals such as the *Astrophysical Journal*.

In compiling this catalog, the first step was to check that an object might be seen through average amateur telescopes. Then I searched for those that might show some features. Many objects within range of small telescopes were left out because they are so near the limits that most would just look like another fuzzy patch. The present list still has many of these (depending on your telescope size and sky quality), but many others clearly show significant detail and have much text devoted to them in books like Burnham's.

When going out for an observing session, one wants to know which objects will be visible that night. Some books (e.g. Burnham's) are organized alphabetically by constellation, so a search through the entire work is required in order to find out what there is to see. This list, on the other hand, is ordered by right ascension. For any given night, one can start at some point on the list and work downward.

The following data are tabulated:

ID: The NGC number of the object, or other catalog number if the object has no NGC designation. The Messier number is also given if appropriate. Before the ID, several codes may appear. An asterisk (*) means the object is discussed in the book and a drawing and photograph are presented in Chapter 7.

An **S** indicates the object is a star cluster with star magnitudes given in Appendix B. Exclamation points indicate the object is an especially fine-looking one, with four (!!!!) being the most spectacular visually.

Positions: Positions in right ascension (**RA**) and declination (**DEC**) are tabulated for equinoxes 1950.0 and 2000.0. Both are given because star charts for both equinoxes are common. The RA is given in hours and decimal minutes of time, and the DEC is given in degrees and arc-minutes. The objects are sorted by increasing 1950.0 RA. The positions in the catalog have been checked against several other catalogs, and as many discrepancies as possible have been eliminated. One good reference is *Sky Catalog 2000.0, Volume 2*, by Hirshfeld and Sinnott (1985). The 2000.0 positions in this list have all been checked against Hirshfeld and Sinnott. Many positions disagree in the last digit between the two catalogs, but the differences are not significant and can be attributed to round-off error, or to where the exact center of a large object was chosen (a center may not be obvious in open clusters and nebulae in particular). When the position disagreement was significant, I investigated and chose the best position. In some cases that meant measuring photographs. Several small galaxies in the couple-of-arc-minute size range had positions disagreeing by about an arc minute or two. However, the coordinates are always accurate enough to find an object with setting circles or by plotting on star charts, even at high power.

Con: The constellation in which the object is located. The following standard abbreviations are used:

283

Andromeda	And	Leo	Leo
Antlia	Ant	Leo Minor	LMi
Apus	Aps	Lepus	Lep
Aquarius	Aqr	Libra	Lib
Aquila	Aql	Lupus	Lup
Ara	Ara	Lynx	Lyn
Aries	Ari	Lyra	Lyr
Auriga	Aur	Mensa	Men
Bootes	Boo	Microscopium	Mic
Caelum	Cae	Monoceros	Mon
Camelopardalis	Cam	Musca	Mus
Cancer	Cnc	Norma	Nor
Canes Venatici	CVn	Octans	Oct
Canis Major	CMa	Ophiuchus	Oph
Canis Minor	CMi	Orion	Ori
Capricornus	Cap	Pavo	Pav
Carina	Car	Pegasus	Peg
Cassiopeia	Cas	Perseus	Per
Centaurus	Cen	Phoenix	Phe
Cepheus	Cep	Pictor	Pic
Cetus	Cet	Pisces	Psc
Chamaeleon	Cha	Piscis Austrinus	PsA
Circinus	Cir	Puppis	Pup
Columba	Col	Pyxis	Pyx
Coma Berenices	Com	Reticulum	Ret
Corona Australis	CrA	Sagitta	Sge
Corona Borealis	CrB	Sagittarius	Sgr
Corvus	Crv	Scorpius	Sco
Crater	Crt	Sculptor	Scl
Crux	Cru	Scutum	Sct
Cygnus	Cyg	Serpens	Ser
Delphinus	Del	Sextans	Sex
Dorado	Dor	Taurus	Tau
Draco	Dra	Telescopium	Tel
Equuleus	Equ	Triangulum	Tri
Eridanus	Eri	Triangulum Australe	TrA
Fornax	For	Tucana	Tuc
Gemini	Gem	Ursa Major	UMa
Grus	Gru	Ursa Minor	UMi
Hercules	Her	Vela	Vel
Horologium	Hor	Virgo	Vir
Hydra	Hya	Volans	Vol
Hydrus	Hyi	Vulpecula	Vul
Indus	Ind		
Lacerta	Lac		

The number of listed objects in each constellation is as follows:

7 And	1 Ant	1 Aps	5 Aql	5 Aqr	5 Ara	2 Ari
8 Aur	3 Boo	7 CMa	26 CVn	7 Cam	2 Cap	5 Car
12 Cas	8 Cen	11 Cep	21 Cet	1 Cir	3 Cnc	2 Col
34 Com	3 CrA	4 Crt	4 Cru	6 Crv	18 Cyg	3 Del
5 Dor	9 Dra	15 Eri	20 For	5 Gem	5 Gru	5 Her
2 Hor	9 Hya	5 LMi	1 Lac	14 Leo	3 Lep	1 Lib
3 Lup	3 Lyn	3 Lyr	6 Mon	3 Mus	3 Nor	9 Oph
12 Ori	2 Pav	7 Peg	12 Per	4 Psc	7 Pup	3 Pyx
2 Ret	9 Scl	19 Sco	5 Sct	5 Ser	4 Sex	32 Sgr
9 Tau	1 Tel	1 TrA	4 Tri	3 Tuc	27 UMa	10 Vel
74 Vir	6 Vul					

Type: The type of object:

Plan:	planetary nebula
Glob:	globular cluster.
Neb:	diffuse nebula.
DNeb:	dark nebula.
OpCl:	open cluster.
S0:	a galaxy with no discernible spiral structure, resembling an elliptical galaxy but may have a barely detectable flattened central plane or disk.
S0p:	peculiar S0 galaxy.
Sa, Sb, Sc, Sd:	spiral galaxies type a, b, c, and d. In type a, the central bulge dominates the spiral arms. The central bulge decreases to type d, where it is almost non-existent. The spiral arms are tightly wound at type a, very loose and open at type d.
SB:	barred spiral galaxy, followed by type a, b, or c, as defined above.
En:	elliptical galaxy, where n is a number from 0 to 8 describing the ellipticity, 0 being circular, 8 most elongated. A "p" indicates peculiar
DwEl:	dwarf elliptical galaxy.
Irr:	irregular galaxy.
Pec:	peculiar galaxy.
QSO:	quasi-stellar object, also called quasar.
Star:	The only star in the catalog is "Messier 40".

This catalog lists 340 galaxies, 108 open clusters, 65 globular clusters, 49 bright nebulae, 38 planetary nebulae, 9 dark nebulae, 1 QSO, and 1 star. The galaxys' types are: 237 spiral (S), 76 elliptical (E), 19 irregular, and 8 peculiar.

Charts: The chart number in an atlas where the object can be found:

2000:	*Sky Atlas 2000.0* by Wil Tirion.
VPSA:	Vehrenberg's *Photographic Star Atlas.*
Uran:	*Uranometria 2000.0* by W. Tirion, B. Rappaport and G. Lovi.

APPENDIX E: A CATALOG OF DEEP-SKY OBJECTS

Magnitude: For an extended object this is difficult to determine, and references often disagree. Furthermore, some objects only have a photographic magnitude determined, which may differ from the visual. A magnitude measurement also depends on how much of the object is included; some objects have very large, very dim outlying areas that cannot be seen visually but contribute substantially to the listed total magnitudes. Thus, the values cannot be relied upon to better than about a half magnitude at best.

Size: The visual size of the object in arc-minutes. This is approximately the size that should be detectable in a telescope by eye. The values are consistently smaller than those in *Sky Catalogue 2000.0* because, as described in that work, the largest sizes that can be determined were listed – and those are often larger than can be seen visually in any telescope.

S.B. The surface brightness in magnitudes per square arc second. This value is only a rough approximation computed from the visual magnitude and the object's size. The S.B. can be found by the equation:

S.B. = v mag + 2.5 log(2827ab)

where a and b are the object's major and minor dimensions in arc-minutes (an elliptical shape is assumed). The constant equals $\pi(60 \text{ arc-sec/arc-minute})^2/4$.

C1: The logarithm of the contrast between the mean surface brightness (column S.B.) and a sky background of 24.25 magnitudes per square arc-second. This sky background corresponds to a dark country sky, as described in Chapter 6. The greater this value, the higher the contrast and the easier the object is to detect. We use a logarithmic parameter here because the eye's response to contrast follows a log scale. To compute *C1*, equation 6.1 was used, except that the sky surface brightness (M_o) was added to the object's surface brightness (M) to give the equivalent value seen by the eye. This must be done in linear units:

if $B = 10^{-0.4M}$ and $B_o = 10^{-0.4M_o}$

then $B' = B + B_o$, and the contrast is

$C = (B' - B_o)/B_o$, or

$C = B/B_o$,

which, when converted to magnitudes is:

$C = 10^{-0.4(M-M_o)}$, or

$C1 = \log(C)$
$= -0.4(M-M_o)$

MDM: The Minimum Optimum Detection Magnification. This is the magnification that enlarges the object to the optimum visual angle (shown in Figure 2.7b) if the telescope size is such that this is its lowest usable power.

If an object is very difficult to detect, then the MDM divided by about 3.5 (representing a telescope's minimum useful magnification per inch of aperture) gives the optimum telescope aperture in inches to view the object at this power! Of course if the contrast is high enough, then the object is easily detected at many magnifications and with many telescopes.

The MDM was computed using the smaller of the two size dimensions and the object's mean surface brightness. It also assumes that the object is viewed through optics with a transmission factor of 0.7, reducing the surface brightness by 0.38 magnitudes at the telescope's lowest usable magnification.

Note that the MDM is not the optimum detection magnification (ODM) unless it is also the telescope's lowest usable power. The ODM for other circumstances is tabulated in Appendix F. If the magnification listed is greater than the telescope's minimum (3.5 or 4 times the aperture in inches), then the best magnification is greater than the MDM. The MDM is only an approximate lower limit for detection. It is approximate because the object's mean surface brightness is used, not its peak brightness.

Comments: General notes that further describe the object (including the Messier number) or its neighbors.

Notes: Additional comments too long to fit in the table. They are listed at the end of the catalog.

	ID		1950.0 RA DEC	2000.0 RA DEC	Con	Type	Charts 2000	Uran	VPSA
	NGC 7814		00 00.7 15 51	00 03.3 16 08	Peg	Sa	10	170	124
	NGC 40		00 10.2 72 15	00 12.9 72 32	Cep	Plan	1	3	10
	NGC 45		00 11.4 -23 27	00 13.9 -23 10	Cet	S	18	305	268
	NGC 55		00 12.5 -39 30	00 15.0 -39 13	Scl	Irr	18	386	336
!!!	NGC 104		00 21.9 -72 21	00 24.1 -72 04	Tuc	Glob	24	440	406
	NGC 128		00 26.7 02 35	00 29.3 02 52	Psc	E8	10	216	196
	NGC 134		00 27.9 -33 32	00 30.4 -33 15	Scl	Sb	18	351	304
	NGC 147		00 30.4 48 14	00 33.1 48 31	Cas	E4	4	60	40
	NGC 150		00 31.8 -28 05	00 34.3 -27 48	Scl	SB	18	306	304
	NGC 157		00 32.3 -08 40	00 34.8 -08 23	Cet	Sc	10	261	232
	NGC 175		00 34.9 -20 12	00 37.4 -19 55	Cet	SBb	10	306	268
	NGC 185		00 36.1 48 04	00 38.9 48 21	Cas	E1	4	60	40
*	NGC 205	M110	00 37.6 41 25	00 40.3 41 41	And	E6	4	60	64
	NGC 188		00 39.4 85 03	00 44.2 85 19	Cep	OpCl	1	1	1
!!!!*	NGC 224	M31	00 40.0 41 00	00 42.7 41 16	And	Sb	4	60	64
*	NGC 221	M32	00 40.0 40 36	00 42.7 40 52	And	E2	4	60	64
S	NGC 225		00 40.9 61 30	00 43.8 61 46	Cas	OpCl	1	16	22
*	NGC 246		00 44.6 -12 09	00 47.1 -11 53	Cet	Plan	10	261	233
	NGC 247		00 44.6 -21 01	00 47.1 -20 45	Cet	Sc	18	306	269
!!*	NGC 253		00 45.1 -25 34	00 47.6 -25 18	Scl	Sc	18	306	305
	NGC 255		00 45.2 -11 45	00 47.7 -11 29	Cet	Sb	10	261	233
	NGC 278		00 49.2 47 18	00 52.0 47 34	Cas	Sc	4	60	40
!!!	NGC 292		00 50.0 -73 30	00 51.7 -73 14	Tuc	Irr	24	460	406
	NGC 288		00 50.2 -26 52	00 52.6 -26 36	Scl	Glob	18	307	305
	NGC 281		00 50.4 56 19	00 53.3 56 35	Cas	Neb	1	36	22
	NGC 300		00 52.6 -37 58	00 55.0 -37 42	Scl	Sc	18	351	337
	NGC 309		00 54.0 -10 13	00 56.5 -09 57	Cet	Sc	10	262	233
	Scl-Sys		00 57.0 -34 00	00 59.4 -33 44	Scl	DwEl	18	351	305
	NGC 337		00 57.3 -07 51	00 59.8 -07 35	Cet	SBc	10	262	233
!	NGC 362		01 00.6 -71 07	01 02.3 -70 51	Tuc	Glob	24	441	406
	I.1613		01 02.5 01 52	01 05.1 02 08	Cet	Irr	10	217	197
	NGC 404		01 06.6 35 27	01 09.4 35 43	And	E0	4	91	65
	NGC 428		01 10.4 00 43	01 13.0 00 59	Cet	Sc	10	217	197
	NGC 457		01 15.9 58 04	01 19.0 58 20	Cas	OpCl	1	36	22
	NGC 488		01 19.1 05 00	01 21.7 05 16	Psc	Sb	10	218	161
	NGC 520		01 22.0 03 32	01 24.6 03 48	Psc	Pec	10	218	198
	NGC 581	M103	01 29.9 60 27	01 33.2 60 42	Cas	OpCl	1	37	23
!!*	NGC 598	M33	01 31.1 30 24	01 33.9 30 39	Tri	Sc	4	91	94
*	NGC 604		01 30.7 30 31	01 34.5 30 46	Tri	Neb	4	91	94
	NGC 613		01 32.0 -29 40	01 34.3 -29 25	Scl	SBc	18	352	306

APPENDIX E: A CATALOG OF DEEP-SKY OBJECTS

ID	v mag	Size arc-min	S.B.	Cl	MDM	Comments
NGC 7814	12.0	1.0x5.0	22.4	0.7	77	edge-on, with dust lane
NGC 40	10.5	1.0x0.7	18.7	2.2	110	central star mag 11.5
NGC 45	12.1	8.0x5.5	24.8	-0.2	14	
NGC 55	7.8	25x40	23.9	0.1	3	
NGC 104	4.5	25x25	20.1	1.7	3	47 Tucanae
NGC 128	12.8	2.2x0.4	21.3	1.2	192	box shape
NGC 134	11.4	5.0x1.0	21.8	1.0	77	
NGC 147	12.1	6.5x3.8	24.2	0.0	20	companion to M31
NGC 150	12.2	2.0x1.0	21.6	1.1	77	
NGC 157	11.1	2.8x2.1	21.7	1.0	37	
NGC 175	12.8	1.5x1.3	22.2	0.8	59	theta shape
NGC 185	11.8	3.5x2.8	22.9	0.5	27	companion to M31
NGC 205	10.8	8.0x3.0	22.9	0.5	26	M110
NGC 188	10.	15x15	24.5	-0.1	5	oldest open cluster
NGC 224	4.0	150x50	22.3	0.8	2	M31, Great Galaxy in Andromeda
NGC 221	9.5	3.6x3.1	20.7	1.4	25	M32
NGC 225	8.	14x14	22.4	0.8	5	20 stars mag 9+
NGC 246	8.5	4.0x2.5	19.6	1.8	31	central star mag 12
NGC 247	10.7	18.0x5.0	24.2	0.0	15	
NGC 253	7.0	22x6	20.9	1.3	13	
NGC 255	12.8	1.5x1.5	22.3	0.8	51	15' from NGC 246
NGC 278	11.6	1.2x1.2	20.6	1.5	64	compact spiral with dust lanes
NGC 292	1.5	210x210	21.7	1.0	1	Small Magellanic Cloud
NGC 288	7.2	10x10	20.8	1.4	8	
NGC 281	8.0	23x27	23.6	0.3	3	
NGC 300	11.3	21x14	26.1	-0.7	5	
NGC 309	12.5	2.4x2.1	22.9	0.5	37	
Scl-Sys	8.8	60x60	26.3	-0.8	1	Sculptor System (see Notes)
NGC 337	12.8	3.0x0.5	21.9	1.0	154	distorted
NGC 362	6.	10x10	19.6	1.8	8	
I.1613	12.0	11x9	25.6	-0.5	9	member of Local Group
NGC 404	11.9	1.3x1.3	21.1	1.3	59	
NGC 428	11.7	3.9x3.5	23.2	0.4	22	
NGC 457	7.	10x10	20.6	1.4	8	100 stars mag 8+
NGC 488	11.2	3.5x3.0	22.4	0.7	26	
NGC 520	12.4	3.0x0.7	21.8	1.0	110	
NGC 581	8.	8x8	21.1	1.2	10	M103, 40 stars mag 8-12
NGC 598	5.3	60x60	22.8	0.6	1	M33, Triangulum Galaxy
NGC 604	?	1x1				HII region in M33
NGC 613	11.1	3.0x2.0	21.7	1.0	38	

	ID		1950.0 RA DEC	2000.0 RA DEC	Con	Type	Charts 2000	Uran	VPSA
	NGC 615		01 32.6 -07 35	01 35.1 -07 20	Cet	Sb	10	263	234
*	NGC 628	M74	01 34.0 15 32	01 36.7 15 47	Psc	Sc	10	173	126
!*	NGC 650-1	M76	01 38.8 51 19	01 41.9 51 34	Per	Plan	1	37	41
	NGC 663		01 42.6 61 01	01 46.0 61 16	Cas	OpCl	1	16	23
	NGC 672		01 45.0 27 11	01 47.8 27 26	Tri	SBc	4	128	126
	NGC 681		01 46.7 -10 40	01 49.2 -10 25	Cet	Sa	10	263	234
	NGC 752		01 54.8 37 26	01 57.8 37 41	And	OpCl	4	92	66
	NGC 779		01 57.2 -06 12	01 59.7 -05 57	Cet	Sb	10	264	234
!!!!	NGC 869		02 15.5 56 55	02 19.0 57 09	Per	OpCl	1	37	23
!!!!	NGC 884		02 18.9 56 53	02 22.4 57 07	Per	OpCl	1	37	23
*	NGC 891		02 19.3 42 07	02 22.4 42 21	And	Sb	4	62	66
	NGC 908		02 20.8 -21 27	02 23.1 -21 13	Cet	Sc	18	309	271
	I.1805		02 29.6 61 13	02 33.4 61 26	Cas	Neb	1	17	23
	NGC 925		02 24.3 33 22	02 27.3 33 36	Tri	Sb	4	93	95
	NGC 936		02 25.1 -01 22	02 27.6 -01 09	Cet	SBa	10	220	199
	NGC 941		02 26.0 -01 22	02 28.5 -01 09	Cet	Sc	10	220	199
	NGC 972		02 31.3 29 06	02 34.2 29 19	Ari	Sc	4	93	95
	For.Sys		02 37.0 -34 40	02 39.1 -34 27	For	Irr	18	354	307
	NGC 1023		02 37.2 38 52	02 40.3 39 05	Per	E7	4	62	67
	NGC 1049		02 37.7 -34 29	02 39.8 -34 16	For	Glob	18	354	307
!	NGC 1039	M34	02 38.8 42 34	02 42.0 42 47	Per	OpCl	4	62	67
	NGC 1055		02 39.2 00 16	02 41.8 00 29	Cet	Sb	10	220	199
*	NGC 1068	M77	02 40.1 -00 14	02 42.7 -00 01	Cet	Sb	10	220	199
	NGC 1073		02 41.2 01 10	02 43.8 01 23	Cet	Sbc	10	220	200
	NGC 1084		02 43.5 -07 47	02 46.0 -07 34	Eri	Sc	10	265	236
	NGC 1087		02 43.9 -00 42	02 46.5 -00 29	Cet	Sc	10	220	200
	NGC 1090		02 44.0 -00 27	02 46.6 -00 14	Cet	Sb	10	220	200
	NGC 1097		02 44.3 -30 29	02 46.4 -30 16	For	SBb	18	354	307
	NGC 1156		02 56.7 25 03	02 59.6 25 15	Ari	Irr	4	131	95
	NGC 1187		03 00.4 -23 04	03 02.6 -22 52	Eri	SBc	18	311	272
	NGC 1232		03 07.5 -20 46	03 09.8 -20 35	Eri	Sc	18	311	272
	NGC 1261		03 10.9 -55 25	03 12.3 -55 14	Hor	Glob	24	419	390
	NGC 1245		03 11.2 47 03	03 14.6 47 14	Per	OpCl	4	63	43
	NGC 1291		03 15.5 -41 17	03 17.3 -41 06	Eri	SB	18	390	339
	NGC 1300		03 17.5 -19 35	03 19.8 -19 24	Eri	SBb	10	311	272
	NGC 1313		03 17.6 -66 40	03 18.2 -66 29	Ret	SB	24	443	407
	NGC 1316		03 20.7 -37 25	03 22.6 -37 14	For	S0p	18	355	339
	NGC 1317		03 20.8 -37 17	03 22.7 -37 06	For	Sb	18	355	339
	NGC 1326		03 22.0 -36 39	03 23.9 -36 28	For	Sb	18	355	339
	NGC 1341		03 26.1 -37 19	03 28.0 -37 09	For	Sba	18	355	340

APPENDIX E: A CATALOG OF DEEP-SKY OBJECTS

```
             Size
ID       v mag  arc-min    S.B.   Cl  MDM     Comments
--------|-----|---------|------|----|---|------------------------------
NGC 615   12.6  2.7x0.8   22.1   0.9   96
NGC 628    9.0  9.0x9.0   22.4   0.7    9  M74
NGC 650-1 10.   1.5x0.7   18.7   2.2  110  M76
NGC 663    7.    11x11    20.8   1.4    7  80 stars mag 9+
NGC 672   11.6  4.5x1.7   22.4   0.7   45

NGC 681   12.9  1.3x1.2   22.0   0.9   64  similar to M104
NGC 752    7.5   45x45    24.4  -0.1    2  70 stars mag 9+
NGC 779   11.8  3.0x0.5   20.9   1.4  154  edge-on spiral
NGC 869    4.4   35x35    20.7   1.4    2  the Double Cluster
NGC 884    4.7   35x35    21.0   1.3    2  the Double Cluster

NGC 891   12.2 12.0x1.0   23.5   0.3   77
NGC 908   11.1  4.0x1.3   21.5   1.1   59
I.1805     9.    60x60    26.5  -0.9    1  Running Dog Nebula w/I.1795
NGC 925   12.0  9.4x9.4   25.5  -0.5    8
NGC 936   11.3  3.0x2.0   21.9   1.0   38  near NGC 941

NGC 941   12.9  1.9x1.3   22.5   0.7   59  near NGC 936
NGC 972   12.3  2.7x1.0   22.0   0.9   77
For.Sys    9.    65x65    26.7  -1.0    1  Fornax System
NGC 1023  11.0  4.5x1.3   21.5   1.1   59
NGC 1049  13.   0.4x0.4   19.6   1.8  192  in Fornax System

NGC 1039   6.    20x20    21.1   1.2    4  M34
NGC 1055  12.0  5.0x1.0   22.4   0.7   77  similar to M104
NGC 1068  10.0  2.5x1.7   20.2   1.6   45  M77
NGC 1073  12.0  4.0x4.0   23.6   0.2   19
NGC 1084  11.1  2.1x1.0   20.5   1.5   77

NGC 1087  11.2  2.3x2.3   21.6   1.0   33
NGC 1090  12.8  4.0x1.5   23.4   0.4   51  near NGC 1087
NGC 1097  10.6  9.0x5.5   23.5   0.3   14
NGC 1156  12.5  2.0x1.5   22.3   0.8   51
NGC 1187  11.3  5.5x4.0   23.3   0.4   19

NGC 1232  10.7  7.0x6.0   23.4   0.3   13
NGC 1261  12.5  2.4x1.0   22.1   0.9   77
NGC 1245   9.    20x20    24.1   0.0    4
NGC 1291  10.2  5.0x2.0   21.3   1.2   38
NGC 1300  11.3  6.0x3.2   23.1   0.4   24  nice barred spiral

NGC 1313  10.8  5.0x3.2   22.4   0.7   24
NGC 1316  10.1  3.5x2.5   21.1   1.3   31  Member of Fornax Cluster of Galaxies
NGC 1317  12.2  0.7x0.6   19.9   1.7  128  Member of Fornax Cluster of Galaxies
NGC 1326  11.8  3.0x2.5   22.6   0.7   31  Member of Fornax Cluster of Galaxies
NGC 1341  13.1  0.8x0.8   21.2   1.2   96  Member of Fornax Cluster of Galaxies
```

	ID		1950.0 RA DEC	2000.0 RA DEC	Con	Type	Charts 2000	Uran	VPSA
	NGC 1351		03 28.6 -35 02	03 30.5 -34 52	For	E4	18	355	340
	NGC 1350		03 29.1 -33 47	03 31.1 -33 37	For	Sb	18	355	308
	NGC 1360		03 31.0 -26 00	03 33.1 -25 50	For	Plan	18	312	308
*	NGC 1365		03 31.8 -36 18	03 33.7 -36 08	For	Sb	18	355	340
	NGC 1374		03 33.4 -35 24	03 35.3 -35 14	For	E0	18	355	340
	NGC 1379		03 34.2 -35 37	03 36.1 -35 27	For	E0	18	355	340
	NGC 1380		03 34.6 -35 09	03 36.5 -34 59	For	E7	18	355	340
	NGC 1381		03 34.7 -35 28	03 36.6 -35 18	For	E7	18	355	340
	NGC 1386		03 35.0 -36 10	03 36.9 -36 00	Eri	S	18	355	340
	NGC 1387		03 35.1 -35 41	03 37.0 -35 31	For	S0	18	355	340
	NGC 1389		03 35.3 -35 55	03 37.2 -35 45	Eri	E	18	355	340
	NGC 1395		03 36.3 -23 11	03 38.5 -23 01	Eri	E3	18	312	273
	NGC 1399		03 36.6 -35 37	03 38.5 -35 27	For	E0	18	355	340
	NGC 1398		03 36.8 -26 30	03 38.9 -26 20	For	SBb	18	312	308
	NGC 1404		03 37.0 -35 45	03 38.9 -35 35	For	E1	18	355	340
	NGC 1400		03 37.2 -18 51	03 39.5 -18 41	Eri	E1	11	312	273
	NGC 1407		03 37.9 -18 44	03 40.2 -18 34	Eri	E0	11	312	273
	NGC 1427		03 40.4 -35 34	03 42.3 -35 24	For	E3	18	356	340
	NGC 1433		03 40.4 -47 24	03 42.0 -47 14	Hor	SBa	18	391	367
	I.342		03 41.9 67 57	03 46.7 68 06	Cam	Sc	1	18	11
	NGC 1437		03 41.7 -36 01	03 43.6 -35 52	Eri	S	18	356	340
*	NGC 1435		03 43.2 23 36	03 46.2 23 45	Tau	Neb	4	132	129
!!!!*	M45	M45	03 43.9 23 58	03 46.9 24 07	Tau	OpCl	4	132	129
	NGC 1491		03 59.5 51 10	04 03.3 51 18	Per	Neb	1	39	43
	NGC 1499		04 00.1 36 17	04 03.4 36 25	Per	Neb	5	95	68
	NGC 1501		04 02.6 60 47	04 06.9 60 55	Cam	Plan	1	39	25
	NGC 1502		04 03.0 62 11	04 07.4 62 19	Cam	OpCl	1	18	25
	NGC 1514		04 06.1 30 38	04 09.2 30 46	Tau	Plan	5	95	97
	NGC 1513		04 06.2 49 23	04 09.9 49 31	Per	OpCl	5	64	44
	NGC 1531		04 10.1 -32 59	04 12.0 -32 51	Eri	E3	19	356	309
	NGC 1532		04 10.2 -33 00	04 12.1 -32 52	Eri	Sb	19	356	309
	NGC 1528		04 11.4 51 07	04 15.2 51 15	Per	OpCl	1	39	44
	NGC 1535		04 12.1 -12 52	04 14.4 -12 44	Eri	Plan	11	268	238
	NGC 1549		04 14.7 -55 42	04 15.8 -55 35	Dor	E1	24	420	391
	NGC 1553		04 15.2 -55 54	04 16.3 -55 47	Dor	S0	24	420	391
	NGC 1559		04 17.0 -62 55	04 17.6 -62 48	Ret	SB	24	444	391
	NGC 1566		04 18.9 -55 04	04 20.0 -54 57	Dor	Sb	24	420	391
	NGC 1555		04 19.0 19 25	04 21.9 19 32	Tau	Neb	11	133	130
	NGC 1579		04 26.9 35 10	04 30.2 35 17	Per	Neb	5	96	69
	NGC 1624		04 36.5 50 21	04 40.3 50 27	Per	Neb	1	40	44

APPENDIX E: A CATALOG OF DEEP-SKY OBJECTS

ID	v mag	Size arc-min	S.B.	Cl	MDM	Comments
NGC 1351	12.8	0.8x0.6	20.6	1.4	128	Member of Fornax Cluster of Galaxies
NGC 1350	11.8	3.0x1.5	22.1	0.9	51	Member of Fornax Cluster of Galaxies
NGC 1360	?	6.0x4.5				central star mag 9
NGC 1365	11.2	8.0x3.5	23.4	0.3	22	Member of Fornax Cluster of Galaxies
NGC 1374	12.4	0.8x0.8	20.5	1.5	96	Member of Fornax Cluster of Galaxies
NGC 1379	12.3	0.6x0.6	19.8	1.8	128	Member of Fornax Cluster of Galaxies
NGC 1380	11.4	3.0x1.0	21.2	1.2	77	Member of Fornax Cluster of Galaxies
NGC 1381	12.6	2.0x0.5	21.2	1.2	154	Member of Fornax Cluster of Galaxies
NGC 1386	12.4	2.5x1.0	22.0	0.9	77	Member of Fornax Cluster of Galaxies
NGC 1387	12.1	1.0x0.9	20.6	1.5	85	Member of Fornax Cluster of Galaxies
NGC 1389	12.8	1.0x1.8	22.1	0.9	77	Member of Fornax Cluster of Galaxies
NGC 1395	10.2	5.0x2.0	21.3	1.2	38	
NGC 1399	10.9	1.4x1.4	20.3	1.6	55	Member of Fornax Cluster of Galaxies
NGC 1398	10.7	4.5x3.8	22.4	0.7	20	
NGC 1404	11.5	1.0x1.0	20.1	1.6	77	Member of Fornax Cluster of Galaxies
NGC 1400	12.4	0.7x0.7	20.3	1.6	110	near NGC 1407
NGC 1407	11.4	0.8x0.8	19.5	1.9	96	near NGC 1400
NGC 1427	12.4	1.4x1.0	21.4	1.1	77	Member of Fornax Cluster of Galaxies
NGC 1433	11.4	7.0x6.0	24.1	0.1	13	
I.342	12.0	15x15	26.5	-0.9	5	
NGC 1437	12.9	2.0x1.5	22.7	0.6	51	Member of Fornax Cluster of Galaxies
NGC 1435	6.8	15x20	21.6	1.1	5	Merope Nebula in M45
M45	1.4	100x100	20.0	1.7	1	M45, the Pleiades
NGC 1491	?	3x3	?			
NGC 1499	6.	145x40	24.0	0.1	2	California Nebula, see Notes
NGC 1501	12.	0.9x0.8	20.3	1.6	96	central star mag 13.5
NGC 1502	7.	8x8	20.1	1.6	10	25 stars mag 8+
NGC 1514	11.	2.0x2.0	21.1	1.2	38	central star mag 10
NGC 1513	9.	12x12	23.0	0.5	6	
NGC 1531	13.0	0.5x0.3	19.6	1.9	256	near NGC 1532
NGC 1532	11.8	5.0x1.0	22.2	0.8	77	near NGC 1531
NGC 1528	6.	25x25	21.6	1.1	3	about 80 stars
NGC 1535	9.	0.3x0.3	15.0	3.7	256	central star mag 11.5
NGC 1549	11.0	2.8x2.5	21.7	1.0	31	near NGC 1553
NGC 1553	10.2	3.1x2.3	21.0	1.3	33	
NGC 1559	11.1	3.0x1.5	21.4	1.2	51	
NGC 1566	10.5	5.0x4.0	22.4	0.7	19	
NGC 1555	var	0.5x0.5	-			Hind's Variable Nebula
NGC 1579	?	8x12	?			
NGC 1624	?	3x3	?			nebula around 7 stars

VISUAL ASTRONOMY OF THE DEEP SKY

	ID		1950.0 RA	DEC	2000.0 RA	DEC	Con	Type	Charts 2000	Uran	VPSA
	NGC 1637		04 38.9	-02 56	04 41.4	-02 50	Eri	Sc	11	224	202
S	NGC 1647		04 43.2	19 00	04 46.1	19 05	Tau	OpCl	11	134	131
	NGC 1746		05 00.6	23 44	05 03.6	23 48	Tau	OpCl	5	134	131
	NGC 1792		05 03.5	-38 04	05 05.2	-38 00	Col	Sc	19	358	341
	NGC 1807		05 07.8	16 28	05 10.7	16 32	Tau	OpCl	11	180	131
	NGC 1817		05 09.2	16 38	05 12.1	16 42	Tau	OpCl	11	180	131
	NGC 1832		05 10.0	-15 47	05 12.3	-15 43	Lep	Sc	11	270	275
	NGC 1851		05 12.4	-40 05	05 14.0	-40 02	Col	Glob	19	393	342
	I.405		05 13.0	34 16	05 16.3	34 19	Aur	Neb	5	97	99
	NGC 1857		05 16.6	39 18	05 20.1	39 21	Aur	OpCl	5	66	70
	I.410		05 19.3	33 28	05 22.6	33 31	Aur	Neb	5	97	99
	NGC 1893		05 19.3	33 21	05 22.6	33 24	Aur	OpCl	5	97	99
!!!!	LMC		05 20.0	-69 00	05 19.7	-68 57	Dor	Irr	24	444	409
	NGC 1904	M79	05 22.2	-24 34	05 24.3	-24 31	Lep	Glob	19	315	276
	NGC 1912	M38	05 25.3	35 48	05 28.7	35 51	Aur	OpCl	5	97	70
	NGC 1964		05 31.2	-21 59	05 33.3	-21 57	Lep	Sb	19	315	276
!*	NGC 1952	M1	05 31.5	21 59	05 34.5	22 01	Tau	Neb	5	135	132
	NGC 1960	M36	05 32.9	34 07	05 36.2	34 09	Aur	OpCl	5	97	99
!!!!*	NGC 1976	M42	05 32.9	-05 25	05 35.4	-05 23	Ori	Neb	11	225	240
	NGC 1977		05 33.0	-04 54	05 35.5	-04 52	Ori	Neb	11	225	204
!!*	NGC 1982	M43	05 33.1	-05 18	05 35.6	-05 16	Ori	Neb	11	225	240
	NGC 1999		05 34.1	-06 45	05 36.5	-06 43	Ori	Neb	11	271	240
*	B33		05 38.4	-02 30	05 40.9	-02 28	Ori	DNeb	11	226	204
*	I.434		05 38.6	-02 26	05 41.1	-02 24	Ori	Neb	11	226	204
*	NGC 2023		05 39.2	-02 15	05 41.7	-02 13	Ori	Neb	11	226	204
!*	NGC 2024		05 39.4	-01 52	05 41.9	-01 50	Ori	Neb	11	226	204
!!!	NGC 2070		05 39.9	-69 04	05 39.6	-69 03	Dor	Neb	24	445	409
!*	NGC 2068	M78	05 44.2	00 02	05 46.8	00 03	Ori	Neb	11	226	204
*	NGC 2071		05 44.6	00 17	05 47.2	00 18	Ori	Neb	11	226	204
!	NGC 2099	M37	05 49.0	32 33	05 52.3	32 34	Aur	OpCl	5	98	99
	Barn. Loop		05 52.5	-01 31	05 55.0	-01 30	Ori	Neb	11	226	204
S	NGC 2129		05 58.1	23 18	06 01.1	23 18	Gem	OpCl	5	136	132
	NGC 2158		06 04.3	24 06	06 07.4	24 06	Gem	OpCl	5	136	133
!	NGC 2168	M35	06 05.7	24 20	06 08.8	24 20	Gem	OpCl	5	137	133
	NGC 2174		06 06.7	20 30	06 09.7	20 30	Ori	Neb	5	137	133
	I.443		06 13.9	22 48	06 16.9	22 47	Gem	Neb	5	137	133
	NGC 2207		06 14.3	-21 21	06 16.4	-21 22	CMa	Sc	19	317	277
	I.2165		06 19.6	-12 57	06 21.9	-12 58	CMa	Plan	11	272	241
	NGC 2237-9		06 29.6	05 05	06 32.3	05 03	Mon	Neb	12	227	169
!	NGC 2244		06 29.7	04 54	06 32.4	04 52	Mon	OpCl	12	227	205

APPENDIX E: A CATALOG OF DEEP-SKY OBJECTS

```
                   Size
    ID      v mag  arc-min    S.B.   Cl   MDM    Comments
--------  |-----|----------|-----|-----|-----|-------------------------------
NGC 1637    11.4   2.7x2.0   21.9   1.0   38
NGC 1647     6.5   40x40     23.1   0.4    2   25 stars mag 8 to 13
NGC 1746     6.    45x45     22.9   0.5    2   50 stars mag 8+
NGC 1792    10.7   3.0x1.0   20.5   1.5   77
NGC 1807     7.5   10x10     21.1   1.2    8   15 stars mag 8 to 9

NGC 1817     8.    15x15     22.5   0.7    5   50 stars mag 10+
NGC 1832    12.3   2.1x1.1   21.8   1.0   70
NGC 1851     7.    5x5       19.1   2.1   15
I.405       10.    18x30     25.5  -0.5    4   nebula with AE Aurigae
NGC 1857     7.    9x9       20.4   1.5    9   45 stars mag 8+

I.410        9.    20x20     24.1   0.0    4
NGC 1893     8.    12x12     22.0   0.9    6   near I.410, 20 stars mag 9 to 12
LMC          1.0   360x360   22.4   0.7    1   Large Magellanic Cloud
NGC 1904     8.4   7.5x7.5   21.4   1.1   10   M79
NGC 1912     6.2   20x20     21.3   1.2    4   M38, 150 stars mag 8+

NGC 1964    11.8   5.0x1.6   22.7   0.6   48
NGC 1952     9.    5x3       20.6   1.5   26   M1, the Crab Nebula
NGC 1960     6.3   12x12     20.3   1.6    6   M36, 60 stars mag 9+
NGC 1976     4.    65x65     21.7   1.0    1   M42, the Great Nebula in Orion
NGC 1977     7.    40x25     23.1   0.4    3   includes NGC 1973, 1975

NGC 1982     8.    7x5       20.5   1.5   15   M43
NGC 1999     9.    2x2       19.1   2.0   38   near M43
B33          -     6x4        -     -          Horsehead Nebula
I.434        ?     60x12      ?                contains Horsehead Nebula, B33
NGC 2023     ?     10x10      ?

NGC 2024     ?     20x20      ?                beautiful complex detail
NGC 2070    ~5.?   20x20    ~20.1   1.7    4   Tarantula Nebula
NGC 2068     8.    8x6       20.8   1.4   13   M78
NGC 2071     ?     4x3        ?                15' NE from NGC 2068
NGC 2099     6.2   20x20     21.3   1.2    4   M37, 150 stars mag 9+

Barn. Loop  6.0    840x60    26.4  -0.9    1   Barnard's Loop: faint arc
NGC 2129     7.    5x5       19.1   2.1   15   about 50 stars
NGC 2158    11.    4x4       22.6   0.6   19   near M35
NGC 2168     5.5   30x30     21.5   1.1    3   M35, 120 stars mag 8+
NGC 2174     ?     40x30                       Ape Man Nebula; includes NGC 2175

I.443        ?     50x40                       supernova remnant
NGC 2207    12.3   2.5x1.5   22.4   0.8   51   interacting pair of galaxies
I.2165      12.5   0.13x0.13 16.7   3.0  591
NGC 2237     9.    60x80     26.8  -1.0    1   Rosette Nebula
NGC 2244     5.5   40x40     22.1   0.8    2   cluster surrounding NGC 2237
```

VISUAL ASTRONOMY OF THE DEEP SKY

	ID		1950.0 RA DEC	2000.0 RA DEC	Con Type	Charts 2000	Uran	VPSA
!*	NGC 2261		06 36.4 08 46	06 39.1 08 43	Mon Neb	12	182	169
	NGC 2264		06 38.4 09 56	06 41.2 09 53	Mon OpCl	12	183	169
!	NGC 2287	M41	06 44.9 -20 42	06 47.0 -20 45	CMa OpCl	19	318	278
	NGC 2281		06 45.8 41 07	06 49.3 41 04	Aur OpCl	5	68	71
	NGC 2323	M50	07 00.5 -08 16	07 02.9 -08 20	Mon OpCl	12	273	242
	NGC 2346		07 06.8 -00 44	07 09.3 -00 49	Mon Plan	12	228	206
	NGC 2276		07 11.0 85 52	07 28.2 85 47	Cep Sc	1	1	1
	NGC 2359		07 15.4 -13 07	07 17.7 -13 12	CMa Neb	12	274	242
	NGC 2360		07 15.4 -15 33	07 17.7 -15 38	CMa OpCl	12	274	278
	NGC 2300		07 16.5 85 50	07 33.5 85 45	Cep E2	1	1	1
	NGC 2362		07 16.6 -24 52	07 18.7 -24 57	CMa OpCl	19	319	278
	NGC 2383		07 22.6 -20 50	07 24.8 -20 56	CMa OpCl	19	319	279
	NGC 2366		07 23.6 69 08	07 28.9 69 02	Cam Irr	1	21	13
!	NGC 2392		07 26.2 21 01	07 29.2 20 55	Gem Plan	5	139	135
	NGC 2403		07 32.0 65 43	07 36.8 65 36	Cam Sc	1	21	13
	NGC 2421		07 34.1 -20 30	07 36.3 -20 37	Pup OpCl	19	319	279
S	NGC 2422	M47	07 34.3 -14 23	07 36.6 -14 30	Pup OpCl	12	274	243
	NGC 2419		07 34.8 39 00	07 38.2 38 53	Lyn Glob	5	100	72
!*	NGC 2437	M46	07 39.6 -14 42	07 41.9 -14 49	Pup OpCl	12	274	243
*	NGC 2438		07 39.6 -14 36	07 41.9 -14 43	Pup Plan	12	274	243
	NGC 2447	M93	07 42.4 -23 45	07 44.5 -23 52	Pup OpCl	19	320	279
	NGC 2477		07 50.5 -38 25	07 52.3 -38 33	Pup OpCl	19	362	345
	NGC 2516		07 59.7 -60 44	08 00.5 -60 52	Car OpCl	25	424	393
	NGC 2547		08 08.9 -49 07	08 10.4 -49 16	Vel OpCl	20	396	372
	NGC 2523		08 09.2 73 45	08 15.0 73 36	Cam SBb	2	7	14
	NGC 2546		08 10.6 -37 29	08 12.4 -37 38	Pup OpCl	20	362	345
	NGC 2548	M48	08 11.2 -05 38	08 13.7 -05 47	Hya OpCl	12	275	244
	NGC 2613		08 31.2 -22 48	08 33.4 -22 58	Pyx Sb	20	321	280
	NGC 2627		08 35.2 -29 46	08 37.3 -29 56	Pyx OpCl	20	363	315
	NGC 2623		08 35.4 25 56	08 38.4 25 46	Cnc Pec	6	141	103
!!	NGC 2632	M44	08 37.5 19 52	08 40.4 19 41	Cnc OpCl	12	141	136
	NGC 2659		08 40.9 -44 46	08 42.6 -44 57	Vel OpCl	20	397	346
	H3		08 44.6 -52 36	08 46.1 -52 47	Vel OpCl	25	425	372
*	NGC 2682	M67	08 48.3 12 00	08 51.0 11 49	Cnc OpCl	12	187	173
	NGC 2683		08 49.6 33 38	08 52.7 33 27	Lyn Sb	6	102	103
	NGC 2681		08 50.0 51 31	08 53.6 51 20	UMa S0	2	44	48
	NGC 2782		09 10.9 40 19	09 14.1 40 07	Lyn Sb	6	71	74
	NGC 2818		09 14.0 -36 24	09 16.0 -36 37	Pyx OpCl	20	364	346
	NGC 2841		09 18.6 51 12	09 22.1 50 59	UMa Sb	2	44	49
	I.2488		09 25.7 -56 45	09 27.2 -56 58	Vel OpCl	25	425	395

APPENDIX E: A CATALOG OF DEEP-SKY OBJECTS

```
                    Size
ID        v mag   arc-min    S.B.   Cl    MDM    Comments
--------|-------|----------|------|-----|-----|---------------------------------
NGC 2261   10.     2x2       20.1   1.6   38   Hubble's Variable Nebula
NGC 2264   5.      15x26     20.1   1.7    5   cluster + Cone Nebula
NGC 2287   6.      30x30     22.0   0.9    3   M41
NGC 2281   6.      15x15     20.5   1.5    5   about 30 stars mag 7+
NGC 2323   6.      10x10     19.6   1.8    8   M50

NGC 2346   10.     1.0x0.9   18.5   2.3   85   central star variable mag 11+
NGC 2276   12.4    2.5x2.0   22.8   0.6   38
NGC 2359   ?       10x5                        complex nebula around 10.4 mag star
NGC 2360   9.      10x10     22.6   0.6    8   50 stars mag 9 to 12
NGC 2300   12.2    1.0x1.7   21.4   1.1   77   NGC 2276 is 6' NW

NGC 2362   4.      6x6       16.5   3.1   13   40 stars & 4th mag 30 CMa
NGC 2383   11.     2x2       21.1   1.2   38   50 stars mag 12+
NGC 2366   12.6    6.0x3.0   24.4   0.0   26
NGC 2392   8.3     0.7x0.7   16.2   3.2  110   Eskimo Nebula
NGC 2403   8.8     16x10     22.9   0.5    8

NGC 2421   10.     8x8       23.1   0.4   10   about 60 stars
NGC 2422   5.      20x20     20.1   1.6    4   M47
NGC 2419   11.5    2x2       21.6   1.0   38   most distant globular in our Galaxy
NGC 2437   8.      25x25     23.6   0.3    3   M46, with planetary NGC 2438
NGC 2438   11.     1.1x1.1   19.8   1.8   70   in M46

NGC 2447   7.      18x18     21.9   0.9    4   M93
NGC 2477   7.      25x25     22.6   0.7    3   300 stars mag 11+
NGC 2516   6.      60x60     23.5   0.3    1   100 stars mag 7 to 13
NGC 2547   5.5     15x15     20.0   1.7    5   50 stars mag 7+
NGC 2523   12.7    1.8x1.4   22.3   0.8   55   theta shape

NGC 2546   8.      25x25     23.6   0.3    3   50 stars mag 9+
NGC 2548   5.5     40x40     22.1   0.8    2   M48
NGC 2613   10.9    6.4x1.5   22.0   0.9   51
NGC 2627   9.      8x8       22.1   0.8   10   70 stars mag 11 to 13
NGC 2623   14.0    2.0x0.4   22.4   0.7  192   extending filaments

NGC 2632   4.5     80x80     22.6   0.6    1   M44, Praesepe
NGC 2659   9.5     10x10     23.1   0.4    8   50 stars mag 11+
H3         6.      7x7       18.9   2.2   11   35 stars mag 10+
NGC 2682   7.      15x15     21.5   1.1    5   M67, stars mag 10+
NGC 2683   10.6    9.0x1.3   21.9   0.9   59

NGC 2681   11.3    2.8x2.5   22.0   0.9   31
NGC 2782   12.4    1.8x1.6   22.2   0.8   48   distorted, extended filament
NGC 2818   11.     9x9       24.4  -0.1    9   contains 13th mag planetary nebula
NGC 2841   10.3    6.2x2.0   21.7   1.0   38
I.2488     7.      20x20     22.1   0.8    4   50 stars mag 11+
```

	ID		1950.0 RA DEC	2000.0 RA DEC	Con	Type	Charts 2000	Uran	VPSA
	NGC 2910		09 28.4 -52 41	09 30.1 -52 54	Vel	OpCl	25	426	373
*	NGC 2903		09 29.3 21 44	09 32.1 21 31	Leo	Sb	6	143	138
	NGC 2925		09 31.9 -53 13	09 33.6 -53 26	Vel	OpCl	25	426	373
	NGC 2976		09 43.2 68 08	09 47.3 67 54	UMa	Sc	2	23	14
	NGC 2997		09 43.5 -30 58	09 45.7 -31 12	Ant	Sc	20	365	317
!!*	NGC 3031	M81	09 51.5 69 18	09 55.6 69 04	UMa	Sa	2	23	14
!!*	NGC 3034	M82	09 51.9 69 56	09 56.1 69 42	UMa	Irr	2	23	14
	NGC 3077		09 59.4 68 58	10 03.4 68 44	UMa	E2	2	23	14
	NGC 3109		10 00.8 -25 55	10 03.1 -26 10	Hya	Irr	20	324	317
	NGC 3114		10 01.1 -59 53	10 02.7 -60 08	Car	OpCl	25	426	395
	NGC 3115		10 02.8 -07 28	10 05.3 -07 43	Sex	E7	13	279	247
	NGC 3132		10 04.9 -40 11	10 07.0 -40 26	Vel	Plan	20	399	347
	NGC 3145		10 07.7 -12 10	10 10.1 -12 25	Hya	SBb	13	279	247
	NGC 3166		10 11.2 03 40	10 13.8 03 25	Sex	Sa	13	234	211
	NGC 3169		10 11.7 03 43	10 14.3 03 28	Sex	Sb	13	234	211
	NGC 3147		10 12.8 73 39	10 17.1 73 24	Dra	Sb	2	8	15
	NGC 3185		10 14.9 21 56	10 17.7 21 41	Leo	SBa	6	144	139
	NGC 3187		10 15.0 22 08	10 17.8 21 53	Leo	SBc	6	144	139
	NGC 3184		10 15.2 41 40	10 18.2 41 25	UMa	Sc	6	72	75
	NGC 3190		10 15.4 22 05	10 18.2 21 50	Leo	Sb	6	144	139
	NGC 3201		10 15.5 -46 09	10 17.6 -46 24	Vel	Glob	20	399	374
	NGC 3193		10 15.7 22 09	10 18.5 21 54	Leo	E0	6	144	139
	NGC 3198		10 16.7 45 49	10 19.8 45 34	UMa	Sc	6	72	50
	NGC 3228		10 19.7 -51 29	10 21.7 -51 44	Vel	OpCl	25	426	374
	NGC 3242		10 22.4 -18 23	10 24.8 -18 38	Hya	Plan	13	325	283
	I.2574		10 25.0 68 43	10 28.7 68 28	UMa	Irr	2	24	15
	NGC 3256		10 25.7 -43 38	10 27.8 -43 53	Vel	Pec	20	399	348
	NGC 3293		10 31.5 -57 58	10 33.4 -58 13	Car	OpCl	25	427	395
	NGC 3310		10 35.7 53 46	10 38.8 53 30	UMa	Irr	2	46	50
	NGC 3319		10 36.4 41 56	10 39.3 41 40	UMa	SBc	6	72	76
	NGC 3344		10 40.7 25 11	10 43.4 24 55	LMi	Sc	6	145	106
	NGC 3351	M95	10 41.3 11 58	10 43.9 11 42	Leo	SBb	13	190	176
!!!!	NGC 3372		10 43.1 -59 25	10 45.0 -59 41	Car	Neb	25	427	396
	NGC 3359		10 43.4 63 30	10 46.7 63 14	UMa	SBc	2	24	30
*	NGC 3368	M96	10 44.2 12 05	10 46.8 11 49	Leo	Sb	13	190	176
*	NGC 3379	M105	10 45.2 12 51	10 47.8 12 35	Leo	E1	13	190	176
*	NGC 3384		10 45.7 12 54	10 48.3 12 38	Leo	E7	13	190	176
*	NGC 3389		10 45.8 12 48	10 48.4 12 32	Leo	Sc	13	190	176
	NGC 3395		10 47.1 33 15	10 49.9 32 59	LMi	Sc	6	105	106
	NGC 3396		10 47.2 33 16	10 50.0 33 00	LMi	Irr	6	105	106

APPENDIX E: A CATALOG OF DEEP-SKY OBJECTS

```
            Size
ID       v mag  arc-min    S.B.  Cl   MDM  Comments
--------|-----|----------|----|----|---|-------------------------
NGC 2910  8.      6x6     20.5  1.5   13  30 stars mag 10+
NGC 2903  9.7  11.0x4.7   22.6  0.7   16
NGC 2925  8.     11x11    21.8  1.0    7  about 30 stars
NGC 2976 10.8   3.4x1.9   21.5  1.1   40  member of M81 group
NGC 2997 11.0   6.0x5.0   23.3  0.4   15

NGC 3031  8.0    18x10    22.3  0.8    8  M81 (near M82)
NGC 3034  9.2   8.0x3.0   21.3  1.2   26  M82 (near M81)
NGC 3077 11.0   2.6x1.9   21.4  1.2   40  member of M81 group
NGC 3109 11.2  11.0x2.0   23.2  0.4   38
NGC 3114  8.0    40x40    24.6 -0.2    2  100 stars mag 9 to 13

NGC 3115 10.0   4.0x1.0   20.1  1.6   77  Spindle Nebula
NGC 3132  8.2   1.4x1.4   17.6  2.7   55  central star mag 10
NGC 3145 12.5   2.4x1.0   22.1  0.9   77
NGC 3166 11.5   4.0x1.5   22.1  0.9   51  near NGC 3169
NGC 3169 11.4   3.9x1.7   22.1  0.9   45  near NGC 3166

NGC 3147 11.3   3.0x2.3   22.0  0.9   33
NGC 3185 12.7   1.5x0.9   21.7  1.0   85  group with NGC 3187, 3193, 3190
NGC 3187 13.0   1.0x1.3   21.9  0.9   77
NGC 3184 10.5   5.5x5.5   22.8  0.6   14
NGC 3190 12.0   3.0x1.0   21.8  1.0   77

NGC 3201  8.5    10x10    22.1  0.8    8
NGC 3193 12.0   0.9x0.9   20.4  1.5   85
NGC 3198 11.0   9.0x3.0   23.2  0.4   26
NGC 3228  6.5    20x20    21.6  1.0    4
NGC 3242  8.9   0.7x0.7   16.8  3.0  110

I.2574   13.0   9.0x4.0   25.5 -0.5   19  member of M81 group
NGC 3256 12.1   2.0x1.5   21.9  0.9   51
NGC 3293  5.      8x8     18.1  2.4   10  50 stars mag 6 to 13
NGC 3310 11.0   3.0x2.0   21.6  1.1   38
NGC 3319 11.8   6.0x2.8   23.5  0.3   27

NGC 3344 11.0   6.0x5.1   23.3  0.4   15
NGC 3351 11.0   4.0x3.0   22.3  0.8   26  M95, theta shape
NGC 3372  ?     85x80      ?                Eta Carinae Nebula
NGC 3359 11.0   6.0x3.0   22.8  0.6   26
NGC 3368 10.2   6.0x4.0   22.3  0.8   19  M96

NGC 3379 10.6   2.1x2.0   20.8  1.4   38  M105
NGC 3384 11.0   4.0x2.0   21.9  0.9   38  group with M105
NGC 3389 12.2   2.2x1.0   21.7  1.0   77  group with M105
NGC 3395 12.4   1.4x0.8   21.2  1.2   96
NGC 3396 12.8   1.0x0.5   20.7  1.4  154  pair with NGC 3395
```

	ID		1950.0 RA DEC	2000.0 RA DEC	Con	Type	Charts 2000	Uran	VPSA
	NGC 3423		10 48.7 06 07	10 51.3 05 51	Sex	Sc	13	190	176
	NGC 3486		10 57.8 29 15	11 00.5 28 59	LMi	Sc	6	106	106
	NGC 3504		11 00.5 28 15	11 03.2 27 59	LMi	Sb	6	146	106
	NGC 3511		11 00.8 -22 50	11 03.2 -23 06	Crt	Sc	20	326	284
	NGC 3513		11 01.1 -22 58	11 03.5 -23 14	Crt	SBc	20	326	284
	NGC 3521		11 03.2 00 14	11 05.8 -00 02	Leo	Sb	13	236	212
	NGC 3532		11 03.4 -58 24	11 05.5 -58 40	Car	OpCl	25	427	396
*	NGC 3556	M108	11 08.7 55 57	11 11.6 55 41	UMa	Sc	2	46	30
*	NGC 3587	M97	11 12.0 55 18	11 14.9 55 02	UMa	Plan	2	46	30
	NGC 3610		11 15.6 59 04	11 18.5 58 48	UMa	E4	2	46	30
	NGC 3613		11 15.7 58 17	11 18.6 58 01	UMa	E5	2	46	30
	NGC 3621		11 15.9 -32 32	11 18.3 -32 48	Hya	Sc	20	367	319
!*	NGC 3623	M65	11 16.3 13 23	11 18.9 13 07	Leo	Sa	13	191	176
!*	NGC 3627	M66	11 17.6 13 17	11 20.2 13 01	Leo	Sb	13	191	176
*	NGC 3628		11 17.7 13 53	11 20.3 13 37	Leo	Sb	13	191	176
	NGC 3631		11 18.3 53 28	11 21.1 53 12	UMa	Sc	2	46	51
	NGC 3672		11 22.5 -09 32	11 25.0 -09 48	Crt	Sb	13	281	249
	NGC 3675		11 23.5 43 52	11 26.2 43 36	UMa	Sb	6	73	77
	NGC 3718		11 29.9 53 21	11 32.7 53 04	UMa	SBa	2	47	51
	NGC 3726		11 30.7 47 19	11 33.4 47 02	UMa	Sc	6	73	51
	NGC 3887		11 44.6 -16 35	11 47.1 -16 52	Crt	Sc	13	282	285
	NGC 3893		11 46.1 49 00	11 48.7 48 43	UMa	Sc	6	74	51
	NGC 3918		11 47.8 -56 54	11 50.3 -57 11	Cen	Plan	25	428	396
*	NGC 3992	M109·	11 55.0 53 39	11 57.6 53 22	UMa	SBb	2	47	51
	NGC 3998		11 55.3 55 44	11 57.9 55 27	UMa	E2	2	47	30
	NGC 4026		11 56.9 51 14	11 59.5 50 57	UMa	E8	2	47	51
	NGC 4027		11 57.0 -18 59	11 59.6 -19 16	Crv	Pec	13	327	285
	NGC 4030		11 57.8 -00 49	12 00.4 -01 06	Vir	Sb	13	238	213
	NGC 4036		11 58.9 62 10	12 01.5 61 53	UMa	E6	2	25	30
!*	NGC 4038		11 59.3 -18 35	12 01.9 -18 52	Crv	Pec	13	328	285
!*	NGC 4039		11 59.3 -18 36	12 01.9 -18 53	Crv	Pec	13	328	285
	NGC 4041		11 59.7 62 25	12 02.3 62 08	UMa	Sc	2	25	30
	NGC 4051		12 00.6 44 48	12 03.2 44 31	UMa	Sb	7	74	78
	NGC 4085		12 02.8 50 38	12 05.3 50 21	UMa	Sb	2	47	52
	NGC 4088		12 03.0 50 49	12 05.5 50 32	UMa	Sb	2	47	52
	NGC 4096		12 03.5 47 45	12 06.0 47 28	UMa	Sc	7	74	52
	NGC 4100		12 03.6 49 51	12 06.1 49 34	UMa	Sb	7	74	52
	NGC 4105		12 04.1 -29 30	12 06.7 -29 47	Hya	E3	21	368	320
	NGC 4106		12 04.2 -29 31	12 06.8 -29 48	Hya	E2	21	368	320
	NGC 4111		12 04.5 43 21	12 07.0 43 04	CVn	E7	7	74	78

APPENDIX E: A CATALOG OF DEEP-SKY OBJECTS

```
                  Size
   ID     v mag  arc-min   S.B.   Cl   MDM           Comments
--------|-----|----------|----|----|---|--------------------------------
NGC 3423  11.7  3.5x3.0   22.9  0.5   26
NGC 3486  11.2  5.5x4.2   23.2  0.4   18
NGC 3504  11.6  2.0x1.8   21.6  1.1   43
NGC 3511  11.9  4.2x1.5   22.5  0.7   51
NGC 3513  12.0  2.0x1.6   21.9  0.9   48

NGC 3521  10.2  6.0x4.0   22.3  0.8   19
NGC 3532   7.   60x60     24.5 -0.1    1  150 stars mag 8 to 12
NGC 3556  10.8  7.8x1.4   22.0  0.9   55  M108, near M97
NGC 3587  11.0  2.5x2.5   21.6  1.1   31  M97, Owl Nebula
NGC 3610  11.6  1.4x0.9   20.5  1.5   85

NGC 3613  11.7  1.7x0.8   20.7  1.4   96
NGC 3621  11.8  3.5x1.4   22.2  0.8   55
NGC 3623  10.3  7.8x1.6   21.7  1.0   48  M65
NGC 3627   9.7  8.0x2.5   21.6  1.1   31  M66
NGC 3628  10.3 12.0x2.0   22.4  0.7   38  near M65, M66

NGC 3631  11.5  4.5x4.0   23.3  0.4   19
NGC 3672  11.8  3.5x1.4   22.2  0.8   55
NGC 3675  11.4  3.5x1.3   21.7  1.0   59
NGC 3718  11.8  3.0x3.0   22.8  0.6   26
NGC 3726  11.3  5.0x3.4   23.0  0.5   23

NGC 3887  11.6  2.8x2.0   22.1  0.9   38
NGC 3893  11.0  3.9x2.5   22.1  0.9   31
NGC 3918   8.   0.17x0.17 12.8  4.6  452
NGC 3992  10.9  6.4x3.5   22.9  0.5   22  M109
NGC 3998  11.6  1.6x1.2   20.9  1.3   64

NGC 4026  11.9  3.3x0.7   21.4  1.1  110
NGC 4027  11.6  2.0x1.7   21.6  1.1   45  near NGC 4038
NGC 4030  11.2  3.3x2.4   22.1  0.9   32
NGC 4036  11.6  3.0x1.0   21.4  1.1   77  near NGC 4041
NGC 4038  11.0  2.5x2.5   21.6  1.1   31  Ringtail Galaxy

NGC 4039  12.   2.5x2.0   22.4  0.7   38  Ringtail Galaxy
NGC 4041  11.7  2.2x1.9   21.9  0.9   40  16' from NGC 4036
NGC 4051  11.2  4.2x3.0   22.6  0.7   26
NGC 4085  12.8  2.2x0.5   21.5  1.1  154  11' from NGC 4088
NGC 4088  11.1  4.7x1.5   21.8  1.0   51  near NGC 4085

NGC 4096  11.5  5.8x1.0   22.0  0.9   77
NGC 4100  11.9  4.5x1.1   22.3  0.8   70
NGC 4105  12.0  1.5x1.5   21.5  1.1   51  interacting with NGC 4106
NGC 4106  12.5  1.0x1.8   21.8  1.0   77  interacting with NGC 4105
NGC 4111  11.6  3.4x0.8   21.3  1.2   96  lenticular shape
```

ID		1950.0 RA	DEC	2000.0 RA	DEC	Con	Type	Charts 2000	Uran	VPSA
NGC 4116		12 05.1	02 58	12 07.7	02 41	Vir	SBc	13	238	214
NGC 4123		12 05.6	03 09	12 08.2	02 52	Vir	SBb	13	238	214
NGC 4147		12 07.6	18 49	12 10.1	18 32	Com	Glob	13	148	142
NGC 4192	M98	12 11.3	15 11	12 13.8	14 54	Com	Sb	13	193	142
NGC 4214		12 13.1	36 36	12 15.6	36 19	CVn	Irr	7	107	78
NGC 4216		12 13.4	13 25	12 15.9	13 08	Vir	Sb	13	193	178
NGC 4236		12 14.3	69 45	12 16.7	69 28	Dra	SB	2	25	16
NGC 4244		12 15.0	38 05	12 17.5	37 48	CVn	Sb	7	107	78
NGC 4251		12 15.7	28 27	12 18.2	28 10	Com	Sa	7	107	108
* NGC 4254	M99	12 16.3	14 42	12 18.8	14 25	Com	Sc	13	193	178
* NGC 4258	M106	12 16.5	47 35	12 19.0	47 18	CVn	Sb	7	74	52
NGC 4260		12 16.8	06 23	12 19.4	06 06	Vir	SBc	13	193	178
NGC 4261		12 16.8	06 06	12 19.4	05 49	Vir	E2	13	193	178
NGC 4273		12 17.4	05 37	12 20.0	05 20	Vir	Sc	13	238	178
NGC 4274		12 17.4	29 53	12 18.8	29 36	Com	Sb	7	107	108
NGC 4281		12 17.8	05 40	12 20.4	05 23	Vir	E5	13	238	178
NGC 4293		12 18.7	18 40	12 21.2	18 23	Com	Sa	13	148	142
NGC 4294		12 18.7	11 47	12 21.2	11 30	Vir	SBc	13	193	178
NGC 4298		12 19.0	14 53	12 21.5	14 36	Com	Sc	13	193	178
NGC 4299		12 19.2	11 47	12 21.7	11 30	Vir	SBc	13	193	178
NGC 4302		12 19.2	14 53	12 21.7	14 36	Com	Sc	13	193	178
NGC 4303	M61	12 19.4	04 45	12 22.0	04 28	Vir	Sc	13	238	214
* NGC 4321	M100	12 20.4	16 06	12 22.9	15 49	Com	Sc	13	193	142
M40	M40	12 20.0	58 22	12 22.4	58 05	UMa	star	2	47	31
NGC 4340		12 21.0	17 00	12 23.5	16 43	Com	SBa	13	193	142
NGC 4349		12 21.4	-61 37	12 24.2	-61 54	Cru	OpCl	25	450	397
NGC 4350		12 21.4	16 58	12 23.9	16 41	Com	E7	13	193	142
NGC 4361		12 21.9	-18 29	12 24.5	-18 46	Crv	Plan	13	328	286
NGC 4365		12 22.0	07 36	12 24.5	07 19	Vir	E2	13	193	178
* NGC 4374	M84	12 22.6	13 10	12 25.1	12 53	Vir	E1	13	193	178
NGC 4382	M85	12 22.8	18 28	12 25.3	18 11	Com	S0	13	148	142
NGC 4372		12 23.0	-72 24	12 25.9	-72 41	Mus	Glob	25	466	413
* NGC 4387		12 23.2	13 06	12 25.7	12 49	Vir	E5	13	193	178
* NGC 4388		12 23.3	12 56	12 25.8	12 39	Vir	SBc	13	193	178
NGC 4394		12 23.4	18 29	12 25.9	18 12	Com	SBb	13	148	142
NGC 4395		12 23.4	33 49	12 25.9	33 32	CVn	S	7	108	108
* NGC 4402		12 23.6	13 24	12 26.1	13 07	Vir	Sb	13	193	178
* NGC 4406	M86	12 23.7	13 13	12 26.2	12 56	Vir	E3	13	193	178
* NGC 4413		12 24.0	12 53	12 26.5	12 36	Vir	SBa	13	193	178
NGC 4414		12 24.0	31 30	12 26.5	31 13	Com	Sc	7	108	108

APPENDIX E: A CATALOG OF DEEP-SKY OBJECTS

```
                    Size
    ID     v mag   arc-min    S.B.   Cl   MDM    Comments
--------|-----|----------|----|----|---|------------------------------
NGC 4116  12.3   3.3x1.4   22.6   0.7   55  near NGC 4123
NGC 4123  12.0   3.5x2.4   22.9   0.5   32  near NGC 4116
NGC 4147  11.    4x4       22.6   0.6   19
NGC 4192  11.0   8.2x2.0   22.7   0.6   38  M98
NGC 4214  10.5   7.0x4.5   22.9   0.6   17

NGC 4216  10.9   7.2x1.0   21.7   1.0   77
NGC 4236  10.7   22x5      24.4  -0.1   15
NGC 4244  10.7  13.0x1.0   22.1   0.9   77  edge-on spiral
NGC 4251  11.3   2.0x0.8   20.4   1.5   96
NGC 4254  10.4   4.5x4.0   22.2   0.8   19  M99

NGC 4258   9.0  19.5x6.5   22.9   0.5   12  M106
NGC 4260  12.7   2.3x0.9   22.1   0.9   85
NGC 4261  11.7   2.0x1.7   21.7   1.0   45  near NGC 4260
NGC 4273  12.3   1.7x1.2   21.7   1.0   64
NGC 4274  11.5   5.0x1.2   22.1   0.9   64

NGC 4281  12.2   1.5x0.8   21.0   1.3   96  near NGC 4270, 4273
NGC 4293  11.7   4.8x1.8   22.7   0.6   43  heavy dust lanes
NGC 4294  12.6   2.4x0.9   22.1   0.9   85  near NGC 4299
NGC 4298  11.9   2.7x1.1   21.7   1.0   70  near NGC 4307
NGC 4299  12.9   1.1x0.9   21.5   1.1   85  near NGC 4294

NGC 4302  12.9   4.5x0.5   22.4   0.7  154  edge-on, with dust lane
NGC 4303  10.2   5.7x5.5   22.6   0.7   14  M61
NGC 4321  10.4   5.2x5.0   22.6   0.7   15  M100
M40        9.0      -                        M40: 2 stars separated by 49"
NGC 4340  13.0   2.2x1.4   22.8   0.6   55  theta shape

NGC 4349  10.    15x15     24.5  -0.1    5  100 stars mag 12-14
NGC 4350  11.9   1.9x0.5   20.5   1.5  154  lenticular shape
NGC 4361  10.5   1.3x1.3   19.7   1.8   59  central star mag 13
NGC 4365  11.0   2.0x1.3   20.7   1.4   59
NGC 4374  10.5   2.0x1.8   20.5   1.5   43  M84

NGC 4382  10.5   3.0x2.0   21.1   1.3   38  M85
NGC 4372   8.    18x18     22.9   0.5    4
NGC 4387  12.0   1.9x1.1   21.4   1.1   70
NGC 4388  12.0   5.0x1.0   22.4   0.7   77  near M84
NGC 4394  12.0   3.0x3.0   23.0   0.5   26  near M85

NGC 4395  11.0  10.0x8.0   24.4  -0.1   10  3-branch spiral
NGC 4402  13.0   2.0x0.8   22.1   0.8   96  near M86
NGC 4406  10.5   3.0x2.8   21.4   1.1   27  M86
NGC 4413  13.2   1.1x0.7   21.5   1.1  110  near NGC 4388
NGC 4414  11.0   3.1x1.5   21.3   1.2   51
```

	ID		1950.0 RA DEC	2000.0 RA DEC	Con	Type	Charts 2000	Uran	VPSA
	NGC 4417		12 24.3 09 52	12 26.8 09 35	Vir	E7	13	193	178
	NGC 4424		12 24.6 09 42	12 27.1 09 25	Vir	SBa	13	193	178
*	NGC 4425		12 24.7 13 01	12 27.2 12 44	Vir	S0	13	193	178
*	NGC 4435		12 25.2 13 21	12 27.7 13 04	Vir	E4	13	193	178
*	NGC 4438		12 25.3 13 17	12 27.8 13 00	Vir	Sa	13	193	178
	NGC 4448		12 25.8 28 54	12 28.3 28 37	Com	Sb	7	108	108
*	NGC 4449		12 25.8 44 22	12 28.2 44 05	CVn	Irr	7	75	78
	NGC 4450		12 25.9 17 21	12 28.4 17 04	Com	Sb	13	148	142
*	NGC 4458		12 26.5 13 32	12 29.0 13 15	Vir	E0	13	193	178
*	NGC 4459		12 26.5 14 15	12 29.0 13 58	Com	S0	13	193	178
	3C273		12 26.6 02 19	12 29.2 02 02	Vir	QSO	13	238	214
*	NGC 4461		12 26.6 13 28	12 29.1 13 11	Vir	S0	13	193	178
	NGC 4472	M49	12 27.3 08 16	12 29.8 07 59	Vir	E3	13	193	178
*	NGC 4473		12 27.3 13 42	12 29.8 13 25	Vir	E4	13	193	178
*	NGC 4476		12 27.5 12 37	12 30.0 12 20	Vir	E4	14	193	178
*	NGC 4477		12 27.5 13 55	12 30.0 13 38	Com	SBa	14	193	178
*	NGC 4478		12 27.8 12 36	12 30.3 12 19	Vir	E1	14	193	178
*	NGC 4479		12 27.8 13 52	12 30.3 13 35	Com	Sb	14	193	178
	NGC 4485		12 28.2 41 58	12 30.6 41 41	CVn	E	7	75	78
!*	NGC 4486	M87	12 28.3 12 40	12 30.8 12 23	Vir	E1	14	193	178
	NGC 4490		12 28.3 41 55	12 30.7 41 38	CVn	Sc	7	75	78
	NGC 4494		12 28.9 26 03	12 31.4 25 46	Com	E1	7	148	108
	NGC 4501	M88	12 29.5 14 42	12 32.0 14 25	Com	Sb	14	194	178
	NGC 4517		12 30.2 00 23	12 32.8 00 06	Vir	Sc	14	239	214
	NGC 4526		12 31.6 07 58	12 34.1 07 41	Vir	E7	14	194	178
	NGC 4527		12 31.6 02 56	12 34.2 02 39	Vir	Sb	14	239	214
	NGC 4535		12 31.8 08 28	12 34.3 08 11	Vir	SBc	14	194	178
	NGC 4536		12 31.9 02 28	12 34.5 02 11	Vir	Sc	14	239	214
	I.3568		12 32.4 82 51	12 33.7 82 34	Cam	Plan	2	9	6
	NGC 4548	M91	12 32.9 14 46	12 35.4 14 29	Com	SBb	14	194	178
	NGC 4552	M89	12 33.1 12 50	12 35.6 12 33	Vir	E0	14	194	178
	NGC 4559		12 33.5 28 14	12 36.0 27 57	Com	Sc	7	149	108
	NGC 4564		12 33.9 11 43	12 36.4 11 26	Vir	E6	14	194	178
!!!*	NGC 4565		12 33.9 26 16	12 36.4 25 59	Com	Sb	7	149	108
!	NGC 4567		12 34.0 11 32	12 36.5 11 15	Vir	Sb	14	194	178
	NGC 4568		12 34.1 11 31	12 36.6 11 14	Vir	Sb	14	194	178
*	NGC 4569	M90	12 34.3 13 26	12 36.8 13 09	Vir	Sb	14	194	178
	NGC 4579	M58	12 35.1 12 05	12 37.6 11 48	Vir	Sb	14	194	178
	H7		12 36.8 -60 20	12 39.7 -60 36	Cru	OpCl	25	429	397
	NGC 4590	M68	12 36.8 -26 29	12 39.5 -26 45	Hya	Glob	21	329	320

APPENDIX E: A CATALOG OF DEEP-SKY OBJECTS

```
                Size
ID       v mag  arc-min    S.B.  Cl   MDM  Comments
--------|-----|----------|----|----|---|------------------------------------
NGC 4417  12.2  2.2x0.8   21.4  1.1   96  near NGC 4424
NGC 4424  12.5  2.5x1.3   22.4  0.7   59  near NGC 4417
NGC 4425  12.9  2.0x0.5   21.5  1.1  154  near M86
NGC 4435  11.8  1.4x0.9   20.7  1.4   85  near NGC 4438
NGC 4438  11.0  4.0x1.5   21.6  1.1   51  near NGC 4435

NGC 4448  11.7  2.9x1.0   21.5  1.1   77
NGC 4449  10.5  4.2x3.0   21.9  0.9   26  rectangular shape
NGC 4450  11.1  3.8x3.0   22.4  0.8   26
NGC 4458  12.0  1.9x1.8   22.0  0.9   43
NGC 4459  11.7  1.5x1.0   20.8  1.4   77  dust lane near nucleus

3C273     12.8  stellar    -              brightest quasar
NGC 4461  12.2  2.0x1.0   21.6  1.1   77
NGC 4472  10.1  4.0x3.4   21.6  1.1   23  M49
NGC 4473  11.3  2.0x1.0   20.7  1.4   77
NGC 4476  13.3  0.7x0.4   20.5  1.5  192  near M87

NGC 4477  10.4  4.0x3.5   21.9  0.9   22
NGC 4478  12.4  1.0x1.8   21.7  1.0   77  near M87
NGC 4479  12.5  1.5x1.5   22.0  0.9   51
NGC 4485  12.5  1.3x0.7   21.0  1.3  110  near NGC 4490
NGC 4486   8.6  3.0x3.0   19.6  1.9   26  M87

NGC 4490  10.1  5.0x2.0   21.2  1.2   38  Cocoon Galaxy
NGC 4494  10.9  1.4x1.4   20.3  1.6   55
NGC 4501  10.5  5.7x2.5   22.0  0.9   31  M88
NGC 4517  11.4  9.0x1.0   22.4  0.7   77  equatorial dust lane
NGC 4526  10.7  4.0x1.0   20.8  1.4   77

NGC 4527  11.3  5.1x1.1   21.8  1.0   70
NGC 4535  10.7  6.0x4.0   22.8  0.6   19
NGC 4536  11.0  7.0x2.0   22.5  0.7   38
I.3568    11.6  0.60x0.60 19.1  2.1  128
NGC 4548  10.9  3.9x3.4   22.3  0.8   23  M91

NGC 4552  11.0  2.0x2.0   21.1  1.2   38  M89
NGC 4559  10.5 10.0x3.0   22.8  0.6   26
NGC 4564  11.1  3.1x1.4   21.3  1.2   55  near NGC 4567, 4568
NGC 4565  10.5 15.0x1.1   22.2  0.8   70  famous edge-on spiral
NGC 4567  12.0  2.4x1.6   22.1  0.9   48  the Siamese Twins

NGC 4568  11.9  3.6x1.8   22.6  0.7   43  near NGC 4567
NGC 4569   9.0  7.0x2.5   20.7  1.4   31  M90
NGC 4579  10.5  4.0x3.5   22.0  0.9   22  M58
H7        10.   8x8       23.1  0.4   10  about 200 stars
NGC 4590   8.   9x9       21.4  1.1    9  M68
```

	ID		1950.0 RA	DEC	2000.0 RA	DEC	Con	Type	Charts 2000	Uran	VPSA
!!*	NGC 4594	M104	12 37.3	-11 21	12 39.9	-11 37	Vir	Sa	14	284	250
	NGC 4605		12 37.8	61 53	12 40.0	61 37	UMa	Sc	2	26	31
	NGC 4618		12 39.2	41 25	12 41.6	41 09	CVn	Sc	7	75	78
	NGC 4621	M59	12 39.5	11 55	12 42.0	11 39	Vir	E3	14	194	178
	NGC 4631		12 39.8	32 49	12 42.2	32 33	CVn	Sc	7	108	108
	NGC 4638		12 40.2	11 43	12 42.7	11 27	Vir	E5	14	194	179
	NGC 4636		12 40.3	02 57	12 42.9	02 41	Vir	E1	14	239	215
	NGC 4639		12 40.3	13 31	12 42.8	13 15	Vir	SBb	14	194	179
	NGC 4643		12 40.8	02 15	12 43.4	01 59	Vir	SBa	14	239	215
	NGC 4647		12 41.0	11 51	12 43.5	11 35	Vir	Sc	14	194	179
	NGC 4649	M60	12 41.1	11 49	12 43.6	11 33	Vir	E1	14	194	179
	NGC 4651		12 41.2	16 40	12 43.7	16 24	Com	Sc	14	194	143
	NGC 4654		12 41.4	13 23	12 43.9	13 07	Vir	Sc	14	194	179
!	NGC 4656		12 41.6	32 26	12 44.0	32 10	CVn	Irr	7	108	108
	NGC 4660		12 42.0	11 26	12 44.5	11 10	Vir	E5	14	194	179
	NGC 4666		12 42.6	-00 12	12 45.2	-00 28	Vir	Sc	14	239	215
	NGC 4668		12 43.0	-00 17	12 45.6	-00 33	Vir	SBc	14	239	215
	NGC 4689		12 45.2	14 01	12 47.7	13 45	Com	Sb	14	194	179
	NGC 4697		12 46.0	-05 32	12 48.6	-05 48	Vir	E5	14	284	251
	NGC 4699		12 46.5	-08 24	12 49.1	-08 40	Vir	Sa	14	284	251
	NGC 4710		12 47.1	15 26	12 49.6	15 10	Com	S0	14	194	143
	NGC 4725		12 48.1	25 46	12 50.5	25 30	Com	SBb	7	149	109
*	NGC 4736	M94	12 48.6	41 23	12 51.0	41 07	CVn	Sb	7	75	78
	NGC 4742		12 49.2	-10 12	12 51.8	-10 28	Vir	E3	14	284	251
	NGC 4747		12 49.4	26 01	12 51.8	25 45	Com	Pec	7	149	109
	NGC 4754		12 49.7	11 35	12 52.2	11 19	Vir	S0	14	194	179
	NGC 4753		12 49.8	-00 55	12 52.4	-01 11	Vir	Irr	14	239	215
	Coalsack		12 50.0	-62 44	12 53.0	-63 00	Cru	DNeb	25	451	397
	NGC 4762		12 50.4	11 31	12 52.9	11 15	Vir	S0	14	194	179
	NGC 4760		12 50.5	-10 13	12 53.1	-10 29	Vir	E1	14	284	251
!!	NGC 4755		12 50.6	-60 05	12 53.6	-60 21	Cru	OpCl	25	429	397
	NGC 4781		12 51.8	-10 16	12 54.4	-10 32	Vir	SBc	14	284	251
	NGC 4782		12 52.0	-12 19	12 54.6	-12 35	Crv	E0	14	284	251
	NGC 4783		12 52.0	-12 18	12 54.6	-12 34	Crv	E0	14	284	251
	NGC 4790		12 52.2	-09 58	12 54.8	-10 14	Vir	SBc	14	284	251
!*	NGC 4826	M64	12 54.3	21 57	12 56.8	21 41	Com	Sa	7	149	143
	NGC 4845		12 55.5	01 51	12 58.1	01 35	Vir	Sa	14	239	215
	NGC 4833		12 56.0	-70 36	12 59.3	-70 52	Mus	Glob	25	451	413
	NGC 4856		12 56.7	-14 46	12 59.3	-15 02	Vir	SBa	14	284	251
	NGC 4874		12 57.2	28 14	12 59.6	27 58	Com	S0	7	149	109

APPENDIX E: A CATALOG OF DEEP-SKY OBJECTS

```
                     Size
   ID      v mag   arc-min    S.B.   Cl   MDM    Comments
--------|-------|-----------|------|-----|-----|----------------------------
NGC 4594   8.2    7.0x1.5    19.4   1.9   51   M104, the Sombrero Galaxy
NGC 4605  10.9    4.0x1.2    21.2   1.2   64
NGC 4618  11.2    3.0x2.5    22.0   0.9   31   ringtail shape
NGC 4621  11.0    2.0x1.5    20.8   1.4   51   M59
NGC 4631   9.7   12.5x1.2    21.3   1.2   64   edge-on spiral

NGC 4638  12.2    1.1x0.5    20.2   1.6  154   near M59 & M60
NGC 4636  11.0    7.0x2.0    22.5   0.7   38
NGC 4639  12.2    2.0x1.3    21.9   1.0   59
NGC 4643  11.6    1.7x0.8    20.6   1.5   96
NGC 4647  12.0    2.3x1.8    22.2   0.8   43   near M60

NGC 4649  10.0    3.0x2.5    20.8   1.4   31   M60
NGC 4651  11.4    3.0x2.5    22.2   0.8   31
NGC 4654  11.2    4.5x2.5    22.5   0.7   31   near NGC 4639
NGC 4656  11.0   19.5x2.0    23.6   0.3   38   stretched S shape
NGC 4660  12.2    1.3x0.6    20.6   1.5  128   near M60

NGC 4666  11.4    3.9x0.7    21.1   1.3  110   near NGC 4668
NGC 4668  13.4    0.8x0.6    21.2   1.2  128   near NGC 4666
NGC 4689  11.7    2.8x2.0    22.2   0.8   38
NGC 4697  10.5    2.5x1.3    20.4   1.5   59
NGC 4699  10.3    3.0x2.0    20.9   1.4   38

NGC 4710  12.0    3.4x0.5    21.2   1.2  154   edge-on, dust lane
NGC 4725  10.5    7.5x4.8    23.0   0.5   16
NGC 4736   8.9    5.0x3.5    20.6   1.4   22   M94
NGC 4742  12.0    1.0x0.6    20.1   1.7  128   near NGC 4760, 4781
NGC 4747  12.8    3.0x0.5    21.9   1.0  154   distorted

NGC 4754  11.8    2.5x1.0    21.4   1.1   77   near NGC 4762
NGC 4753  10.6    2.8x2.0    21.1   1.3   38
Coalsack    -    400x300      -     -           next to Southern Cross
NGC 4762  11.5    3.7x0.4    20.6   1.5  192   near NGC 4754
NGC 4760  12.5    0.6x0.5    19.8   1.8  154   near NGC 4742, 4781

NGC 4755   4.2    10x10      17.8   2.6    8   the Jewel Box
NGC 4781  11.7    2.6x1.1    21.5   1.1   70   near NGC 4742, 4760
NGC 4782  12.9    0.5x0.5    20.0   1.7  154   touching NGC 4783
NGC 4783  12.9    0.5x0.5    20.0   1.7  154   touching NGC 4782
NGC 4790  12.5    1.4x1.0    21.5   1.1   77   near NGC 4781

NGC 4826   8.6    7.5x3.5    20.8   1.4   22   M64, the Black Eye Galaxy
NGC 4845  12.6    4.0x0.8    22.5   0.7   96   equatorial dust lane
NGC 4833   8.5    6x6        21.0   1.3   13
NGC 4856  11.4    2.5x0.7    20.6   1.4  110   theta shape
NGC 4874  13.5    1.0x1.0    22.1   0.8   77   near NGC 4889
```

VISUAL ASTRONOMY OF THE DEEP SKY

	ID		1950.0 RA	1950.0 DEC	2000.0 RA	2000.0 DEC	Con	Type	Charts 2000	Uran	VPSA
	NGC 4889		12 57.7	28 15	13 00.1	27 59	Com	E4	7	149	109
	NGC 4902		12 58.3	-14 15	13 00.9	-14 31	Vir	SBb	14	284	251
	NGC 4945		13 02.4	-49 13	13 05.3	-49 29	Cen	Sc	21	402	377
	NGC 5005		13 08.5	37 19	13 10.8	37 03	CVn	Sb	7	109	79
	NGC 5024	M53	13 10.5	18 26	13 12.9	18 10	Com	Glob	14	150	143
	NGC 5033		13 11.2	36 51	13 13.5	36 35	CVn	Sb	7	109	79
*	NGC 5055	M63	13 13.5	42 17	13 15.7	42 01	CVn	Sb	7	76	79
	NGC 5053		13 13.9	17 57	13 16.3	17 41	Com	Glob	14	150	143
	NGC 5102		13 19.1	-36 23	13 21.9	-36 39	Cen	S0	21	370	351
!!!*	NGC 5128		13 22.4	-42 45	13 25.3	-43 01	Cen	S0	21	403	351
!!!!*	NGC 5139		13 23.8	-47 13	13 26.8	-47 29	Cen	Glob	21	403	377
!!!*	NGC 5194	M51	13 27.8	47 27	13 29.9	47 12	CVn	Sc	7	76	53
*	NGC 5195		13 27.9	47 31	13 30.0	47 16	CVn	Pec	7	76	53
	NGC 5189		13 29.9	-65 43	13 33.4	-65 58	Mus	Neb	25	451	413
!!*	NGC 5236	M83	13 34.3	-29 37	13 37.1	-29 52	Hya	Sc	21	370	322
	NGC 5248		13 35.1	09 08	13 37.6	08 53	Boo	Sc	14	196	180
	NGC 5253		13 37.1	-31 24	13 39.9	-31 39	Cen	E	21	370	322
!	NGC 5272	M3	13 39.9	28 38	13 42.2	28 23	CVn	Glob	7	110	110
	NGC 5322		13 47.6	60 26	13 49.3	60 11	UMa	E2	2	49	32
	NGC 5350		13 51.3	40 37	13 53.4	40 22	CVn	Sb	7	76	80
	NGC 5353		13 51.3	40 31	13 53.4	40 16	CVn	E5	7	76	80
	NGC 5354		13 51.3	40 32	13 53.4	40 17	CVn	E3	7	76	80
	NGC 5363		13 53.6	05 29	13 56.1	05 14	Vir	Irr	14	241	180
	NGC 5371		13 53.6	40 43	13 55.7	40 28	CVn	Sb	7	76	80
	NGC 5364		13 53.7	05 15	13 56.2	05 00	Vir	Sb	14	241	180
	NGC 5377		13 54.3	47 27	13 56.3	47 12	CVn	Sa	7	76	53
	NGC 5367		13 54.7	-39 44	13 57.7	-39 59	Cen	Neb	21	403	352
	NGC 5383		13 55.0	42 05	13 57.1	41 50	CVn	SBb	7	76	80
	NGC 5394		13 56.5	37 40	13 58.6	37 25	CVn	Sb	7	110	80
	NGC 5395		13 56.5	37 39	13 58.6	37 24	CVn	Sb	7	110	80
	NGC 5426		14 00.8	-05 49	14 03.4	-06 03	Vir	Sc	14	286	253
	NGC 5427		14 00.8	-05 47	14 03.4	-06 01	Vir	Sc	14	286	253
!	NGC 5457	M101	14 01.4	54 35	14 03.2	54 21	UMa	Sc	2	49	54
	NGC 5466		14 03.2	28 46	14 05.4	28 32	Boo	Glob	7	110	110
	NGC 5474		14 03.2	53 54	14 05.0	53 40	UMa	Irr	2	49	54
	NGC 5585		14 18.0	56 57	14 19.6	56 43	UMa	Sc	2	49	32
	NGC 5614		14 22.0	35 05	14 24.1	34 51	Boo	Sa	7	111	80
	NGC 5617		14 26.0	-60 30	14 29.7	-60 43	Cen	OpCl	25	430	398
	NGC 5643		14 29.4	-43 59	14 32.6	-44 12	Lup	Sb	21	404	352
	NGC 5740		14 41.9	01 54	14 44.4	01 41	Vir	Sb	14	243	218

APPENDIX E: A CATALOG OF DEEP-SKY OBJECTS

ID	v mag	Size arc-min	S.B.	Cl	MDM	Comments
NGC 4889	13.2	1.0x0.6	21.3	1.2	128	center of Coma Cluster of Galaxies
NGC 4902	11.6	2.0x2.0	21.7	1.0	38	
NGC 4945	?	?	?			
NGC 5005	10.8	4.7x1.6	21.6	1.1	48	
NGC 5024	8.	10x10	21.6	1.0	8	M53
NGC 5033	11.0	8.0x4.0	23.4	0.3	19	
NGC 5055	9.8	9.0x4.0	22.3	0.8	19	M63
NGC 5053	10.5	8x8	23.6	0.2	10	near M53
NGC 5102	10.8	6.0x2.5	22.4	0.8	31	
NGC 5128	7.2	10x8	20.6	1.5	10	peculiar, with dark lane
NGC 5139	3.6	30x30	19.6	1.9	3	Omega Centauri: spectacular globular
NGC 5194	8.1	10.0x5.5	21.1	1.3	14	M51
NGC 5195	11.0	2.0x1.5	20.8	1.4	51	companion to NGC 5194
NGC 5189	?	185x130	?			
NGC 5236	8.0	10.0x8.0	21.4	1.1	10	M83
NGC 5248	11.0	3.2x1.4	21.3	1.2	55	
NGC 5253	10.8	4.0x1.5	21.4	1.2	51	
NGC 5272	6.	18x18	20.9	1.3	4	M3
NGC 5322	11.3	1.4x1.0	20.3	1.6	77	
NGC 5350	11.4	3.2x1.4	21.7	1.0	55	near NGC 5353
NGC 5353	12.3	1.1x0.4	20.0	1.7	192	near NGC 5354
NGC 5354	13.0	0.9x0.7	21.1	1.2	110	near NGC 5353
NGC 5363	11.1	1.7x1.5	20.7	1.4	51	near NGC 5364
NGC 5371	11.5	3.7x3.0	22.7	0.6	26	
NGC 5364	11.5	5.0x4.0	23.4	0.3	19	near NGC 5363
NGC 5377	12.0	3.0x0.6	21.3	1.2	128	outer ring
NGC 5367	10.	1.3x1.0	18.9	2.1	77	double nucleus
NGC 5383	12.7	2.2x2.0	22.9	0.5	38	
NGC 5394	13.5	0.5x0.5	20.6	1.5	154	
NGC 5395	12.7	2.1x1.0	22.1	0.8	77	interacting with NGC 5394?
NGC 5426	12.7	1.5x1.1	21.9	1.0	70	near NGC 5427
NGC 5427	12.0	2.0x1.7	22.0	0.9	45	near NGC 5426
NGC 5457	9.0	22x20	24.2	0.0	4	M101
NGC 5466	9.0	5x5	21.1	1.3	15	
NGC 5474	11.5	4.0x3.0	22.8	0.6	26	
NGC 5585	11.6	4.5x2.3	22.8	0.6	33	
NGC 5614	12.9	2.1x0.8	22.1	0.9	96	
NGC 5617	8.	15x15	22.5	0.7	5	50 stars mag 8+
NGC 5643	11.4	2.5x2.3	21.9	0.9	33	
NGC 5740	12.6	2.5x1.5	22.7	0.6	51	near NGC 5746

	ID		1950.0 RA DEC	2000.0 RA DEC	Con	Type	Charts 2000	Uran	VPSA
	NGC 5746		14 42.3　02 10	14 44.8　01 57	Vir	Sb	14	243	218
	NGC 5824		15 00.9 -32 53	15 04.0 -33 05	Lup	Glob	21	373	324
	NGC 5823		15 01.9 -55 24	15 05.6 -55 36	Cir	OpCl	25	431	399
	NGC 5846		15 04.0　01 48	15 06.5　01 36	Vir	E0	14	243	218
	NGC 5850		15 04.6　01 44	15 07.1　01 32	Vir	SBb	14	243	218
	NGC 5866		15 05.1　55 57	15 06.5　55 45	Dra	E6	2	50	33
	NGC 5905		15 14.1　55 42	15 15.4　55 31	Dra	SBb	2	50	33
	NGC 5897		15 14.5 -20 50	15 17.4 -21 01	Lib	Glob	21	334	290
	NGC 5907		15 14.6　56 31	15 15.9　56 20	Dra	Sb	2	50	33
	NGC 5908		15 15.4　55 36	15 16.7　55 25	Dra	Sb	2	50	33
!	NGC 5904	M5	15 16.0　02 16	15 18.5　02 05	Ser	Glob	14	244	218
	NGC 5962		15 34.2　16 46	15 36.5　16 36	Ser	Sc	15	199	147
	NGC 5985		15 38.6　59 30	15 39.6　59 20	Dra	Sb	2	51	33
	NGC 5986		15 42.8 -37 37	15 46.1 -37 46	Lup	Glob	21	374	354
	SP1		15 47.4 -51 21	15 51.1 -51 30	Nor	Plan	25	432	379
	NGC 6027		15 57.0　20 55	15 59.2　20 46	Ser	E	7	155	147
	NGC 6025		15 59.4 -60 22	16 03.7 -60 30	TrA	OpCl	26	432	399
	NGC 6067		16 09.3 -54 05	16 13.2 -54 13	Nor	OpCl	26	432	380
	NGC 6093	M80	16 14.1 -22 52	16 17.1 -22 59	Sco	Glob	22	336	292
	NGC 6101		16 20.0 -72 06	16 25.7 -72 13	Aps	Glob	26	454	415
!*	NGC 6121	M4	16 20.6 -26 24	16 23.7 -26 31	Sco	Glob	22	336	325
	NGC 6124		16 22.2 -40 35	16 25.6 -40 42	Sco	OpCl	22	407	355
	NGC 6152		16 28.8 -52 31	16 32.7 -52 37	Nor	OpCl	26	433	380
	NGC 6144		16 24.2 -25 56	16 27.3 -26 03	Sco	Glob	22	336	325
	NGC 6171	M107	16 29.7 -12 57	16 32.5 -13 03	Oph	Glob	15	291	256
	NGC 6153		16 28.0 -40 08	16 31.4 -40 15	Sco	Plan	22	407	355
	NGC 6188		16 35.9 -48 55	16 39.7 -49 01	Ara	Neb	22	407	380
	NGC 6192		16 36.8 -43 17	16 40.3 -43 23	Sco	OpCl	22	407	355
	NGC 6193		16 37.6 -48 40	16 41.3 -48 46	Ara	OpCl	22	407	380
!!!*	NGC 6205	M13	16 39.9　36 33	16 41.7　36 27	Her	Glob	8	114	83
	NGC 6207		16 41.3　36 56	16 43.1　36 50	Her	Sc	8	114	83
	NGC 6210		16 42.5　23 53	16 44.6　23 47	Her	Plan	8	156	149
	NGC 6218	M12	16 44.6 -01 52	16 47.2 -01 57	Oph	Glob	15	246	221
	NGC 6229		16 45.6　47 37	16 47.0　47 32	Her	Glob	8	80	56
	NGC 6231		16 50.7 -41 43	16 54.2 -41 48	Sco	OpCl	22	407	355
	H12		16 52.7 -40 38	16 56.2 -40 43	Sco	OpCl	22	407	355
	NGC 6254	M10	16 54.5 -04 02	16 57.1 -04 07	Oph	Glob	15	247	221
	NGC 6259		16 57.1 -44 36	17 00.7 -44 41	Sco	OpCl	22	407	355
	NGC 6266	M62	16 58.1 -30 03	17 01.3 -30 07	Sco	Glob	22	376	326
	NGC 6273	M19	16 59.5 -26 11	17 02.6 -26 15	Oph	Glob	22	337	326

APPENDIX E: A CATALOG OF DEEP-SKY OBJECTS

```
                    Size
   ID      v mag   arc-min    S.B.   Cl   MDM   Comments
--------|-------|-----------|------|-----|-----|------------------------------
NGC 5746  11.7   6.5x0.8    22.1   0.9    96  near NGC 5740
NGC 5824   9.5   3x3        20.5   1.5    26
NGC 5823  11.    9x9        24.4  -0.1     9  80 stars mag 13+
NGC 5846  11.5   1.0x1.0    20.1   1.6    77  near NGC 5850
NGC 5850  12.0   2.6x2.1    22.5   0.7    37  theta shape

NGC 5866  11.1   2.9x1.0    20.9   1.3    77  equatorial dust lane
NGC 5905  13.1   4.4x3.2    24.6  -0.1    24  near NGC 5908
NGC 5897   9.    8.5x8.5    22.3   0.8     9
NGC 5907  11.0  11.0x0.6    21.7   1.0   128
NGC 5908  13.0   2.4x0.4    21.6   1.1   192  equatorial dust lane

NGC 5904   6.2   13x13      20.4   1.5     6  M5
NGC 5962  11.9   2.2x1.3    21.7   1.0    59
NGC 5985  12.0   4.3x2.1    23.0   0.5    37
NGC 5986   8.    5x5        20.1   1.7    15
SP1        8.5   1.2x1.2    17.5   2.7    64  ring, 13.5m central star

NGC 6027  14.    1.7x1.3    23.5   0.3    59  5 close galaxies
NGC 6025   6.    10x10      19.6   1.8     8  30 stars mag 7+
NGC 6067   8.    15x15      22.5   0.7     5  100 stars mag 10+
NGC 6093   8.    7x7        20.9   1.4    11  M80
NGC 6101  10.    4x4        21.6   1.0    19

NGC 6121   7.4   20x20      22.5   0.7     4  M4
NGC 6124   8.    25x25      23.6   0.3     3  100 stars mag 9 to 12
NGC 6152   8.    30x30      24.0   0.1     3  60 stars mag 9+
NGC 6144  10.    3x3        21.0   1.3    26
NGC 6171   8.1   3x3        19.1   2.1    26  M107

NGC 6153  11.5   0.33x0.33  17.7   2.6   233
NGC 6188   ?     20x12       ?                bright and dark nebulae
NGC 6192  10.    7x7        22.9   0.6    11
NGC 6193   5.2   15x15      19.7   1.8     5  about 30 stars near NGC 6188
NGC 6205   5.7   23x23      21.1   1.2     3  M13, spectacular

NGC 6207  12.3   2.0x1.0    21.7   1.0    77  near M13
NGC 6210   9.7   0.33x0.27  15.7   3.4   284  central star 12.5
NGC 6218   8.    10x10      21.6   1.0     8  M12
NGC 6229   8.7   3.5x3.5    20.0   1.7    22
NGC 6231   6.    15x15      20.5   1.5     5  100 stars mag 7 to 13

H12        ?     40x40       ?                200 stars
NGC 6254   7.    8x8        20.1   1.6    10  M10
NGC 6259  10.    15x15      24.5  -0.1     5  100 stars mag 11+
NGC 6266   6.5   6x6        19.0   2.1    13  M62
NGC 6273   7.    6x6        19.5   1.9    13  M19
```

	ID		1950.0 RA DEC	2000.0 RA DEC	Con	Type	Charts 2000	Uran	VPSA
	NGC 6302		17 10.5 -37 03	17 13.9 -37 07	Sco	Neb	22	376	356
	NGC 6318		17 14.3 -39 24	17 17.8 -39 27	Sco	OpCl	22	408	356
!	NGC 6341	M92	17 15.6 43 12	17 17.1 43 09	Her	Glob	8	81	84
	NGC 6333	M9	17 16.2 -18 28	17 19.1 -18 31	Oph	Glob	15	337	293
	NGC 6334		17 17.2 -36 01	17 20.6 -36 04	Sco	Neb	22	376	356
	NGC 6337		17 18.9 -38 25	17 22.3 -38 28	Sco	Plan	22	376	356
	B72		17 21.0 -23 35	17 23.5 -23 38	Oph	DNeb	22	338	294
	NGC 6357		17 21.3 -34 07	17 24.6 -34 10	Sco	Neb	22	376	327
	NGC 6352		17 21.6 -48 26	17 25.4 -48 29	Ara	Glob	22	408	381
	NGC 6362		17 26.6 -67 01	17 31.8 -67 03	Ara	Glob	26	455	416
	NGC 6384		17 29.9 07 06	17 32.3 07 04	Oph	Sb	15	203	186
	B78		17 29.9 -25 58	17 33.0 -26 00	Oph	DNeb	22	338	327
	NGC 6388		17 32.6 -44 43	17 36.3 -44 45	Sco	Glob	22	408	356
	NGC 6402	M14	17 35.0 -03 13	17 37.6 -03 15	Oph	Glob	15	248	222
	NGC 6397		17 36.8 -53 39	17 40.9 -53 41	Ara	Glob	26	434	381
!	NGC 6405	M6	17 36.8 -32 11	17 40.1 -32 13	Sco	OpCl	22	377	327
	NGC 6441		17 46.8 -37 02	17 50.2 -37 03	Sco	Glob	22	377	356
!	NGC 6475	M7	17 50.7 -34 48	17 54.0 -34 49	Sco	OpCl	22	377	327
S!	NGC 6494	M23	17 54.0 -19 01	17 56.9 -19 01	Sgr	OpCl	15	339	294
	NGC 6543		17 58.8 66 38	17 58.8 66 38	Dra	Plan	3	30	18
!!!*	NGC 6514	M20	17 58.9 -23 02	18 01.9 -23 02	Sgr	Neb	22	339	294
	B86		18 00.0 -27 50	18 03.1 -27 50	Sgr	DNeb	22	339	328
	NGC 6520		18 00.3 -27 54	18 03.5 -27 54	Sgr	OpCl	22	339	328
	B87		18 01.0 -32 30	18 04.3 -32 30	Sgr	DNeb	22	377	328
!!!*	NGC 6523	M8	18 01.6 -24 20	18 04.7 -24 20	Sgr	Neb	22	339	295
*	NGC 6530		18 01.6 -24 20	18 04.7 -24 20	Sgr	OpCl	22	339	295
	NGC 6531	M21	18 01.8 -22 30	18 04.8 -22 30	Sgr	OpCl	22	339	295
	NGC 6569		18 10.4 -31 50	18 13.7 -31 49	Sgr	Glob	22	377	328
	B92		18 12.7 -18 20	18 15.6 -18 19	Sgr	DNeb	15	339	295
	NGC 6584		18 14.6 -52 14	18 18.6 -52 13	Tel	Glob	26	434	382
!	M24	M24	18 15.5 -18 27	18 18.4 -18 26	Sgr	OpCl	15	339	295
	NGC 6603		18 15.5 -18 27	18 18.4 -18 26	Sgr	OpCl	15	339	295
*	NGC 6611	M16	18 16.0 -13 48	18 18.8 -13 47	Ser	Neb	15	294	259
	NGC 6613	M18	18 17.0 -17 09	18 19.9 -17 08	Sgr	OpCl	15	339	295
!!!*	NGC 6618	M17	18 18.0 -16 12	18 20.9 -16 11	Sgr	Neb	15	294	295
	NGC 6624		18 20.5 -30 23	18 23.7 -30 22	Sgr	Glob	22	378	328
	NGC 6643		18 21.2 74 33	18 19.7 74 35	Dra	Sc	3	12	19
	NGC 6626	M28	18 21.5 -24 54	18 24.6 -24 52	Sgr	Glob	22	340	295
	NGC 6629		18 22.7 -23 14	18 25.7 -23 12	Sgr	Plan	22	340	295
	NGC 6638		18 27.9 -25 32	18 31.0 -25 30	Sgr	Glob	22	340	328

APPENDIX E: A CATALOG OF DEEP-SKY OBJECTS

```
                  Size
   ID    v mag  arc-min   S.B.  Cl   MDM     Comments
--------|-----|---------|----|----|---|------------------------------
NGC 6302   ?      2x1       ?               figure 8 shape
NGC 6318  11.     5x5      23.1  0.5  15    60 stars mag 12 to 14
NGC 6341   6.5    8x8      19.6  1.8  10    M92
NGC 6333   8.     4x4      19.6  1.8  19    M9
NGC 6334   ?     30x30      ?

NGC 6337   ?     0.6x0.5    ?
B72        -    30x30       -    -          dark S nebula: the Snake
NGC 6357   ?     4x1        ?
NGC 6352   9.    8x8       22.1  0.8  10
NGC 6362   8.    9x9       21.4  1.1   9

NGC 6384  12.3   4.0x3.0   23.6  0.2  26
B78        -   200x140      -    -          bowl of Pipe Nebula
NGC 6388   7.    4x4       18.6  2.2  19
NGC 6402   9.    6x6       21.5  1.1  13    M14
NGC 6397   7.   19x19      22.0  0.9   4    may be nearest globular

NGC 6405   6.   25x25      21.6  1.1   3    M6
NGC 6441   8.    3x3       19.0  2.1  26
NGC 6475   5.   60x60      22.5  0.7   1    M7
NGC 6494   7.   25x25      22.6  0.7   3    M23
NGC 6543   8.6  0.4x0.3    14.9  3.7 256    central star mag 10

NGC 6514   8.5  29x27      24.4  0.0   3    M20, Trifid Nebula
B86        -    4.5x3       -    -          dark nebula
NGC 6520   9.    5x5       21.1  1.3  15    25 stars mag 9 to 12
B87        -   12x12        -    -          Parrot's Head
NGC 6523   5.   80x40      22.4  0.7   2    M8, Lagoon Nebula

NGC 6530   6.   10x10      19.6  1.8   8    cluster in M8
NGC 6531   7.   10x10      20.6  1.4   8    M21
NGC 6569  10.    2x2       20.1  1.6  38
B92        -   15x15        -    -          dark nebula
NGC 6584   8.5   6x6       21.0  1.3  13

M24        4.5  60x90      22.5  0.7   1    M24
NGC 6603  11.4   4x4       23.0  0.5  19    within M24
NGC 6611   6.5  25x25      22.1  0.9   3    M16, nebula +60 stars
NGC 6613   8.    7x7       20.9  1.4  11    M18
NGC 6618   6.   45x35      22.6  0.7   2    M17, Omega Nebula

NGC 6624   8.5   3x3       19.5  1.9  26
NGC 6643  12.0  3.0x1.3    22.1  0.9  59    central star mag 10
NGC 6626   8.    6x6       20.5  1.5  13    M28
NGC 6629  10.5 0.25x0.25   16.1  3.3 307    central star mag 13.5
NGC 6638   9.5   2x2       19.6  1.8  38
```

VISUAL ASTRONOMY OF THE DEEP SKY

	ID		1950.0 RA DEC	2000.0 RA DEC	Con	Type	Charts 2000	Uran	VPSA
	NGC 6637	M69	18 28.1 -32 23	18 31.4 -32 21	Sgr	Glob	22	378	328
	I.4725	M25	18 28.8 -19 17	18 31.7 -19 15	Sgr	OpCl	16	340	295
	NGC 6642		18 28.8 -23 31	18 31.8 -23 29	Sgr	Glob	22	340	295
	NGC 6645		18 29.8 -16 56	18 32.7 -16 54	Sgr	OpCl	16	295	295
	NGC 6649		18 30.7 -10 26	18 33.5 -10 24	Sct	OpCl	16	295	259
	NGC 6652		18 32.5 -33 02	18 35.8 -33 00	Sgr	Glob	22	378	328
!!	NGC 6656	M22	18 33.3 -23 58	18 36.4 -23 56	Sgr	Glob	22	340	295
	NGC 6664		18 34.0 -08 16	18 36.7 -08 14	Sct	OpCl	16	295	259
	I.4756		18 36.6 05 26	18 39.1 05 29	Ser	OpCl	16	250	187
	NGC 6681	M70	18 40.0 -32 21	18 43.3 -32 18	Sgr	Glob	22	378	328
	NGC 6694	M26	18 42.5 -09 27	18 45.2 -09 24	Sct	OpCl	16	295	260
!!*	NGC 6705	M11	18 48.4 -06 20	18 51.1 -06 16	Sct	OpCl	16	295	260
	NGC 6709		18 49.1 10 17	18 51.5 10 21	Aql	OpCl	16	205	188
	NGC 6712		18 50.3 -08 47	18 53.0 -08 43	Sct	Glob	16	295	260
!!*	NGC 6720	M57	18 51.7 32 58	18 53.6 33 02	Lyr	Plan	8	117	117
	NGC 6715	M54	18 52.0 -30 32	18 55.2 -30 28	Sgr	Glob	22	378	329
	NGC 6723		18 56.2 -36 42	18 59.6 -36 38	Sgr	Glob	22	378	358
	NGC 6726		18 58.3 -36 57	19 01.7 -36 53	CrA	Neb	22	379	358
	NGC 6727		18 58.3 -36 56	19 01.7 -36 52	CrA	Neb	22	379	358
	NGC 6729		18 58.4 -37 02	19 01.8 -36 58	CrA	Neb	22	379	358
	B133		19 04.5 -06 05	19 07.2 -06 00	Aql	DNeb	16	296	260
	NGC 6744		19 05.0 -63 56	19 09.7 -63 51	Pav	SBc	26	456	402
	NGC 6752		19 06.4 -60 04	19 10.8 -59 59	Pav	Glob	26	435	402
	NGC 6779	M56	19 14.6 30 05	19 16.5 30 10	Lyr	Glob	8	118	117
	NGC 6781		19 16.0 06 26	19 18.4 06 31	Aql	Plan	16	206	188
	NGC 6791		19 19.0 37 40	19 20.8 37 46	Lyr	OpCl	8	118	86
	NGC 6811		19 36.7 46 27	19 38.2 46 34	Cyg	OpCl	8	84	59
	NGC 6809		19 36.9 -31 03	19 40.1 -30 56	Sgr	Glob	22	380	330
	B143		19 38.0 11 00	19 40.4 11 07	Aql	DNeb	16	207	189
	NGC 6819		19 39.6 40 06	19 41.3 40 13	Cyg	OpCl	8	84	86
	NGC 6814		19 39.9 -10 25	19 42.6 -10 18	Aql	Sb	16	297	261
	NGC 6820		19 41.1 23 10	19 43.2 23 17	Vul	Neb	8	162	153
	NGC 6818		19 41.1 -14 17	19 43.9 -14 10	Sgr	Plan	16	297	261
S	NGC 6823		19 41.1 23 12	19 43.2 23 19	Vul	OpCl	8	162	153
	NGC 6822		19 42.1 -14 53	19 44.9 -14 46	Sgr	Irr	16	297	261
	NGC 6826		19 43.4 50 24	19 44.7 50 31	Cyg	Plan	3	55	59
	NGC 6834		19 50.2 29 17	19 52.2 29 25	Vul	OpCl	8	119	118
	NGC 6838	M71	19 51.5 18 39	19 53.7 18 47	Sgr	Glob	16	162	153
	NGC 6842		19 53.0 29 09	19 55.0 29 17	Vul	Plan	8	119	118
!!*	NGC 6853	M27	19 57.4 22 35	19 59.6 22 43	Vul	Plan	8	162	153

APPENDIX E: A CATALOG OF DEEP-SKY OBJECTS

ID	v mag	Size arc-min	S.B.	Cl	MDM	Comments
NGC 6637	7.5	4x4	19.1	2.0	19	M69
I.4725	6.	35x35	22.3	0.8	2	M25
NGC 6642	8.	2x2	18.1	2.4	38	
NGC 6645	9.	10x10	22.6	0.6	8	75 stars mag 11 to 15
NGC 6649	9.	9x9	22.4	0.7	9	
NGC 6652	8.5	2x2	18.6	2.2	38	
NGC 6656	6.	18x18	20.9	1.3	4	M22
NGC 6664	9.	18x18	23.9	0.1	4	
I.4756	6.	70x70	23.9	0.2	1	80 stars mag 7+
NGC 6681	8.	4x4	19.6	1.8	19	M70
NGC 6694	9.5	9x9	22.9	0.5	9	M26
NGC 6705	6.	12x12	20.0	1.7	6	M11
NGC 6709	8.	12x12	22.0	0.9	6	
NGC 6712	9.	3x3	20.0	1.7	26	
NGC 6720	9.	1.3x1.0	17.9	2.5	77	M57, the Ring Nebula
NGC 6715	9.	6x6	21.5	1.1	13	M54
NGC 6723	6.	7x7	18.9	2.2	11	
NGC 6726	?	2x2	?			nebula around var. star
NGC 6727	?	2x2	?			nebula around var. star
NGC 6729	?	1x1	?			nebula around var. star
B133	-	10x5	-	-		dark oval nebula
NGC 6744	10.6	9.0x9.0	24.0	0.1	9	
NGC 6752	7.	15x15	21.5	1.1	5	
NGC 6779	8.	5x5	20.1	1.7	15	M56
NGC 6781	12.5	1.75x1.75	22.3	0.8	44	central star mag 15.5
NGC 6791	11.	20x20	26.1	-0.8	4	couple of hundred stars
NGC 6811	9.	15x15	23.5	0.3	5	50 stars mag 11 to 14
NGC 6809	7.	15x15	21.5	1.1	5	
B143	-	30x30	-	-		dark two-pronged nebula
NGC 6819	10.	6x6	22.5	0.7	13	150 stars mag 11 to 15
NGC 6814	12.2	2.0x2.0	22.3	0.8	38	
NGC 6820	?	20x20	?			in cluster NGC 6823
NGC 6818	10.	0.4x0.4	16.6	3.0	192	
NGC 6823	10.	5x5	22.1	0.9	15	30 stars mag 11+
NGC 6822	10.0	20x10	24.4	-0.1	8	
NGC 6826	8.8	0.4x0.4	15.4	3.5	192	central star mag 11
NGC 6834	10.	4x4	21.6	1.0	19	50 stars mag 11+
NGC 6838	9.	6x6	21.5	1.1	13	M71
NGC 6842	13.	0.8x0.75	21.1	1.3	102	central star mag 14.5
NGC 6853	8.	8x6	20.8	1.4	13	M27, the Dumbbell Nebula

	ID		1950.0 RA DEC	2000.0 RA DEC	Con	Type	Charts 2000	Uran	VPSA
	NGC 6866		20 02.1 44 02	20 03.7 44 10	Cyg	OpCl	9	84	87
	NGC 6864	M75	20 03.2 -22 04	20 06.1 -21 55	Sgr	Glob	23	343	298
	NGC 6884		20 08.8 46 19	20 10.4 46 28	Cyg	Plan	9	84	60
*	NGC 6888		20 10.7 38 16	20 12.5 38 25	Cyg	Neb	9	119	87
	NGC 6891		20 12.8 12 35	20 15.2 12 44	Del	Plan	16	208	190
S	NGC 6910		20 21.2 40 38	20 23.0 40 48	Cyg	OpCl	9	84	87
	NGC 6907		20 22.1 -24 58	20 25.1 -24 48	Cap	SBb	23	343	298
	NGC 6913	M29	20 22.2 38 21	20 24.0 38 31	Cyg	OpCl	9	120	87
	NGC 6939		20 30.4 60 28	20 31.4 60 38	Cep	OpCl	3	56	37
	NGC 6934		20 31.7 07 14	20 34.1 07 24	Del	Glob	16	209	190
	NGC 6940		20 32.5 28 08	20 34.6 28 18	Vul	OpCl	9	120	119
!*	NGC 6946		20 33.9 59 58	20 35.0 60 08	Cep	Sc	3	56	37
	NGC 6951		20 36.5 65 56	20 37.1 66 07	Cep	SB	3	32	20
!!*	NGC 6960		20 43.6 30 32	20 45.7 30 43	Cyg	Neb	9	120	119
	NGC 6981	M72	20 50.8 -12 44	20 53.5 -12 33	Aqr	Glob	16	299	263
!!*	NGC 6992-5		20 54.3 31 30	20 56.4 31 42	Cyg	Neb	9	120	119
	NGC 6997		20 54.7 44 27	20 56.5 44 39	Cyg	OpCl	9	85	88
	NGC 6994	M73	20 56.2 -12 51	20 58.9 -12 39	Aqr	OpCl	16	299	263
!*	NGC 7000		20 57.0 44 08	20 58.8 44 20	Cyg	Neb	9	85	88
	NGC 7006		20 59.1 16 00	21 01.4 16 12	Del	Glob	16	209	155
	NGC 7008		20 59.1 54 21	21 00.6 54 33	Cyg	Plan	3	56	60
	NGC 7009		21 01.4 -11 34	21 04.1 -11 22	Aqr	Plan	16	300	263
	NGC 7023		21 01.4 67 58	21 02.0 68 10	Cep	Neb	3	33	20
	NGC 7026		21 04.6 47 39	21 06.3 47 51	Cyg	Plan	9	85	61
	NGC 7027		21 05.1 42 02	21 07.0 42 14	Cyg	Plan	9	85	88
S	NGC 7031		21 07.9 50 40	21 07.3 50 50	Cep	OpCl	3	56	61
	NGC 7048		21 12.6 46 04	21 14.4 46 16	Cyg	Plan	9	86	61
!!*	NGC 7078	M15	21 27.6 11 57	21 30.0 12 10	Peg	Glob	17	210	192
	NGC 7086		21 29.8 51 22	21 31.5 51 35	Cyg	OpCl	3	57	61
	NGC 7092	M39	21 30.4 48 13	21 32.2 48 26	Cyg	OpCl	9	86	61
*	NGC 7089	M2	21 30.9 -01 03	21 33.5 -00 50	Aqr	Glob	17	255	228
	NGC 7099	M30	21 37.5 -23 25	21 40.3 -23 11	Cap	Glob	23	346	300
	NGC 7217		22 05.6 31 07	22 07.8 31 22	Peg	Sb	9	122	121
S	NGC 7235		22 10.7 57 00	22 12.5 57 15	Cep	OpCl	3	57	38
	NGC 7243		22 13.2 49 38	22 15.2 49 53	Lac	OpCl	9	87	62
!!*	NGC 7293		22 27.0 -21 06	22 29.7 -20 51	Aqr	Plan	23	347	301
*	NGC 7331		22 34.8 34 10	22 37.1 34 26	Peg	Sb	9	123	122
	NGC 7410		22 52.1 -39 56	22 54.9 -39 40	Gru	SB	23	415	362
	NGC 7418		22 53.8 -37 17	22 56.6 -37 01	Gru	SBc	23	384	362
	NGC 7479		23 02.4 12 03	23 04.9 12 19	Peg	SBb	17	213	194

APPENDIX E: A CATALOG OF DEEP-SKY OBJECTS

```
                    Size
    ID     v mag  arc-min    S.B.   Cl   MDM     Comments
---------|-----|-----------|----|----|---|---------------------------------
NGC 6866   8.      8x8      21.1  1.2   10  50 stars mag 10+
NGC 6864   8.      3x3      19.0  2.1   26  M75
NGC 6884  12.5    0.1x0.1   16.1  3.2  768
NGC 6888   ?      18x12      ?               Crescent Nebula
NGC 6891  10.     0.2x0.1   14.4  3.9  768

NGC 6910   6.5     8x8      19.6  1.8   10  40 stars mag 10+
NGC 6907  12.1    2.5x2.0   22.5  0.7   38  S-shaped spiral
NGC 6913   7.      7x7      19.9  1.8   11  M29: 20 stars mag 8+
NGC 6939  10.      8x8      23.1  0.4   10  100 stars mag 12 to 16
NGC 6934   9.      2x2      19.1  2.0   38

NGC 6940   8.     20x20     23.1  0.4    4  100 stars mag 9+
NGC 6946  11.1   8.0x8.0    24.2  0.0   10  near NGC 6939
NGC 6951  12.3   3.5x3.5    23.6  0.2   22
NGC 6960   8.     70x6      23.2  0.4   13  Veil Nebula
NGC 6981   8.6    3x3       19.6  1.9   26  M72

NGC 6992-5 8.     78x8      23.6  0.3   10  Veil Nebula
NGC 6997  10.      7x7      22.9  0.6   11  cluster in NGC 7000
NGC 6994  10.      1x1      18.6  2.2   77  M73 (four stars)
NGC 7000   5.    100x100    23.6  0.2    1  North America Nebula
NGC 7006  11.5     1x1      20.1  1.6   77

NGC 7008  12.     1.4x1.2   21.2  1.2   64  central star mag 13
NGC 7009   8.     0.4x0.4   14.6  3.8  192  Saturn Nebula
NGC 7023   ?      18x18      ?               nebula around 7 mag star
NGC 7026  12.     0.4x0.4   18.6  2.2  192  central star mag 15
NGC 7027   9.     0.3x0.2   14.6  3.9  384

NGC 7031  10.      6x6      22.5  0.7   13  50 stars mag 11+
NGC 7048  11.     1.0x0.9   19.5  1.9   85  central star mag 18
NGC 7078   6.5    10x10     20.1  1.6    8  M15
NGC 7086   9.      8x8      22.1  0.8   10  50 stars mag 11+
NGC 7092   5.     30x30     21.0  1.3    3  M39, 25 stars mag 7+

NGC 7089   6.0     7x7      18.9  2.2   11  M2
NGC 7099   8.      6x6      20.5  1.5   13  M30
NGC 7217  11.3   2.7x2.4    22.0  0.9   32
NGC 7235   9.      4x4      20.6  1.4   19  about 25 stars
NGC 7243   8.     20x20     23.1  0.4    4

NGC 7293   6.5   12x16      20.8  1.4    6  the Helix Nebula
NGC 7331  10.4  10.0x2.4    22.5  0.7   32  Stephan's Quintet 0.5 deg SSW
NGC 7410  11.8   4.0x1.1    22.0  0.9   70
NGC 7418  11.8   2.8x2.5    22.5  0.7   31
NGC 7479  11.8   3.2x3.5    23.1  0.5   24  S-shape
```

ID		1950.0 RA	1950.0 DEC	2000.0 RA	2000.0 DEC	Con	Type	Charts 2000	Uran	VPSA
NGC 7510		23 09.2	60 18	23 11.3	60 34	Cep	OpCl	3	58	39
NGC 7582		23 15.8	-42 38	23 18.6	-42 22	Gru	SBb	23	415	363
NGC 7590		23 16.3	-42 31	23 19.1	-42 15	Gru	Sb	23	415	363
NGC 7599		23 16.7	-42 32	23 19.5	-42 16	Gru	SBc	23	415	363
NGC 7619		23 17.8	07 55	23 20.3	08 11	Peg	E1	17	214	194
NGC 7626		23 18.2	07 56	23 20.7	08 12	Peg	E2	17	214	194
NGC 7635		23 18.5	60 54	23 20.7	61 10	Cas	Neb	3	34	39
NGC 7654	M52	23 22.0	61 20	23 24.2	61 36	Cas	OpCl	3	15	39
* NGC 7662		23 23.5	42 14	23 25.9	42 33	And	Plan	9	88	91
NGC 7789		23 54.5	56 26	23 57.0	56 43	Cas	OpCl	3	35	39
NGC 7793		23 55.3	-32 51	23 57.9	-32 34	Scl	Sd	23	350	335

APPENDIX E: A CATALOG OF DEEP-SKY OBJECTS

```
                  Size
   ID     v mag  arc-min   S.B.   Cl  MDM    Comments
--------|------|---------|------|----|---|------------------------------
NGC 7510   9.     3x3      20.0  1.7  26  30 stars mag 10+
NGC 7582  11.8   3.0x3.0   22.8  0.6  26  group with NGC 7590, 7599
NGC 7590  11.9   2.2x0.8   21.1  1.2  96  group with NGC 7582, 7599
NGC 7599  12.0   3.8x1.2   22.3  0.8  64  group with NGC 7582, 7590
NGC 7619  12.6   0.8x0.6   20.4  1.5 128  near NGC 7826

NGC 7626  12.7   0.9x0.7   20.8  1.4 110  near NGC 7619
NGC 7635  11.    10x5      23.9  0.1  15  the Bubble Nebula
NGC 7654   7.    12x12     21.0  1.3   6  M52, 120 stars mag 9+
NGC 7662   8.5   0.5x0.5   15.6  3.5 154  central star mag 14
NGC 7789  10.    20x20     25.1 -0.4   4  900 stars mag 11+

NGC 7793   9.7   6.0x4.0   21.8  1.0  19
```

NOTES

Scl-Sys: The Sculptor System is one of the most difficult objects in this catalog to detect in any telescope. The average surface brightness is low at 26.3 magnitude per square arc-second, and even the maximum at its center is a faint 23.9. The brightness distribution is similar to some globular star clusters and probably only the central 20 arc-minutes is detectable visually in the darkest skies.

NGC 1435: The Merope Nebula in the Pleiades. The surface brightness was obtained directly by O'Dell (1965), so the value listed is among the best for any of the nebulae in this book. The total visual magnitude was computed from the size and surface brightness.

NGC 1499: The California Nebula is extremely low in surface brightness but can be seen under good conditions. See *Sky & Telescope*, December, 1982, p. 612, for a discussion of different successful observations.

M24: M24 is a detached portion of the Milky Way and not an open cluster. See *The Messier Album* (Mallas and Kreimer, 1978) for details.

Appendix F
Optimum detection magnifications for deep-sky objects

Appendix E listed the *minimum* optimum detection magnification (MDM) for each of the 611 deep-sky objects catalogued. But that magnification is the best for an object only if it is the lowest your telescope can give. No one will be viewing each object at this power optimally, unless they happen to be using a whole battery of different-sized telescopes.

For your particular instrument, what is the best power for detecting each galaxy, cluster, or nebula? This appendix suggests answers. It lists the *optimum detection magnification*, which we will call the ODM, for each catalogued object in a range of telescopes under a dark country sky.

Of course, if an object is bright enough and the sky dark enough, it will be easy to see at many magnifications and no calculation is needed. But if the object is a challenging one near the limit of detection, it's very helpful to know the ODM.

HOW THE COMPUTATION IS DONE

For those who want to know how these values were calculated, or who wish to perform their own calculations for light-polluted conditions or for objects not listed in the table, the next few pages explain the method. Other readers may wish to skip this section.

The computation of the optimum detection magnification is a complex iterative procedure ideal for a computer. The goal is to find the power that magnifies the object to the optimum magnified visual angle. This sounds simple, but each time you change the telescope's magnification, the surface brightness of everything you are viewing also changes. So to find the correct magnification you simply make a guess and then compute if that is correct. If not, you make a new guess and try again. By following some simple rules the solution can be found quickly.

First check that the object's total magnitude is within reach of your telescope. Use Table 4.1 or equation 4.2 to find your telescope's limiting magnitude. If the object should be detectable at all, then you can do a computation to find the ODM.

The ODM depends on three things. First is the surface brightness of the background (M_o). As in Appendix E we will use $M_o = 24.25$ magnitudes per square arc-second for a dark country sky. Second is the size of the object you wish to study. Use the smaller dimension that describes the object. For example, if you are examining an elliptical galaxy, use the minor axis size. If you are examining a wisp of nebulosity, use its width, not its length. Third is the reduction in surface brightness due to the telescope's transmission factor and magnification.

The first step is to make a guess at the ODM. You can use any number you wish here. To generate the entries in this appendix I simply made a first guess of 100×. You could use the MDM value from Appendix E. To get close you might try using Figure 2.7b in the following way. Add about one to the background surface brightness M_o (for example for a dark country sky try 25), then find the angle corresponding to this value from Figure 2.7b (at $M_o = 25$ we find about 75 arc-minutes). This angle divided by the object's size gives the first guess.

Now that we have made a guess at the magnification, we need to see whether it is close. We must determine what effect the magnification has on the object's brightness. Recall from Chapter 4 that magnification dims the apparent surface brightness of anything seen through a telescope. To get a particular surface brightness reduction M_b in a

APPENDIX F: OPTIMUM DETECTION MAGNIFICATIONS FOR DEEP-SKY OBJECTS

Table F.1. *Magnification causing a given surface brightness reduction*

Telescope aperture		Reduction in magnitudes/sq. arc-sec					
		0.38	2.12	3.62	4.73	6.24	7.12
Inches	mm	Magnification per inch of aperture					
		3.4	7.5	15.0	25.0	50.0	75.0
2	51	7	15	30	50	100	150
4	102	14	30	60	100	200	300
6	152	20	45	90	150	300	450
8	203	27	60	120	200	400	600
10	254	34	75	150	250	500	750
12	305	41	90	180	300	600	900
14	356	47	105	210	350	700	1050
16	406	54	120	240	400	800	1200
18	457	61	135	270	450	900	1350
20	508	68	150	300	500	1000	1500
24	610	81	180	360	600	1200	1800
30	762	101	225	450	750	1500	2250
36	914	121	270	540	900	1800	2700
40	1016	135	300	600	1000	2000	3000

given telescope means using a particular magnification m. The relation can be easily derived by rearranging equation 4.3:

$$m = 0.1116\, D\, 10^{(M_b/5)}, \text{ or} \quad \text{(equation F.1)}$$

$$M_b = 5 \log(m / (0.1116\, D)), \quad \text{(equation F.2)}$$

where D is the telescope aperture in millimeters and M_b is in magnitudes per square arc-second. (If you wish to give D in inches, change the constant 0.1116 to 2.833.) Table F.1 uses this formula to list the magnifications that yield certain surface-brightness reductions in various telescopes.

Now compute the background surface brightness as viewed by the eye through the telescope at this power. It is found by adding the surface brightness reduction M_b from equation F.2 to the initial background M_o:

$$B_o = M_o + M_b. \quad \text{(equation F.3)}$$

Now read the OMVA from Figure 2.7b using this new B_o as the background surface brightness. (If you extrapolate the OMVA beyond 27.0 mag. per square arc-sec., do not let it become greater than 360 arc-minutes, because the object is then spread over too large an area in the eye.)

A new guess for the optimum detection magnification, which for now we will simply call m, is given by

$$m = \text{OMVA/size} \quad \text{(equation F.4)}$$

where "size" is the object's smallest dimension, as discussed above. If m turns out to be the same magnification that was used to compute M_b, then m is the optimum detection magnification (ODM), and you're done. More likely m will be somewhat different. In this case use the m just computed and try again until the two results match within reason. With a computer, a few iterations can quickly narrow in on a good solution.

For those wishing to program this procedure into a computer you must be able to compute the curve in Figure 2.7b. Table F.2 lists specific values along that curve. Other values can be found by interpolation. If you do interpolation, you should interpolate the log(OMVA) values because the line curves too much using normal OMVA values.

Let's work through an example. Suppose we intend to hunt for NGC 134 in an 8-inch telescope. What is the best magnification?

Look at Table F.3, which shows computations for this object. At iteration 1 an initial

Table F.2. *The optimum magnified visual angle (in arc-minutes) for various surface brightnesses*

S.B.	OMVA	Log(OMVA)	S.B.	OMVA	Log(OMVA)	S.B.	OMVA	Log(OMVA)
4.0	9	0.97	16.0	16	1.20	22.0	62	1.79
9.0	10	1.01	17.0	18	1.26	23.0	68	1.83
11.0	11	1.04	18.0	23	1.36	24.0	72	1.86
13.0	12	1.07	19.0	31	1.49	25.0	79	1.90
14.0	13	1.11	20.0	42	1.62	26.0	91	1.96
15.0	14	1.14	21.0	51	1.71	27.0	117	2.07

S.B. is the surface brightness in magnitudes per square arc-second.

guess for the magnification was made. In this case, 100× was selected (for no other reason than it is a starting point). Now compute the background surface brightness B_o through the 8-inch telescope at 100× using equations F.2 and F.3. We find that B_o is 27.5. Next estimate the OMVA from Figure 2.7b or interpolate it using Table F.2. In the table the last entry is at 27, so the OMVA must be found by extrapolation. After doing this we find the OMVA should be 132.4 arc-minutes.

Using this guess for the optimum size, compute the magnification that would give that size using equation F.3. Because NGC 134 has a smaller dimension of 1.0 arc-minutes we find the optimum magnification would be 132.4/1.0 or 132.4×. But we guessed 100×. These do not match very well so use 132.4× as a new guess and start over. Examining each iteration in Table F.3 for the 8-inch we see that the guess and the computed magnification are getting closer. By iteration 9 the difference is less than 1, so 186× is very close. At iteration 12 the difference is only 0.1× so 187× is certainly better than one needs. In practice the final magnification is only approximate for reasons discussed below. Any value between 180 and 190 is accurate enough.

Once the two magnifications agree you have converged on the ODM!

Try the same exercise using a 24-inch telescope. Only 5 iterations are needed to converge on a value of 76×. Because 81× is already the lowest useful magnification on a 24-inch, a lower power could not be beneficially used, so 81× should be considered the ODM.

WHAT THE ODM MEANS

By examining the entries in the catalog you will note that the optimum magnification on the large telescope is less than on the small one! The reason is that the larger a telescope is, the greater an object's surface brightness is at a given magnification, so it can be detected at a smaller apparent size.

Examining the solutions for NGC 134 in the catalog, we see that the ODM on the 8-inch is 187×, and on the 24-inch 81×. Notice that for very small telescopes, the magnification must be higher still. A 4-inch needs about twice the power of an 8-inch to show NGC 134 best.

Once an object has been found, and if it is not at the very threshold of detection, try raising the power to enlarge any internal detail toward *its* ODM. Recall that the optimum magnifications in the catalog are only approximate, because the entire object's mean surface brightness is used. Many objects show considerable variation in brightness across their surfaces.

Spiral galaxies, for instance, typically have a bright central region and faint arms. The fainter parts may not be seen at all, while the bright inner area is much smaller (tending to raise the ODM) and much brighter (tending to lower it, but usually not as much as the increase caused by the smaller size). This catalog is only a guide; if an object can't be detected at the listed ODM, try both higher and lower powers.

In the catalog, when the ODM was computed to be less than the minimum useful power of the telescope, the telescope's mini-

APPENDIX F: OPTIMUM DETECTION MAGNIFICATIONS FOR DEEP-SKY OBJECTS

Table F.3. *Sample iterative calculation for NGC 134*

Iteration #	Telescope size (inches)	Guess magnification	B_0	Possible OMVA (arc-min)	Computed magnification
1	8.0	100.0	27.5	132.4	132.4
2	8.0	132.4	28.1	154.6	154.6
3	8.0	154.6	28.4	168.3	168.3
4	8.0	168.3	28.6	176.3	176.3
5	8.0	176.3	28.7	180.9	180.9
6	8.0	180.9	28.8	183.5	183.5
7	8.0	183.5	28.8	184.9	184.9
8	8.0	184.9	28.8	185.7	185.7
9	8.0	185.7	28.8	186.2	186.2
10	8.0	186.2	28.8	186.4	186.4
11	8.0	186.4	28.8	186.5	186.5
12	8.0	186.5	28.8	186.6	186.6
1	24.0	100.0	25.1	80.4	80.4
2	24.0	80.4	24.6	76.7	76.7
3	24.0	76.7	24.5	75.9	75.9
4	24.0	75.9	24.5	75.8	75.8
5	24.0	75.8	24.5	75.8	75.8

mum was listed. If the total magnitude of the object is simply too faint for the telescope under any conditions whatsoever, the entry for that telescope is left blank.

THE ROLE OF CONTRAST

Even where an optimum magnification is listed, it does not necessarily mean the object will be seen at that power. The contrast must also be high enough, as described in Chapter 6. To check on the contrast, find the apparent size of the object's minimum dimension at the ODM, as well as its apparent surface brightness, as described earlier.

For example: in a 12-inch telescope NGC 134 has an ODM of about 114×, at which a dark country sky background is reduced 2.63 magnitudes to a value of 24.25 + 2.63 = 26.88 magnitudes per square arc-second. We can round this off to 27. The size of NGC 134's smaller dimension is 1.0 arc-minutes, which when magnified 114× is 114 arc-minutes. This is the object's OMVA.

The smallest detectable contrast can be found from Figure 2.6 or interpolating the values in Table F.4. The Table lists the log threshold contrast for different values of background surface brightness and OMVA. The entries were used to plot the curves in Figure 2.6. Again you must use log contrast values for any interpolations.

Figure 2.6 shows that, at an apparent size of 114 arc-minutes on the curve for background brightness 27, a contrast of about 1 is needed for detection. This is a log contrast of 0.

The Cl column gives a log contrast for NGC 134 of 1.0 (which equals 10^1 or a contrast of 10), well above the threshold. Thus, NGC 134 should be easily detectable. I have used the logarithm of the contrast because, as for brightness, the eye responds to contrast on a logarithmic scale. Just think of the Cl column as a contrast scale with a higher number meaning more contrast.

The contrast of each object in the catalog was evaluated for each telescope to determine whether it is high enough for detection. Where the contrast is too low, the letter "u" (for undetectable) appears in front of the magnification entry. Once again, this does not necessarily mean all of the object is invisible, because the calculation was done using the mean surface brightness. If the object has a brighter region, like the nucleus of a galaxy,

Table F.4. *Log threshold contrast as a function of angle and surface brightness*

Background surface brightness (magnitudes per sq. arc-second)	Angle (arc-minutes)						
	0.595	3.60	9.68	18.2	55.2	121	360
	Log angle (arc-minutes)						
	−0.2255	0.5563	0.9859	1.260	1.742	2.083	2.556
4	−0.3769	−1.8064	−2.3368	−2.4601	−2.5469	−2.5610	−2.5660
5	−0.3315	−1.7747	−2.3337	−2.4608	−2.5465	−2.5607	−2.5658
6	−0.2682	−1.7345	−2.3310	−2.4605	−2.5467	−2.5608	−2.5658
7	−0.1982	−1.6851	−2.3140	−2.4572	−2.5481	−2.5615	−2.5665
8	−0.1238	−1.6252	−2.2791	−2.4462	−2.5463	−2.5597	−2.5646
9	−0.0424	−1.5529	−2.2297	−2.4214	−2.5343	−2.5501	−2.5552
10	0.0498	−1.4655	−2.1659	−2.3763	−2.5047	−2.5269	−2.5333
11	0.1596	−1.3581	−2.0810	−2.3036	−2.4499	−2.4823	−2.4937
12	0.2934	−1.2256	−1.9674	−2.1965	−2.3631	−2.4092	−2.4318
13	0.4557	−1.0673	−1.8186	−2.0531	−2.2445	−2.3083	−2.3491
14	0.6500	−0.8841	−1.6292	−1.8741	−2.0989	−2.1848	−2.2505
15	0.8808	−0.6687	−1.3967	−1.6611	−1.9284	−2.0411	−2.1375
16	1.1558	−0.3952	−1.1264	−1.4176	−1.7300	−1.8727	−2.0034
17	1.4822	−0.0419	−0.8243	−1.1475	−1.5021	−1.6768	−1.8420
18	1.8559	0.3458	−0.4924	−0.8561	−1.2661	−1.4721	−1.6624
19	2.2669	0.6960	−0.1315	−0.5510	−1.0562	−1.2892	−1.4827
20	2.6760	1.0880	0.2060	−0.3210	−0.8800	−1.1370	−1.3620
21	2.7766	1.2065	0.3467	−0.1377	−0.7361	−0.9964	−1.2439
22	2.9304	1.3821	0.5353	0.0328	−0.5605	−0.8606	−1.1187
23	3.1634	1.6107	0.7708	0.2531	−0.3895	−0.7030	−0.9681
24	3.4643	1.9034	1.0338	0.4943	−0.2033	−0.5259	−0.8288
25	3.8211	2.2564	1.3265	0.7605	0.0172	−0.2992	−0.6394
26	4.2210	2.6320	1.6990	1.1320	0.2860	−0.0510	−0.4080
27	4.6100	3.0660	2.1320	1.5850	0.6520	0.2410	−0.1210

you still might find it. Of all the objects for which magnitude data could be found, all can be detected in at least one of the telescope sizes listed.

In some cases, the listed optimum detection magnification is extremely high because the object is very small. A good example is the tiny planetary nebula NGC 3918. However, its contrast is so high ($Cl = 4.6$, for a contrast of nearly 40 000!) that it can easily be seen at any power in any of the telescopes. Again, the ODM is a guide for when you're having difficulty detecting an object; it's not an absolute rule.

Finally, a computer program in FORTRAN that computes the ODM for any object, and indicates whether or not the contrast is high enough to see it in a dark country sky, is given in Table F.5.

APPENDIX F: OPTIMUM DETECTION MAGNIFICATIONS FOR DEEP-SKY OBJECTS

		size			Optimum Detection Magnification for a given telescope aperture					
ID	v mag	arc-min	S.B.	C1	2	4	6	8	12	16	24
NGC 7814	12.0	1.0x5.0	22.4	0.7	360	360	265	187	114	87	81
NGC 40	10.5	1.0x0.7	18.7	2.2	514	514	514	412	251	177	122
NGC 45	12.1	8.0x5.5	24.8	-0.2	23	u14	20	27	41	54	81
NGC 55	7.8	25x40	23.9	0.1	7	14	20	27	41	54	81
NGC 104	4.5	25x25	20.1	1.7	7	14	20	27	41	54	81
NGC 128	12.8	2.2x0.4	21.3	1.2	-	900	900	900	871	613	373
NGC 134	11.4	5.0x1.0	21.8	1.0	360	360	265	187	114	87	81
NGC 147	12.1	6.5x3.8	24.2	0.0	52	23	20	27	41	54	81
NGC 150	12.2	2.0x1.0	21.6	1.1	360	360	265	187	114	87	81
NGC 157	11.1	2.8x2.1	21.7	1.0	171	84	51	41	41	54	81
NGC 175	12.8	1.5x1.3	22.2	0.8	-	243	148	104	68	60	81
NGC 185	11.8	3.5x2.8	22.9	0.5	103	44	30	28	41	54	81
NGC 205	10.8	8.0x3.0	22.9	0.5	89	38	28	27	41	54	81
NGC 188	10.	15x15	24.5	-0.1	7	14	20	27	41	54	81
NGC 224	4.0	150x50	22.3	0.8	7	14	20	27	41	54	81
NGC 221	9.5	3.6x3.1	20.7	1.4	82	35	26	27	41	54	81
NGC 225	8.	14x14	22.4	0.8	7	14	20	27	41	54	81
NGC 246	8.5	4.0x2.5	19.6	1.8	133	57	36	32	41	54	81
NGC 247	10.7	18.0x5.0	24.2	0.0	29	16	20	27	41	54	81
NGC 253	7.0	22x6	20.9	1.3	19	14	20	27	41	54	81
NGC 255	12.8	1.5x1.5	22.3	0.8	-	177	108	76	55	54	81
NGC 278	11.6	1.2x1.2	20.6	1.5	300	290	177	124	76	67	81
NGC 292	1.5	210x210	21.7	1.0	7	14	20	27	41	54	81
NGC 288	7.2	10x10	20.8	1.4	8	14	20	27	41	54	81
NGC 281	8.0	23x27	23.6	0.3	7	14	20	27	41	54	81
NGC 300	11.3	21x14	26.1	-0.7	u7	u14	u20	27	41	54	81
NGC 309	12.5	2.4x2.1	22.9	0.5	-	84	51	41	41	54	81
Scl-Sys	8.8	60x60	26.3	-0.8	u7	14	20	27	41	54	81
NGC 337	12.8	3.0x0.5	21.9	1.0	-	720	720	720	531	373	227
NGC 362	6.	10x10	19.6	1.8	8	14	20	27	41	54	81
I.1613	12.0	11x9	25.6	-0.5	u9	u14	20	27	41	54	81
NGC 404	11.9	1.3x1.3	21.1	1.3	277	243	148	104	68	60	81
NGC 428	11.7	3.9x3.5	23.2	0.4	63	27	22	27	41	54	81
NGC 457	7.	10x10	20.6	1.4	8	14	20	27	41	54	81
NGC 488	11.2	3.5x3.0	22.4	0.7	89	38	28	27	41	54	81
NGC 520	12.4	3.0x0.7	21.8	1.0	-	514	514	412	251	177	122
NGC 581	8.	8x8	21.1	1.2	11	14	20	27	41	54	81
NGC 598	5.3	60x60	22.8	0.6	7	14	20	27	41	54	81
NGC 604	?	1x1									
NGC 613	11.1	3.0x2.0	21.7	1.0	180	93	57	44	41	54	81

VISUAL ASTRONOMY OF THE DEEP SKY

					\multicolumn{7}{c}{Optimum Detection Magnification for a given telescope aperture}
ID	v mag	size arc-min	S.B.	Cl	2	4	6	8	12	16	24
NGC 615	12.6	2.7x0.8	22.1	0.9	-	450	436	306	187	131	101
NGC 628	9.0	9.0x9.0	22.4	0.7	9	14	20	27	41	54	81
NGC 650-1	10.	1.5x0.7	18.7	2.2	514	514	514	412	251	177	122
NGC 663	7.	11x11	20.8	1.4	7	14	20	27	41	54	81
NGC 672	11.6	4.5x1.7	22.4	0.7	212	134	82	57	46	54	81
NGC 681	12.9	1.3x1.2	22.0	0.9	-	290	177	124	76	67	81
NGC 752	7.5	45x45	24.4	-0.1	7	14	20	27	41	54	81
NGC 779	11.8	3.0x0.5	20.9	1.4	720	720	720	720	531	373	227
NGC 869	4.4	35x35	20.7	1.4	7	14	20	27	41	54	81
NGC 884	4.7	35x35	21.0	1.3	7	14	20	27	41	54	81
NGC 891	12.2	12.0x1.0	23.5	0.3	360	360	265	187	114	87	81
NGC 908	11.1	4.0x1.3	21.5	1.1	277	243	148	104	68	60	81
I.1805	9.	60x60	26.5	-0.9	u7	14	20	27	41	54	81
NGC 925	12.0	9.4x9.4	25.5	-0.5	u9	u14	20	27	41	54	81
NGC 936	11.3	3.0x2.0	21.9	1.0	180	93	57	44	41	54	81
NGC 941	12.9	1.9x1.3	22.5	0.7	-	243	148	104	68	60	81
NGC 972	12.3	2.7x1.0	22.0	0.9	-	360	265	187	114	87	81
For.Sys	9.	65x65	26.7	-1.0	u7	u14	20	27	41	54	81
NGC 1023	11.0	4.5x1.3	21.5	1.1	277	243	148	104	68	60	81
NGC 1049	13.	0.4x0.4	19.6	1.8	-	900	900	900	871	613	373
NGC 1039	6.	20x20	21.1	1.2	7	14	20	27	41	54	81
NGC 1055	12.0	5.0x1.0	22.4	0.7	360	360	265	187	114	87	81
NGC 1068	10.0	2.5x1.7	20.2	1.6	212	134	82	57	46	54	81
NGC 1073	12.0	4.0x4.0	23.6	0.2	47	22	20	27	41	54	81
NGC 1084	11.1	2.1x1.0	20.5	1.5	360	360	265	187	114	87	81
NGC 1087	11.2	2.3x2.3	21.6	1.0	157	69	42	36	41	54	81
NGC 1090	12.8	4.0x1.5	23.4	0.4	-	177	108	76	55	54	81
NGC 1097	10.6	9.0x5.5	23.5	0.3	23	14	20	27	41	54	81
NGC 1156	12.5	2.0x1.5	22.3	0.8	-	177	108	76	55	54	81
NGC 1187	11.3	5.5x4.0	23.3	0.4	47	22	20	27	41	54	81
NGC 1232	10.7	7.0x6.0	23.4	0.3	19	14	20	27	41	54	81
NGC 1261	12.5	2.4x1.0	22.1	0.9	-	360	265	187	114	87	81
NGC 1245	9.	20x20	24.1	0.0	7	14	20	27	41	54	81
NGC 1291	10.2	5.0x2.0	21.3	1.2	180	93	57	44	41	54	81
NGC 1300	11.3	6.0x3.2	23.1	0.4	77	33	25	27	41	54	81
NGC 1313	10.8	5.0x3.2	22.4	0.7	77	33	25	27	41	54	81
NGC 1316	10.1	3.5x2.5	21.1	1.3	133	57	36	32	41	54	81
NGC 1317	12.2	0.7x0.6	19.9	1.7	600	600	600	581	354	249	152
NGC 1326	11.8	3.0x2.5	22.6	0.7	133	57	36	32	41	54	81
NGC 1341	13.1	0.8x0.8	21.2	1.2	-	450	436	306	187	131	101

APPENDIX F: OPTIMUM DETECTION MAGNIFICATIONS FOR DEEP-SKY OBJECTS

```
                                Optimum Detection Magnification
                    size        for a given telescope aperture
   ID      v mag   arc-min   S.B.   Cl    2      4      6      8     12     16     24
--------|-------|----------|------|------|------|------|------|------|------|------|------
NGC 1351   12.8   0.8x0.6   20.6   1.4     -    600    600    581    354    249    152
NGC 1350   11.8   3.0x1.5   22.1   0.9   240    177    108     76     55     54     81
NGC 1360    ?     6.0x4.5
NGC 1365   11.2   8.0x3.5   23.4   0.3    63     27     22     27     41     54     81
NGC 1374   12.4   0.8x0.8   20.5   1.5     -    450    436    306    187    131    101

NGC 1379   12.3   0.6x0.6   19.8   1.8     -    600    600    581    354    249    152
NGC 1380   11.4   3.0x1.0   21.2   1.2   360    360    265    187    114     87     81
NGC 1381   12.6   2.0x0.5   21.2   1.2     -    720    720    720    531    373    227
NGC 1386   12.4   2.5x1.0   22.0   0.9     -    360    265    187    114     87     81
NGC 1387   12.1   1.0x0.9   20.6   1.5   400    400    335    236    144    101     86

NGC 1389   12.8   1.0x1.8   22.1   0.9     -    360    265    187    114     87     81
NGC 1395   10.2   5.0x2.0   21.3   1.2   180     93     57     44     41     54     81
NGC 1399   10.9   1.4x1.4   20.3   1.6   257    206    126     88     61     55     81
NGC 1398   10.7   4.5x3.8   22.4   0.7    52     23     20     27     41     54     81
NGC 1404   11.5   1.0x1.0   20.1   1.6   360    360    265    187    114     87     81

NGC 1400   12.4   0.7x0.7   20.3   1.6     -    514    514    412    251    177    122
NGC 1407   11.4   0.8x0.8   19.5   1.9   450    450    436    306    187    131    101
NGC 1427   12.4   1.4x1.0   21.4   1.1     -    360    265    187    114     87     81
NGC 1433   11.4   7.0x6.0   24.1   0.1    19     14     20     27     41     54     81
I.342      12.0   15x15     26.5  -0.9    u7    u14    u20    u27    u41     54     81

NGC 1437   12.9   2.0x1.5   22.7   0.6     -    177    108     76     55     54     81
NGC 1435    6.8   15x20     21.6   1.1     7     14     20     27     41     54     81
M45         1.4   100x100   20.0   1.7     7     14     20     27     41     54     81
NGC 1491    ?     3x3        ?
NGC 1499    6.    145x40    24.0   0.1     7     14     20     27     41     54     81

NGC 1501   12.    0.9x0.8   20.3   1.6   450    450    436    306    187    131    101
NGC 1502    7.    8x8       20.1   1.6    11     14     20     27     41     54     81
NGC 1514   11.    2.0x2.0   21.1   1.2   180     93     57     44     41     54     81
NGC 1513    9.    12x12     23.0   0.5     7     14     20     27     41     54     81
NGC 1531   13.0   0.5x0.3   19.6   1.9     -   1200   1200   1200   1200   1162    708

NGC 1532   11.8   5.0x1.0   22.2   0.8   360    360    265    187    114     87     81
NGC 1528    6.    25x25     21.6   1.1     7     14     20     27     41     54     81
NGC 1535    9.    0.3x0.3   15.0   3.7  1200   1200   1200   1200   1200   1162    708
NGC 1549   11.0   2.8x2.5   21.7   1.0   133     57     36     32     41     54     81
NGC 1553   10.2   3.1x2.3   21.0   1.3   157     69     42     36     41     54     81

NGC 1559   11.1   3.0x1.5   21.4   1.2   240    177    108     76     55     54     81
NGC 1566   10.5   5.0x4.0   22.4   0.7    47     22     20     27     41     54     81
NGC 1555   var    0.5x0.5    -
NGC 1579    ?     8x12       ?
NGC 1624    ?     3x3        ?
```

					Optimum Detection Magnification for a given telescope aperture					
ID	v mag	size arc-min	S.B.	Cl	2	4	6	8	12	16	24
NGC 1637	11.4	2.7x2.0	21.9	1.0	180	93	57	44	41	54	81
NGC 1647	6.5	40x40	23.1	0.4	7	14	20	27	41	54	81
NGC 1746	6.	45x45	22.9	0.5	7	14	20	27	41	54	81
NGC 1792	10.7	3.0x1.0	20.5	1.5	360	360	265	187	114	87	81
NGC 1807	7.5	10x10	21.1	1.2	8	14	20	27	41	54	81
NGC 1817	8.	15x15	22.5	0.7	7	14	20	27	41	54	81
NGC 1832	12.3	2.1x1.1	21.8	1.0	-	327	215	151	92	76	81
NGC 1851	7.	5x5	19.1	2.1	29	16	20	27	41	54	81
I.405	10.	18x30	25.5	-0.5	u7	14	20	27	41	54	81
NGC 1857	7.	9x9	20.4	1.5	9	14	20	27	41	54	81
I.410	9.	20x20	24.1	0.0	7	14	20	27	41	54	81
NGC 1893	8.	12x12	22.0	0.9	7	14	20	27	41	54	81
LMC		1.0	360x360	22.4	0.7	7	14	20	27	41	54	81
NGC 1904	8.4	7.5x7.5	21.4	1.1	12	14	20	27	41	54	81
NGC 1912	6.2	20x20	21.3	1.2	7	14	20	27	41	54	81
NGC 1964	11.8	5.0x1.6	22.7	0.6	225	153	93	66	50	54	81
NGC 1952	9.	5x3	20.6	1.5	89	38	28	27	41	54	81
NGC 1960	6.3	12x12	20.3	1.6	7	14	20	27	41	54	81
NGC 1976	4.	65x65	21.7	1.0	7	14	20	27	41	54	81
NGC 1977	7.	40x25	23.1	0.4	7	14	20	27	41	54	81
NGC 1982	8.	7x5	20.5	1.5	29	16	20	27	41	54	81
NGC 1999	9.	2x2	19.1	2.0	180	93	57	44	41	54	81
B33	-	6x4	-	-							
I.434	?	60x12	?								
NGC 2023	?	10x10	?								
NGC 2024	?	20x20	?								
NGC 2070	~5.?	20x20	~20.1	1.7	7	14	20	27	41	54	81
NGC 2068	8.	8x6	20.8	1.4	19	14	20	27	41	54	81
NGC 2071	?	4x3	?								
NGC 2099	6.2	20x20	21.3	1.2	7	14	20	27	41	54	81
Barn. Loop	6.0	840x60	26.4	-0.9	u7	14	20	27	41	54	81
NGC 2129	7.	5x5	19.1	2.1	29	16	20	27	41	54	81
NGC 2158	11.	4x4	22.6	0.6	47	22	20	27	41	54	81
NGC 2168	5.5	30x30	21.5	1.1	7	14	20	27	41	54	81
NGC 2174	?	40x30									
I.443	?	50x40									
NGC 2207	12.3	2.5x1.5	22.4	0.8	-	177	108	76	55	54	81
I.2165	12.5	0.13x0.13	16.7	3.0	-	2769	2769	2769	2769	2769	2769
NGC 2237	9.	60x80	26.8	-1.0	u7	u14	20	27	41	54	81
NGC 2244	5.5	40x40	22.1	0.8	7	14	20	27	41	54	81

APPENDIX F: OPTIMUM DETECTION MAGNIFICATIONS FOR DEEP-SKY OBJECTS

```
                                     Optimum Detection Magnification
                 size                for a given telescope aperture
   ID     v mag  arc-min   S.B.  Cl   2     4     6     8    12    16    24
--------|------|----------|----|----|----|----|----|----|----|----|----
NGC 2261  10.    2x2       20.1  1.6  180   93    57    44    41    54    81
NGC 2264  5.     15x26     20.1  1.7    7   14    20    27    41    54    81
NGC 2287  6.     30x30     22.0  0.9    7   14    20    27    41    54    81
NGC 2281  6.     15x15     20.5  1.5    7   14    20    27    41    54    81
NGC 2323  6.     10x10     19.6  1.8    8   14    20    27    41    54    81

NGC 2346  10.    1.0x0.9   18.5  2.3  400  400   335   236   144   101    86
NGC 2276  12.4   2.5x2.0   22.8  0.6    -   93    57    44    41    54    81
NGC 2359   ?     10x5
NGC 2360  9.     10x10     22.6  0.6    8   14    20    27    41    54    81
NGC 2300  12.2   1.0x1.7   21.4  1.1  360  360   265   187   114    87    81

NGC 2362  4.     6x6       16.5  3.1   19   14    20    27    41    54    81
NGC 2383  11.    2x2       21.1  1.2  180   93    57    44    41    54    81
NGC 2366  12.6   6.0x3.0   24.4  0.0    -   38    28    27    41    54    81
NGC 2392  8.3    0.7x0.7   16.2  3.2  514  514   514   412   251   177   122
NGC 2403  8.8    16x10     22.9  0.5    8   14    20    27    41    54    81

NGC 2421  10.    8x8       23.1  0.4   11   14    20    27    41    54    81
NGC 2422  5.     20x20     20.1  1.6    7   14    20    27    41    54    81
NGC 2419  11.5   2x2       21.6  1.0  180   93    57    44    41    54    81
NGC 2437  8.     25x25     23.6  0.3    7   14    20    27    41    54    81
NGC 2438  11.    1.1x1.1   19.8  1.8  327  327   215   151    92    76    81

NGC 2447  7.     18x18     21.9  0.9    7   14    20    27    41    54    81
NGC 2477  7.     25x25     22.6  0.7    7   14    20    27    41    54    81
NGC 2516  6.     60x60     23.5  0.3    7   14    20    27    41    54    81
NGC 2547  5.5    15x15     20.0  1.7    7   14    20    27    41    54    81
NGC 2523  12.7   1.8x1.4   22.3  0.8    -  206   126    88    61    55    81

NGC 2546  8.     25x25     23.6  0.3    7   14    20    27    41    54    81
NGC 2548  5.5    40x40     22.1  0.8    7   14    20    27    41    54    81
NGC 2613  10.9   6.4x1.5   22.0  0.9  240  177   108    76    55    54    81
NGC 2627  9.     8x8       22.1  0.8   11   14    20    27    41    54    81
NGC 2623  14.0   2.0x0.4   22.4  0.7    -    -   900   900   871   613   373

NGC 2632  4.5    80x80     22.6  0.6    7   14    20    27    41    54    81
NGC 2659  9.5    10x10     23.1  0.4    8   14    20    27    41    54    81
H3        6.     7x7       18.9  2.2   14   14    20    27    41    54    81
NGC 2682  7.     15x15     21.5  1.1    7   14    20    27    41    54    81
NGC 2683  10.6   9.0x1.3   21.9  0.9  277  243   148   104    68    60    81

NGC 2681  11.3   2.8x2.5   22.0  0.9  133   57    36    32    41    54    81
NGC 2782  12.4   1.8x1.6   22.2  0.8       153    93    66    50    54    81
NGC 2818  11.    9x9       24.4 -0.1    9   14    20    27    41    54    81
NGC 2841  10.3   6.2x2.0   21.7  1.0  180   93    57    44    41    54    81
I.2488    7.     20x20     22.1  0.8    7   14    20    27    41    54    81
```

VISUAL ASTRONOMY OF THE DEEP SKY

		size			Optimum Detection Magnification for a given telescope aperture						
ID	v mag	arc-min	S.B.	Cl	2	4	6	8	12	16	24
NGC 2910	8.	6x6	20.5	1.5	19	14	20	27	41	54	81
NGC 2903	9.7	11.0x4.7	22.6	0.7	33	17	20	27	41	54	81
NGC 2925	8.	11x11	21.8	1.0	7	14	20	27	41	54	81
NGC 2976	10.8	3.4x1.9	21.5	1.1	189	105	64	47	41	54	81
NGC 2997	11.0	6.0x5.0	23.3	0.4	29	16	20	27	41	54	81
NGC 3031	8.0	18x10	22.3	0.8	8	14	20	27	41	54	81
NGC 3034	9.2	8.0x3.0	21.3	1.2	89	38	28	27	41	54	81
NGC 3077	11.0	2.6x1.9	21.4	1.2	189	105	64	47	41	54	81
NGC 3109	11.2	11.0x2.0	23.2	0.4	180	93	57	44	41	54	81
NGC 3114	8.0	40x40	24.6	-0.2	7	14	20	27	41	54	81
NGC 3115	10.0	4.0x1.0	20.1	1.6	360	360	265	187	114	87	81
NGC 3132	8.2	1.4x1.4	17.6	2.7	257	206	126	88	61	55	81
NGC 3145	12.5	2.4x1.0	22.1	0.9	-	360	265	187	114	87	81
NGC 3166	11.5	4.0x1.5	22.1	0.9	240	177	108	76	55	54	81
NGC 3169	11.4	3.9x1.7	22.1	0.9	212	134	82	57	46	54	81
NGC 3147	11.3	3.0x2.3	22.0	0.9	157	69	42	36	41	54	81
NGC 3185	12.7	1.5x0.9	21.7	1.0	-	400	335	236	144	101	86
NGC 3187	13.0	1.0x1.3	21.9	0.9	-	360	265	187	114	87	81
NGC 3184	10.5	5.5x5.5	22.8	0.6	23	14	20	27	41	54	81
NGC 3190	12.0	3.0x1.0	21.8	1.0	360	360	265	187	114	87	81
NGC 3201	8.5	10x10	22.1	0.8	8	14	20	27	41	54	81
NGC 3193	12.0	0.9x0.9	20.4	1.5	400	400	335	236	144	101	86
NGC 3198	11.0	9.0x3.0	23.2	0.4	89	38	28	27	41	54	81
NGC 3228	6.5	20x20	21.6	1.0	7	14	20	27	41	54	81
NGC 3242	8.9	0.7x0.7	16.8	3.0	514	514	514	412	251	177	122
I.2574	13.0	9.0x4.0	25.5	-0.5	-	u22	u20	u27	u41	54	81
NGC 3256	12.1	2.0x1.5	21.9	0.9	240	177	108	76	55	54	81
NGC 3293	5.	8x8	18.1	2.4	11	14	20	27	41	54	81
NGC 3310	11.0	3.0x2.0	21.6	1.1	180	93	57	44	41	54	81
NGC 3319	11.8	6.0x2.8	23.5	0.3	103	44	30	28	41	54	81
NGC 3344	11.0	6.0x5.1	23.3	0.4	27	15	20	27	41	54	81
NGC 3351	11.0	4.0x3.0	22.3	0.8	89	38	28	27	41	54	81
NGC 3372	?	85x80	?								
NGC 3359	11.0	6.0x3.0	22.8	0.6	89	38	28	27	41	54	81
NGC 3368	10.2	6.0x4.0	22.3	0.8	47	22	20	27	41	54	81
NGC 3379	10.6	2.1x2.0	20.8	1.4	180	93	57	44	41	54	81
NGC 3384	11.0	4.0x2.0	21.9	0.9	180	93	57	44	41	54	81
NGC 3389	12.2	2.2x1.0	21.7	1.0	360	360	265	187	114	87	81
NGC 3395	12.4	1.4x0.8	21.2	1.2	-	450	436	306	187	131	101
NGC 3396	12.8	1.0x0.5	20.7	1.4	-	720	720	720	531	373	227

APPENDIX F: OPTIMUM DETECTION MAGNIFICATIONS FOR DEEP-SKY OBJECTS

					\multicolumn{6}{c}{Optimum Detection Magnification for a given telescope aperture}
ID	v mag	size arc-min	S.B.	Cl	2	4	6	8	12	16	24
NGC 3423	11.7	3.5x3.0	22.9	0.5	89	38	28	27	41	54	81
NGC 3486	11.2	5.5x4.2	23.2	0.4	42	20	20	27	41	54	81
NGC 3504	11.6	2.0x1.8	21.6	1.1	200	118	72	51	43	54	81
NGC 3511	11.9	4.2x1.5	22.5	0.7	240	177	108	76	55	54	81
NGC 3513	12.0	2.0x1.6	21.9	0.9	225	153	93	66	50	54	81
NGC 3521	10.2	6.0x4.0	22.3	0.8	47	22	20	27	41	54	81
NGC 3532	7.	60x60	24.5	-0.1	7	14	20	27	41	54	81
NGC 3556	10.8	7.8x1.4	22.0	0.9	257	206	126	88	61	55	81
NGC 3587	11.0	2.5x2.5	21.6	1.1	133	57	36	32	41	54	81
NGC 3610	11.6	1.4x0.9	20.5	1.5	400	400	335	236	144	101	86
NGC 3613	11.7	1.7x0.8	20.7	1.4	450	450	436	306	187	131	101
NGC 3621	11.8	3.5x1.4	22.2	0.8	257	206	126	88	61	55	81
NGC 3623	10.3	7.8x1.6	21.7	1.0	225	153	93	66	50	54	81
NGC 3627	9.7	8.0x2.5	21.6	1.1	133	57	36	32	41	54	81
NGC 3628	10.3	12.0x2.0	22.4	0.7	180	93	57	44	41	54	81
NGC 3631	11.5	4.5x4.0	23.3	0.4	47	22	20	27	41	54	81
NGC 3672	11.8	3.5x1.4	22.2	0.8	257	206	126	88	61	55	81
NGC 3675	11.4	3.5x1.3	21.7	1.0	277	243	148	104	68	60	81
NGC 3718	11.8	3.0x3.0	22.8	0.6	89	38	28	27	41	54	81
NGC 3726	11.3	5.0x3.4	23.0	0.5	67	29	23	27	41	54	81
NGC 3887	11.6	2.8x2.0	22.1	0.9	180	93	57	44	41	54	81
NGC 3893	11.0	3.9x2.5	22.1	0.9	133	57	36	32	41	54	81
NGC 3918	8.	0.17x0.17	12.8	4.6	2118	2118	2118	2118	2118	2118	2118
NGC 3992	10.9	6.4x3.5	22.9	0.5	63	27	22	27	41	54	81
NGC 3998	11.6	1.6x1.2	20.9	1.3	300	290	177	124	76	67	81
NGC 4026	11.9	3.3x0.7	21.4	1.1	514	514	514	412	251	177	122
NGC 4027	11.6	2.0x1.7	21.6	1.1	212	134	82	57	46	54	81
NGC 4030	11.2	3.3x2.4	22.1	0.9	145	62	38	34	41	54	81
NGC 4036	11.6	3.0x1.0	21.4	1.1	360	360	265	187	114	87	81
NGC 4038	11.0	2.5x2.5	21.6	1.1	133	57	36	32	41	54	81
NGC 4039	12.	2.5x2.0	22.4	0.7	180	93	57	44	41	54	81
NGC 4041	11.7	2.2x1.9	21.9	0.9	189	105	64	47	41	54	81
NGC 4051	11.2	4.2x3.0	22.6	0.7	89	38	28	27	41	54	81
NGC 4085	12.8	2.2x0.5	21.5	1.1	-	720	720	720	531	373	227
NGC 4088	11.1	4.7x1.5	21.8	1.0	240	177	108	76	55	54	81
NGC 4096	11.5	5.8x1.0	22.0	0.9	360	360	265	187	114	87	81
NGC 4100	11.9	4.5x1.1	22.3	0.8	327	327	215	151	92	76	81
NGC 4105	12.0	1.5x1.5	21.5	1.1	240	177	108	76	55	54	81
NGC 4106	12.5	1.0x1.8	21.8	1.0	-	360	265	187	114	87	81
NGC 4111	11.6	3.4x0.8	21.3	1.2	450	450	436	306	187	131	101

					Optimum Detection Magnification for a given telescope aperture					
ID	v mag	size arc-min	S.B.	Cl	2	4	6	8	12	16	24
NGC 4116	12.3	3.3x1.4	22.6	0.7	-	206	126	88	61	55	81
NGC 4123	12.0	3.5x2.4	22.9	0.5	145	62	38	34	41	54	81
NGC 4147	11.	4x4	22.6	0.6	47	22	20	27	41	54	81
NGC 4192	11.0	8.2x2.0	22.7	0.6	180	93	57	44	41	54	81
NGC 4214	10.5	7.0x4.5	22.9	0.6	36	18	20	27	41	54	81
NGC 4216	10.9	7.2x1.0	21.7	1.0	360	360	265	187	114	87	81
NGC 4236	10.7	22x5	24.4	-0.1	29	16	20	27	41	54	81
NGC 4244	10.7	13.0x1.0	22.1	0.9	360	360	265	187	114	87	81
NGC 4251	11.3	2.0x0.8	20.4	1.5	450	450	436	306	187	131	101
NGC 4254	10.4	4.5x4.0	22.2	0.8	47	22	20	27	41	54	81
NGC 4258	9.0	19.5x6.5	22.9	0.5	16	14	20	27	41	54	81
NGC 4260	12.7	2.3x0.9	22.1	0.9	-	400	335	236	144	101	86
NGC 4261	11.7	2.0x1.7	21.7	1.0	212	134	82	57	46	54	81
NGC 4273	12.3	1.7x1.2	21.7	1.0	-	290	177	124	76	67	81
NGC 4274	11.5	5.0x1.2	22.1	0.9	300	290	177	124	76	67	81
NGC 4281	12.2	1.5x0.8	21.0	1.3	450	450	436	306	187	131	101
NGC 4293	11.7	4.8x1.8	22.7	0.6	200	118	72	51	43	54	81
NGC 4294	12.6	2.4x0.9	22.1	0.9	-	400	335	236	144	101	86
NGC 4298	11.9	2.7x1.1	21.7	1.0	327	327	215	151	92	76	81
NGC 4299	12.9	1.1x0.9	21.5	1.1	-	400	335	236	144	101	86
NGC 4302	12.9	4.5x0.5	22.4	0.7	-	720	720	720	531	373	227
NGC 4303	10.2	5.7x5.5	22.6	0.7	23	14	20	27	41	54	81
NGC 4321	10.4	5.2x5.0	22.6	0.7	29	16	20	27	41	54	81
M40	9.0	-									
NGC 4340	13.0	2.2x1.4	22.8	0.6	-	206	126	88	61	55	81
NGC 4349	10.	15x15	24.5	-0.1	7	14	20	27	41	54	81
NGC 4350	11.9	1.9x0.5	20.5	1.5	720	720	720	720	531	373	227
NGC 4361	10.5	1.3x1.3	19.7	1.8	277	243	148	104	68	60	81
NGC 4365	11.0	2.0x1.3	20.7	1.4	277	243	148	104	68	60	81
NGC 4374	10.5	2.0x1.8	20.5	1.5	200	118	72	51	43	54	81
NGC 4382	10.5	3.0x2.0	21.1	1.3	180	93	57	44	41	54	81
NGC 4372	8.	18x18	22.9	0.5	7	14	20	27	41	54	81
NGC 4387	12.0	1.9x1.1	21.4	1.1	327	327	215	151	92	76	81
NGC 4388	12.0	5.0x1.0	22.4	0.7	360	360	265	187	114	87	81
NGC 4394	12.0	3.0x3.0	23.0	0.5	89	38	28	27	41	54	81
NGC 4395	11.0	10.0x8.0	24.4	-0.1	11	14	20	27	41	54	81
NGC 4402	13.0	2.0x0.8	22.1	0.8	-	450	436	306	187	131	101
NGC 4406	10.5	3.0x2.8	21.4	1.1	103	44	30	28	41	54	81
NGC 4413	13.2	1.1x0.7	21.5	1.1	-	514	514	412	251	177	122
NGC 4414	11.0	3.1x1.5	21.3	1.2	240	177	108	76	55	54	81

APPENDIX F: OPTIMUM DETECTION MAGNIFICATIONS FOR DEEP-SKY OBJECTS

		size			\multicolumn{6}{c}{Optimum Detection Magnification for a given telescope aperture}				
ID	v mag	arc-min	S.B.	Cl	2	4	6	8	12	16	24
NGC 4417	12.2	2.2x0.8	21.4	1.1	450	450	436	306	187	131	101
NGC 4424	12.5	2.5x1.3	22.4	0.7	-	243	148	104	68	60	81
NGC 4425	12.9	2.0x0.5	21.5	1.1	-	720	720	720	531	373	227
NGC 4435	11.8	1.4x0.9	20.7	1.4	400	400	335	236	144	101	86
NGC 4438	11.0	4.0x1.5	21.6	1.1	240	177	108	76	55	54	81
NGC 4448	11.7	2.9x1.0	21.5	1.1	360	360	265	187	114	87	81
NGC 4449	10.5	4.2x3.0	21.9	0.9	89	38	28	27	41	54	81
NGC 4450	11.1	3.8x3.0	22.4	0.8	89	38	28	27	41	54	81
NGC 4458	12.0	1.9x1.8	22.0	0.9	200	118	72	51	43	54	81
NGC 4459	11.7	1.5x1.0	20.8	1.4	360	360	265	187	114	87	81
3C273	12.8	stellar	-								
NGC 4461	12.2	2.0x1.0	21.6	1.1	360	360	265	187	114	87	81
NGC 4472	10.1	4.0x3.4	21.6	1.1	67	29	23	27	41	54	81
NGC 4473	11.3	2.0x1.0	20.7	1.4	360	360	265	187	114	87	81
NGC 4476	13.3	0.7x0.4	20.5	1.5	-	900	900	900	871	613	373
NGC 4477	10.4	4.0x3.5	21.9	0.9	63	27	22	27	41	54	81
NGC 4478	12.4	1.0x1.8	21.7	1.0	-	360	265	187	114	87	81
NGC 4479	12.5	1.5x1.5	22.0	0.9	-	177	108	76	55	54	81
NGC 4485	12.5	1.3x0.7	21.0	1.3	-	514	514	412	251	177	122
NGC 4486	8.6	3.0x3.0	19.6	1.9	89	38	28	27	41	54	81
NGC 4490	10.1	5.0x2.0	21.2	1.2	180	93	57	44	41	54	81
NGC 4494	10.9	1.4x1.4	20.3	1.6	257	206	126	88	61	55	81
NGC 4501	10.5	5.7x2.5	22.0	0.9	133	57	36	32	41	54	81
NGC 4517	11.4	9.0x1.0	22.4	0.7	360	360	265	187	114	87	81
NGC 4526	10.7	4.0x1.0	20.8	1.4	360	360	265	187	114	87	81
NGC 4527	11.3	5.1x1.1	21.8	1.0	327	327	215	151	92	76	81
NGC 4535	10.7	6.0x4.0	22.8	0.6	47	22	20	27	41	54	81
NGC 4536	11.0	7.0x2.0	22.5	0.7	180	93	57	44	41	54	81
I.3568	11.6	0.60x0.60	19.1	2.1	600	600	600	581	354	249	152
NGC 4548	10.9	3.9x3.4	22.3	0.8	67	29	23	27	41	54	81
NGC 4552	11.0	2.0x2.0	21.1	1.2	180	93	57	44	41	54	81
NGC 4559	10.5	10.0x3.0	22.8	0.6	89	38	28	27	41	54	81
NGC 4564	11.1	3.1x1.4	21.3	1.2	257	206	126	88	61	55	81
NGC 4565	10.5	15.0x1.1	22.2	0.8	327	327	215	151	92	76	81
NGC 4567	12.0	2.4x1.6	22.1	0.9	225	153	93	66	50	54	81
NGC 4568	11.9	3.6x1.8	22.6	0.7	200	118	72	51	43	54	81
NGC 4569	9.0	7.0x2.5	20.7	1.4	133	57	36	32	41	54	81
NGC 4579	10.5	4.0x3.5	22.0	0.9	63	27	22	27	41	54	81
H7	10.	8x8	23.1	0.4	11	14	20	27	41	54	81
NGC 4590	8.	9x9	21.4	1.1	9	14	20	27	41	54	81

		size			Optimum Detection Magnification for a given telescope aperture					
ID	v mag	arc-min	S.B.	Cl	2	4	6	8	12	16	24
NGC 4594	8.2	7.0x1.5	19.4	1.9	240	177	108	76	55	54	81
NGC 4605	10.9	4.0x1.2	21.2	1.2	300	290	177	124	76	67	81
NGC 4618	11.2	3.0x2.5	22.0	0.9	133	57	36	32	41	54	81
NGC 4621	11.0	2.0x1.5	20.8	1.4	240	177	108	76	55	54	81
NGC 4631	9.7	12.5x1.2	21.3	1.2	300	290	177	124	76	67	81
NGC 4638	12.2	1.1x0.5	20.2	1.6	720	720	720	720	531	373	227
NGC 4636	11.0	7.0x2.0	22.5	0.7	180	93	57	44	41	54	81
NGC 4639	12.2	2.0x1.3	21.9	1.0	277	243	148	104	68	60	81
NGC 4643	11.6	1.7x0.8	20.6	1.5	450	450	436	306	187	131	101
NGC 4647	12.0	2.3x1.8	22.2	0.8	200	118	72	51	43	54	81
NGC 4649	10.0	3.0x2.5	20.8	1.4	133	57	36	32	41	54	81
NGC 4651	11.4	3.0x2.5	22.2	0.8	133	57	36	32	41	54	81
NGC 4654	11.2	4.5x2.5	22.5	0.7	133	57	36	32	41	54	81
NGC 4656	11.0	19.5x2.0	23.6	0.3	180	93	57	44	41	54	81
NGC 4660	12.2	1.3x0.6	20.6	1.5	600	600	600	581	354	249	152
NGC 4666	11.4	3.9x0.7	21.1	1.3	514	514	514	412	251	177	122
NGC 4668	13.4	0.8x0.6	21.2	1.2	-	600	600	581	354	249	152
NGC 4689	11.7	2.8x2.0	22.2	0.8	180	93	57	44	41	54	81
NGC 4697	10.5	2.5x1.3	20.4	1.5	277	243	148	104	68	60	81
NGC 4699	10.3	3.0x2.0	20.9	1.4	180	93	57	44	41	54	81
NGC 4710	12.0	3.4x0.5	21.2	1.2	720	720	720	720	531	373	227
NGC 4725	10.5	7.5x4.8	23.0	0.5	31	17	20	27	41	54	81
NGC 4736	8.9	5.0x3.5	20.6	1.4	63	27	22	27	41	54	81
NGC 4742	12.0	1.0x0.6	20.1	1.7	600	600	600	581	354	249	152
NGC 4747	12.8	3.0x0.5	21.9	1.0	-	720	720	720	531	373	227
NGC 4754	11.8	2.5x1.0	21.4	1.1	360	360	265	187	114	87	81
NGC 4753	10.6	2.8x2.0	21.1	1.3	180	93	57	44	41	54	81
Coalsack	-	400x300	-	-							
NGC 4762	11.5	3.7x0.4	20.6	1.5	900	900	900	900	871	613	373
NGC 4760	12.5	0.6x0.5	19.8	1.8	-	720	720	720	531	373	227
NGC 4755	4.2	10x10	17.8	2.6	8	14	20	27	41	54	81
NGC 4781	11.7	2.6x1.1	21.5	1.1	327	327	215	151	92	76	81
NGC 4782	12.9	0.5x0.5	20.0	1.7	-	720	720	720	531	373	227
NGC 4783	12.9	0.5x0.5	20.0	1.7	-	720	720	720	531	373	227
NGC 4790	12.5	1.4x1.0	21.5	1.1	-	360	265	187	114	87	81
NGC 4826	8.6	7.5x3.5	20.8	1.4	63	27	22	27	41	54	81
NGC 4845	12.6	4.0x0.8	22.5	0.7	-	450	436	306	187	131	101
NGC 4833	8.5	6x6	21.0	1.3	19	14	20	27	41	54	81
NGC 4856	11.4	2.5x0.7	20.6	1.4	514	514	514	412	251	177	122
NGC 4874	13.5	1.0x1.0	22.1	0.8	-	360	265	187	114	87	81

APPENDIX F: OPTIMUM DETECTION MAGNIFICATIONS FOR DEEP-SKY OBJECTS

| | | size | | | \multicolumn{6}{c}{Optimum Detection Magnification for a given telescope aperture} |
ID	v mag	arc-min	S.B.	Cl	2	4	6	8	12	16	24
NGC 4889	13.2	1.0x0.6	21.3	1.2	-	600	600	581	354	249	152
NGC 4902	11.6	2.0x2.0	21.7	1.0	180	93	57	44	41	54	81
NGC 4945	?	?	?								
NGC 5005	10.8	4.7x1.6	21.6	1.1	225	153	93	66	50	54	81
NGC 5024	8.	10x10	21.6	1.0	8	14	20	27	41	54	81
NGC 5033	11.0	8.0x4.0	23.4	0.3	47	22	20	27	41	54	81
NGC 5055	9.8	9.0x4.0	22.3	0.8	47	22	20	27	41	54	81
NGC 5053	10.5	8x8	23.6	0.2	11	14	20	27	41	54	81
NGC 5102	10.8	6.0x2.5	22.4	0.8	133	57	36	32	41	54	81
NGC 5128	7.2	10x8	20.6	1.5	11	14	20	27	41	54	81
NGC 5139	3.6	30x30	19.6	1.9	7	14	20	27	41	54	81
NGC 5194	8.1	10.0x5.5	21.1	1.3	23	14	20	27	41	54	81
NGC 5195	11.0	2.0x1.5	20.8	1.4	240	177	108	76	55	54	81
NGC 5189	?	185x130	?								
NGC 5236	8.0	10.0x8.0	21.4	1.1	11	14	20	27	41	54	81
NGC 5248	11.0	3.2x1.4	21.3	1.2	257	206	126	88	61	55	81
NGC 5253	10.8	4.0x1.5	21.4	1.2	240	177	108	76	55	54	81
NGC 5272	6.	18x18	20.9	1.3	7	14	20	27	41	54	81
NGC 5322	11.3	1.4x1.0	20.3	1.6	360	360	265	187	114	87	81
NGC 5350	11.4	3.2x1.4	21.7	1.0	257	206	126	88	61	55	81
NGC 5353	12.3	1.1x0.4	20.0	1.7	-	900	900	900	871	613	373
NGC 5354	13.0	0.9x0.7	21.1	1.2	-	514	514	412	251	177	122
NGC 5363	11.1	1.7x1.5	20.7	1.4	240	177	108	76	55	54	81
NGC 5371	11.5	3.7x3.0	22.7	0.6	89	38	28	27	41	54	81
NGC 5364	11.5	5.0x4.0	23.4	0.3	47	22	20	27	41	54	81
NGC 5377	12.0	3.0x0.6	21.3	1.2	600	600	600	581	354	249	152
NGC 5367	10.	1.3x1.0	18.9	2.1	360	360	265	187	114	87	81
NGC 5383	12.7	2.2x2.0	22.9	0.5	-	93	57	44	41	54	81
NGC 5394	13.5	0.5x0.5	20.6	1.5	-	720	720	720	531	373	227
NGC 5395	12.7	2.1x1.0	22.1	0.8	-	360	265	187	114	87	81
NGC 5426	12.7	1.5x1.1	21.9	1.0	-	327	215	151	92	76	81
NGC 5427	12.0	2.0x1.7	22.0	0.9	212	134	82	57	46	54	81
NGC 5457	9.0	22x20	24.2	0.0	7	14	20	27	41	54	81
NGC 5466	9.0	5x5	21.1	1.3	29	16	20	27	41	54	81
NGC 5474	11.5	4.0x3.0	22.8	0.6	89	38	28	27	41	54	81
NGC 5585	11.6	4.5x2.3	22.8	0.6	157	69	42	36	41	54	81
NGC 5614	12.9	2.1x0.8	22.1	0.9	-	450	436	306	187	131	101
NGC 5617	8.	15x15	22.5	0.7	7	14	20	27	41	54	81
NGC 5643	11.4	2.5x2.3	21.9	0.9	157	69	42	36	41	54	81
NGC 5740	12.6	2.5x1.5	22.7	0.6	-	177	108	76	55	54	81

VISUAL ASTRONOMY OF THE DEEP SKY

		size			Optimum Detection Magnification for a given telescope aperture						
ID	v mag	arc-min	S.B.	Cl	2	4	6	8	12	16	24
NGC 5746	11.7	6.5x0.8	22.1	0.9	450	450	436	306	187	131	101
NGC 5824	9.5	3x3	20.5	1.5	89	38	28	27	41	54	81
NGC 5823	11.	9x9	24.4	-0.1	9	14	20	27	41	54	81
NGC 5846	11.5	1.0x1.0	20.1	1.6	360	360	265	187	114	87	81
NGC 5850	12.0	2.6x2.1	22.5	0.7	171	84	51	41	41	54	81
NGC 5866	11.1	2.9x1.0	20.9	1.3	360	360	265	187	114	87	81
NGC 5905	13.1	4.4x3.2	24.6	-0.1	-	33	25	27	41	54	81
NGC 5897	9.	8.5x8.5	22.3	0.8	10	14	20	27	41	54	81
NGC 5907	11.0	11.0x0.6	21.7	1.0	600	600	600	581	354	249	152
NGC 5908	13.0	2.4x0.4	21.6	1.1	-	900	900	900	871	613	373
NGC 5904	6.2	13x13	20.4	1.5	7	14	20	27	41	54	81
NGC 5962	11.9	2.2x1.3	21.7	1.0	277	243	148	104	68	60	81
NGC 5985	12.0	4.3x2.1	23.0	0.5	171	84	51	41	41	54	81
NGC 5986	8.	5x5	20.1	1.7	29	16	20	27	41	54	81
SP1	8.5	1.2x1.2	17.5	2.7	300	290	177	124	76	67	81
NGC 6027	14.	1.7x1.3	23.5	0.3	-	-	148	104	68	60	81
NGC 6025	6.	10x10	19.6	1.8	8	14	20	27	41	54	81
NGC 6067	8.	15x15	22.5	0.7	7	14	20	27	41	54	81
NGC 6093	8.	7x7	20.9	1.4	14	14	20	27	41	54	81
NGC 6101	10.	4x4	21.6	1.0	47	22	20	27	41	54	81
NGC 6121	7.4	20x20	22.5	0.7	7	14	20	27	41	54	81
NGC 6124	8.	25x25	23.6	0.3	7	14	20	27	41	54	81
NGC 6152	8.	30x30	24.0	0.1	7	14	20	27	41	54	81
NGC 6144	10.	3x3	21.0	1.3	89	38	28	27	41	54	81
NGC 6171	8.1	3x3	19.1	2.1	89	38	28	27	41	54	81
NGC 6153	11.5	0.33x0.33	17.7	2.6	1091	1091	1091	1091	1091	940	573
NGC 6188	?	20x12	?								
NGC 6192	10.	7x7	22.9	0.6	14	14	20	27	41	54	81
NGC 6193	5.2	15x15	19.7	1.8	7	14	20	27	41	54	81
NGC 6205	5.7	23x23	21.1	1.2	7	14	20	27	41	54	81
NGC 6207	12.3	2.0x1.0	21.7	1.0	-	360	265	187	114	87	81
NGC 6210	9.7	0.33x0.27	15.7	3.4	1333	1333	1333	1333	1333	1333	895
NGC 6218	8.	10x10	21.6	1.0	8	14	20	27	41	54	81
NGC 6229	8.7	3.5x3.5	20.0	1.7	63	27	22	27	41	54	81
NGC 6231	6.	15x15	20.5	1.5	7	14	20	27	41	54	81
H12	?	40x40	?								
NGC 6254	7.	8x8	20.1	1.6	11	14	20	27	41	54	81
NGC 6259	10.	15x15	24.5	-0.1	7	14	20	27	41	54	81
NGC 6266	6.5	6x6	19.0	2.1	19	14	20	27	41	54	81
NGC 6273	7.	6x6	19.5	1.9	19	14	20	27	41	54	81

APPENDIX F: OPTIMUM DETECTION MAGNIFICATIONS FOR DEEP-SKY OBJECTS

		size			Optimum Detection Magnification for a given telescope aperture						
ID	v mag	arc-min	S.B.	Cl	2	4	6	8	12	16	24
NGC 6302	?	2x1	?								
NGC 6318	11.	5x5	23.1	0.5	29	16	20	27	41	54	81
NGC 6341	6.5	8x8	19.6	1.8	11	14	20	27	41	54	81
NGC 6333	8.	4x4	19.6	1.8	47	22	20	27	41	54	81
NGC 6334	?	30x30	?								
NGC 6337	?	0.6x0.5	?								
B72	-	30x30	-	-							
NGC 6357	?	4x1	?								
NGC 6352	9.	8x8	22.1	0.8	11	14	20	27	41	54	81
NGC 6362	8.	9x9	21.4	1.1	9	14	20	27	41	54	81
NGC 6384	12.3	4.0x3.0	23.6	0.2	-	38	28	27	41	54	81
B78	-	200x140	-	-							
NGC 6388	7.	4x4	18.6	2.2	47	22	20	27	41	54	81
NGC 6402	9.	6x6	21.5	1.1	19	14	20	27	41	54	81
NGC 6397	7.	19x19	22.0	0.9	7	14	20	27	41	54	81
NGC 6405	6.	25x25	21.6	1.1	7	14	20	27	41	54	81
NGC 6441	8.	3x3	19.0	2.1	89	38	28	27	41	54	81
NGC 6475	5.	60x60	22.5	0.7	7	14	20	27	41	54	81
NGC 6494	7.	25x25	22.6	0.7	7	14	20	27	41	54	81
NGC 6543	8.6	0.4x0.3	14.9	3.7	1200	1200	1200	1200	1200	1162	708
NGC 6514	8.5	29x27	24.4	0.0	7	14	20	27	41	54	81
B86	-	4.5x3	-	-							
NGC 6520	9.	5x5	21.1	1.3	29	16	20	27	41	54	81
B87	-	12x12									
NGC 6523	5.	80x40	22.4	0.7	7	14	20	27	41	54	81
NGC 6530	6.	10x10	19.6	1.8	8	14	20	27	41	54	81
NGC 6531	7.	10x10	20.6	1.4	8	14	20	27	41	54	81
NGC 6569	10.	2x2	20.1	1.6	180	93	57	44	41	54	81
B92	-	15x15	-	-							
NGC 6584	8.5	6x6	21.0	1.3	19	14	20	27	41	54	81
M24	4.5	60x90	22.5	0.7	7	14	20	27	41	54	81
NGC 6603	11.4	4x4	23.0	0.5	47	22	20	27	41	54	81
NGC 6611	6.5	25x25	22.1	0.9	7	14	20	27	41	54	81
NGC 6613	8.	7x7	20.9	1.4	14	14	20	27	41	54	81
NGC 6618	6.	45x35	22.6	0.7	7	14	20	27	41	54	81
NGC 6624	8.5	3x3	19.5	1.9	89	38	28	27	41	54	81
NGC 6643	12.0	3.0x1.3	22.1	0.9	277	243	148	104	68	60	81
NGC 6626	8.	6x6	20.5	1.5	19	14	20	27	41	54	81
NGC 6629	10.5	0.25x0.25	16.1	3.3	1440	1440	1440	1440	1440	1440	1061
NGC 6638	9.5	2x2	19.6	1.8	180	93	57	44	41	54	81

VISUAL ASTRONOMY OF THE DEEP SKY

					\multicolumn{6}{c}{Optimum Detection Magnification for a given telescope aperture}
ID	v mag	size arc-min	S.B.	Cl	2	4	6	8	12	16	24
NGC 6637	7.5	4x4	19.1	2.0	47	22	20	27	41	54	81
I.4725	6.	35x35	22.3	0.8	7	14	20	27	41	54	81
NGC 6642	8.	2x2	18.1	2.4	180	93	57	44	41	54	81
NGC 6645	9.	10x10	22.6	0.6	8	14	20	27	41	54	81
NGC 6649	9.	9x9	22.4	0.7	9	14	20	27	41	54	81
NGC 6652	8.5	2x2	18.6	2.2	180	93	57	44	41	54	81
NGC 6656	6.	18x18	20.9	1.3	7	14	20	27	41	54	81
NGC 6664	9.	18x18	23.9	0.1	7	14	20	27	41	54	81
I.4756	6.	70x70	23.9	0.2	7	14	20	27	41	54	81
NGC 6681	8.	4x4	19.6	1.8	47	22	20	27	41	54	81
NGC 6694	9.5	9x9	22.9	0.5	9	14	20	27	41	54	81
NGC 6705	6.	12x12	20.0	1.7	7	14	20	27	41	54	81
NGC 6709	8.	12x12	22.0	0.9	7	14	20	27	41	54	81
NGC 6712	9.	3x3	20.0	1.7	89	38	28	27	41	54	81
NGC 6720	9.	1.3x1.0	17.9	2.5	360	360	265	187	114	87	81
NGC 6715	9.	6x6	21.5	1.1	19	14	20	27	41	54	81
NGC 6723	6.	7x7	18.9	2.2	14	14	20	27	41	54	81
NGC 6726	?	2x2	?								
NGC 6727	?	2x2	?								
NGC 6729	?	1x1	?								
B133	-	10x5	-	-							
NGC 6744	10.6	9.0x9.0	24.0	0.1	9	14	20	27	41	54	81
NGC 6752	7.	15x15	21.5	1.1	7	14	20	27	41	54	81
NGC 6779	8.	5x5	20.1	1.7	29	16	20	27	41	54	81
NGC 6781	12.5	1.75x1.75	22.3	0.8	-	126	77	54	45	54	81
NGC 6791	11.	20x20	26.1	-0.8	u7	u14	u20	27	41	54	81
NGC 6811	9.	15x15	23.5	0.3	7	14	20	27	41	54	81
NGC 6809	7.	15x15	21.5	1.1	7	14	20	27	41	54	81
B143	-	30x30	-	-							
NGC 6819	10.	6x6	22.5	0.7	19	14	20	27	41	54	81
NGC 6814	12.2	2.0x2.0	22.3	0.8	180	93	57	44	41	54	81
NGC 6820	?	20x20	?								
NGC 6818	10.	0.4x0.4	16.6	3.0	900	900	900	900	871	613	373
NGC 6823	10.	5x5	22.1	0.9	29	16	20	27	41	54	81
NGC 6822	10.0	20x10	24.4	-0.1	8	14	20	27	41	54	81
NGC 6826	8.8	0.4x0.4	15.4	3.5	900	900	900	900	871	613	373
NGC 6834	10.	4x4	21.6	1.0	47	22	20	27	41	54	81
NGC 6838	9.	6x6	21.5	1.1	19	14	20	27	41	54	81
NGC 6842	13.	0.8x0.75	21.1	1.3	-	480	480	354	215	152	110
NGC 6853	8.	8x6	20.8	1.4	19	14	20	27	41	54	81

APPENDIX F: OPTIMUM DETECTION MAGNIFICATIONS FOR DEEP-SKY OBJECTS

		size			Optimum Detection Magnification for a given telescope aperture					
ID	v mag	arc-min	S.B.	Cl	2	4	6	8	12	16	24
NGC 6866	8.	8x8	21.1	1.2	11	14	20	27	41	54	81
NGC 6864	8.	3x3	19.0	2.1	89	38	28	27	41	54	81
NGC 6884	12.5	0.1x0.1	16.1	3.2	-	3600	3600	3600	3600	3600	3600
NGC 6888	?	18x12	?								
NGC 6891	10.	0.2x0.1	14.4	3.9	3600	3600	3600	3600	3600	3600	3600
NGC 6910	6.5	8x8	19.6	1.8	11	14	20	27	41	54	81
NGC 6907	12.1	2.5x2.0	22.5	0.7	180	93	57	44	41	54	81
NGC 6913	7.	7x7	19.9	1.8	14	14	20	27	41	54	81
NGC 6939	10.	8x8	23.1	0.4	11	14	20	27	41	54	81
NGC 6934	9.	2x2	19.1	2.0	180	93	57	44	41	54	81
NGC 6940	8.	20x20	23.1	0.4	7	14	20	27	41	54	81
NGC 6946	11.1	8.0x8.0	24.2	0.0	11	14	20	27	41	54	81
NGC 6951	12.3	3.5x3.5	23.6	0.2	-	27	22	27	41	54	81
NGC 6960	8.	70x6	23.2	0.4	19	14	20	27	41	54	81
NGC 6981	8.6	3x3	19.6	1.9	89	38	28	27	41	54	81
NGC 6992-5	8.	78x8	23.6	0.3	11	14	20	27	41	54	81
NGC 6997	10.	7x7	22.9	0.6	14	14	20	27	41	54	81
NGC 6994	10.	1x1	18.6	2.2	360	360	265	187	114	87	81
NGC 7000	5.	100x100	23.6	0.2	7	14	20	27	41	54	81
NGC 7006	11.5	1x1	20.1	1.6	360	360	265	187	114	87	81
NGC 7008	12.	1.4x1.2	21.2	1.2	300	290	177	124	76	67	81
NGC 7009	8.	0.4x0.4	14.6	3.8	900	900	900	900	871	613	373
NGC 7023	?	18x18	?								
NGC 7026	12.	0.4x0.4	18.6	2.2	900	900	900	900	871	613	373
NGC 7027	9.	0.3x0.2	14.6	3.9	1800	1800	1800	1800	1800	1800	1743
NGC 7031	10.	6x6	22.5	0.7	19	14	20	27	41	54	81
NGC 7048	11.	1.0x0.9	19.5	1.9	400	400	335	236	144	101	86
NGC 7078	6.5	10x10	20.1	1.6	8	14	20	27	41	54	81
NGC 7086	9.	8x8	22.1	0.8	11	14	20	27	41	54	81
NGC 7092	5.	30x30	21.0	1.3	7	14	20	27	41	54	81
NGC 7089	6.0	7x7	18.9	2.2	14	14	20	27	41	54	81
NGC 7099	8.	6x6	20.5	1.5	19	14	20	27	41	54	81
NGC 7217	11.3	2.7x2.4	22.0	0.9	145	62	38	34	41	54	81
NGC 7235	9.	4x4	20.6	1.4	47	22	20	27	41	54	81
NGC 7243	8.	20x20	23.1	0.4	7	14	20	27	41	54	81
NGC 7293	6.5	12x16	20.8	1.4	7	14	20	27	41	54	81
NGC 7331	10.4	10.0x2.4	22.5	0.7	145	62	38	34	41	54	81
NGC 7410	11.8	4.0x1.1	22.0	0.9	327	327	215	151	92	76	81
NGC 7418	11.8	2.8x2.5	22.5	0.7	133	57	36	32	41	54	81
NGC 7479	11.8	3.2x3.5	23.1	0.5	77	33	25	27	41	54	81

		size			Optimum Detection Magnification for a given telescope aperture						
ID	v mag	arc-min	S.B.	Cl	2	4	6	8	12	16	24
NGC 7510	9.	3x3	20.0	1.7	89	38	28	27	41	54	81
NGC 7582	11.8	3.0x3.0	22.8	0.6	89	38	28	27	41	54	81
NGC 7590	11.9	2.2x0.8	21.1	1.2	450	450	436	306	187	131	101
NGC 7599	12.0	3.8x1.2	22.3	0.8	300	290	177	124	76	67	81
NGC 7619	12.6	0.8x0.6	20.4	1.5	-	600	600	581	354	249	152
NGC 7626	12.7	0.9x0.7	20.8	1.4	-	514	514	412	251	177	122
NGC 7635	11.	10x5	23.9	0.1	29	16	20	27	41	54	81
NGC 7654	7.	12x12	21.0	1.3	7	14	20	27	41	54	81
NGC 7662	8.5	0.5x0.5	15.6	3.5	720	720	720	720	531	373	227
NGC 7789	10.	20x20	25.1	-0.4	7	14	20	27	41	54	81
NGC 7793	9.7	6.0x4.0	21.8	1.0	47	22	20	27	41	54	81

APPENDIX F: OPTIMUM DETECTION MAGNIFICATIONS FOR DEEP-SKY OBJECTS

Table F.5. *Sample FORTRAN computer program: TELEODM*

```fortran
c       this program computes the optimum detection
c       magnification (odm) values and generates
c       entries like those in the Appendix F catalog.
c
c       optsiz = optimum size (angle)
c       opower = optimum detection magnification on an object
c       odm    = optimum detection magnification on an object
c                for each telescope size given by tele
c       power  = telescope's power
c       tele   = telescope's aperture in inches
c       tmag   = telescope's limiting magnitude

        implicit integer*4 (i-n)
        integer*4 odm, crtin, crtout
        real*4 tele, mb, tmag
        character*80 in, in2
        character*1 il, ihx, ihblnk

        data ihx, ihblnk /'x', ' '/

        crtin  = 0
        crtout = 0

1       write (crtout, 2)
2       format (1x,'***** Program Teleodm: Telescope Optimum Detection',
     1          ' Magnification *****',//,1x,
     2          'Enter the telescope aperture in inches')
        read (crtin,*) tele
        if (tele .le. 0.01) go to 1
c
        size1 = 0.0
        size2 = 0.0
c
c       sb0 is the background surface brightness. You could add a
c       read statement to input this value to examine other sky
c       conditions
c
        sb0 = 24.25
        sb  = 0.0

c determine faintest star based on equation 4.1 (= tmag):
c dmm = telescope's aperture in millimeters
c dsube = aperture of eye in millimeters
c trans = telescope transmission factor
c       trans = 0.7
c       dmm = tele*25.4
c       dsube = 7.5
c       tmag = 8.5 + 2.5* alog10(dmm*dmm*trans/(dsube*dsube))
```

Table F.5 (cont.)

```
c     which reduces to eqn 4.2:

          dmm = tele*25.4

          tmag = 3.7 + 2.5* alog10(dmm*dmm)
c
25        write (crtout, 26)
26        format (' Enter the total magnitude of the object')
          read (crtin, *) xmag
          if ((xmag .lt. 0.01) .or. (xmag.gt. 25)) go to 25

c         check if total magnitude is bright enough to be seen

          if (tmag .lt. xmag) then
               write (crtout, 28) xmag, tele
28             format (' Magnitude',f6.1,' is too faint to be',
     1                 ' seen in a',f6.1,' inch telescope.',/,
     2                 ' Try again.')
               go to 25
          endif

30        write (crtout, 31)
31        format (' Enter the minimum and maximum sizes of the object',
     1            ' in arc-minutes')
          read (crtin, *) size1, size2
          if ((size1.le.0.0).or.(size2.le.0.0)) go to 30

c         compute surface brightness (see Appendix E)

          sb = xmag + 2.5* alog10(2827.0*size1*size2)

c         compute the cl contrast index.  See Appendix E.

          cl = -0.4 * (sb - sb0)

c         find minimum size

          if (size1.lt.size2) xminsz = size1
          if (size2.le.size1) xminsz = size2

c         find minimum useful magnification for telescope

          ipmin = int(tele*3.375 + 0.5)

c         add sky to object surface brightness

          sbs = -2.5* alog10(10.0**(-0.4*sb) + 10.0**(-0.4*sb0))
c
c ***** find optimum angle : optsiz = opvang (sbm)
c ***** then minimum detection magnification is mdm = optsiz / xminsz

c         start initial guess at 100x:

          power = 100.0
```

APPENDIX F: OPTIMUM DETECTION MAGNIFICATIONS FOR DEEP-SKY OBJECTS

Table F.5 *(cont.)*

```
c        find surface brightness reduction in the telescope
c ***** START of iteration LOOP to find ODM

50       mb = 5.0 * alog10(power/(2.833487 * tele))
         sbm = sb0 + mb
         optsiz = opvang(sbm)
         opower = optsiz/xminsz

c the following is for debug
c        write (6,501) tele,power,opower,sbm,optsiz

c501     format (1x,'D=',f4.0,' inches, guess',f6.1,
c    1            ' new guess',f6.1,' sbm=',f5.1,' optsiz=',f6.1)

         if (opower .lt. ipmin) opower = ipmin
         if (abs(opower - power) .lt. 0.1) then

c              CONVERGED on right solution!
c              round off odm then exit loop

               odm = opower + 0.5
               go to 75
         else
               power = opower
         endif
         go to 50
c ***** END of iteration LOOP

c check if odm < min useful power, and limit max power

75       if (odm .lt. ipmin) odm = ipmin
         if (odm .gt. 9999) odm = 9999

c
c check if object is visible in each telescope and determine if it
c        has a high enough contrast to be detectable.

         angle = odm * xminsz

         xc = alog10( thrcon(angle,sbm) )

c **** start writing output for user

         write (crtout, 100) tele, ipmin, tmag
100      format (///,' Telescope aperture:',f7.1,' inches',/,
     1            ' Minimum usable magnification:',i5,'x',/,
     2            ' Faintest star:',f6.1,/)

c        sbp = surface brightness in telescope at magnification= power

         sbp = sbs + mb
```

341

Table F.5 (cont.)

```
              write (crtout, 102) size1, size2, sb, mb, sbp, sbm
102           format (' Object size:',f8.2,' by',f8.2,' arc minutes',/,
     1                ' surface brightness (magnitudes / sq. arc-sec.):',/,
     2               5x,'object with no telescope:       ',f7.1,/,
     3               5x,'reduction due to ODM:          ',f7.1,/,
     4               5x,'object in telescope at ODM:    ',f7.1,/,
     5               5x,'background in telescope at ODM:',f7.1,
     6               /)

              write (crtout,104) odm
104           format (' ODM:',i6,'x',/)

              write (crtout, 106) cl, xc
106           format (' Log object contrast:    ',f5.1,/,
     1                ' Log threshold contrast:',f5.1)

              if (cl .lt. xc) then
                      write (crtout, 115)
115                   format (' The CONTRAST IS TOO LOW TO DETECT ',
     1                        'the object',/)
              else
                      write (crtout, 120)
120                   format (' The object is detectable',/)
              endif

              go to 1

2000          stop
              end

c------------------------------------------------

              real*4 function thrcon (angle,sb)
              implicit integer*4 (i-n)
c
c             this function computes the contrast threshold for a given angle
c                by two dimensional interpolation.
c
c             angle = angle in arc-minutes
c             sb = surface brightness
c             angx = log10 angles (arc-min) data for 7 values of contrast
c             sbx = log base 10 contrast for angx values for each surface
c                   brightness ranging from 4 to 27 mag/arc-sec
c
c             angx and sbx are used to interpolate to the thrcon given
c                       by ang.
c
              real*4 angx(7), sbx(7,24), angle, sb

              real*4 sbx1(7), sbx2(7), sbx3(7), sbx4(7), sbx5(7), sbx6(7)
              real*4 sbx7(7), sbx8(7), sbx9(7), sbx10(7), sbx11(7), sbx12(7)
              real*4 sbx13(7), sbx14(7), sbx15(7), sbx16(7), sbx17(7)
```

APPENDIX F: OPTIMUM DETECTION MAGNIFICATIONS FOR DEEP-SKY OBJECTS

Table F.5 (cont.)

```
        real*4 sbx18(7), sbx19(7), sbx20(7), sbx21(7), sbx22(7)
        real*4 sbx23(7), sbx24(7)

        equivalence (sbx1(1),sbx(1,1))
        equivalence (sbx2(1),sbx(1,2))
        equivalence (sbx3(1),sbx(1,3))
        equivalence (sbx4(1),sbx(1,4))
        equivalence (sbx5(1),sbx(1,5))
        equivalence (sbx6(1),sbx(1,6))
        equivalence (sbx7(1),sbx(1,7))
        equivalence (sbx8(1),sbx(1,8))
        equivalence (sbx9(1),sbx(1,9))
        equivalence (sbx10(1),sbx(1,10))
        equivalence (sbx11(1),sbx(1,11))
        equivalence (sbx12(1),sbx(1,12))
        equivalence (sbx13(1),sbx(1,13))
        equivalence (sbx14(1),sbx(1,14))
        equivalence (sbx15(1),sbx(1,15))
        equivalence (sbx16(1),sbx(1,16))
        equivalence (sbx17(1),sbx(1,17))
        equivalence (sbx18(1),sbx(1,18))
        equivalence (sbx19(1),sbx(1,19))
        equivalence (sbx20(1),sbx(1,20))
        equivalence (sbx21(1),sbx(1,21))
        equivalence (sbx22(1),sbx(1,22))
        equivalence (sbx23(1),sbx(1,23))
        equivalence (sbx24(1),sbx(1,24))

        data angx /-0.2255,0.5563,0.9859,1.2601,1.7419,2.0828,2.5563/
c:::::::::::::::
c mag/arc-sec =4
c:::::::::::::::
        data sbx1/ -0.376911,-1.806444,-2.336825,-2.460093,
     1             -2.546943,-2.560968,-2.565971/
c:::::::::::::::
c mag/arc-sec =5
c:::::::::::::::
        data sbx2/ -0.331451,-1.774717,-2.333688,-2.460781,
     1             -2.546521,-2.560729,-2.565756/
c:::::::::::::::
c mag/arc-sec =6
c:::::::::::::::
        data sbx3/ -0.268103,-1.734482,-2.331011,-2.460521,
     1             -2.546681,-2.560819,-2.565837/
c:::::::::::::::
c mag/arc-sec =7
c:::::::::::::::
        data sbx4/-0.198177,-1.685101,-2.313951,-2.457228,
     1             -2.548051,-2.561529,-2.566460/
c:::::::::::::::
c mag/arc-sec =8
c:::::::::::::::
        data sbx5/ -0.123815,-1.625192,-2.279119,-2.446245,
     1             -2.546321,-2.559690,-2.564570/
```

Table F.5 (cont.)

```
c:::::::::::::::
c mag/arc-sec =9
c:::::::::::::::
        data sbx6/  -0.042444,-1.552937,-2.229734,-2.421422,
     1             -2.534318,-2.550130,-2.555202/
c:::::::::::::::
c mag/arc-sec =10
c:::::::::::::::
        data sbx7/   .049842,-1.465538,-2.165945,-2.376266,
     1             -2.504717,-2.526875,-2.533309/
c:::::::::::::::
c mag/arc-sec =11
c:::::::::::::::
        data sbx8/   .159558,-1.358054,-2.081004,-2.303572,
     1             -2.449928,-2.482261,-2.493710/
c:::::::::::::::
c mag/arc-sec =12
c:::::::::::::::
        data sbx9/   .293354,-1.225640,-1.967364,-2.196521,
     1             -2.363052,-2.409170,-2.431807/
c:::::::::::::::
c mag/arc-sec =13
c:::::::::::::::
        data sbx10/ 0.455739,-1.067267,-1.818647,-2.053094,
     1             -2.244490,-2.308296,-2.349087/
c:::::::::::::::
c mag/arc-sec =14
c:::::::::::::::
        data sbx11/ 0.649990,-0.884068,-1.629165,-1.874059,
     1             -2.098851,-2.184839,-2.250534/
c:::::::::::::::
c mag/arc-sec =15
c:::::::::::::::
        data sbx12/ 0.880777,-0.668706,-1.396670,-1.661059,
     1             -1.928370,-2.041067,-2.137519/
c:::::::::::::::
c mag/arc-sec =16
c:::::::::::::::
        data sbx13/ 1.155840,-0.395243,-1.126426,-1.417576,
     1             -1.729969,-1.872722,-2.003406/
c:::::::::::::::
c mag/arc-sec =17
c:::::::::::::::
        data sbx14/ 1.482228,-0.041944,-0.824314,-1.147481,
     1             -1.502126,-1.676828,-1.842016/
c:::::::::::::::
c mag/arc-sec =18
c:::::::::::::::
        data sbx15/ 1.855923,0.345791,-0.492425,-0.856105,
     1             -1.266128,-1.472054,-1.662449/
c:::::::::::::::
c mag/arc-sec =19
c:::::::::::::::
```

APPENDIX F: OPTIMUM DETECTION MAGNIFICATIONS FOR DEEP-SKY OBJECTS

Table F.5 (cont.)

```
        data sbx16/  2.266949,0.696000,-0.131546,-0.550963,
     1               -1.056193,-1.289210,-1.482659/
c::::::::::::::
c mag/arc-sec =20
c::::::::::::::
        data sbx17/  2.676000,1.088000,0.206000,-0.321000,
     1               -0.880000,-1.137000,-1.362000/
c::::::::::::::
c mag/arc-sec =21
c::::::::::::::
        data sbx18/  2.776553,1.206474,0.346695,-0.137740,
     1               -0.736142,-0.996400,-1.243900/
c::::::::::::::
c mag/arc-sec =22
c::::::::::::::
        data sbx19/  2.930431,1.382138,0.535251,0.032813,
     1               -0.560534,-0.860600,-1.118700/
c::::::::::::::
c mag/arc-sec =23
c::::::::::::::
        data sbx20/  3.163406,1.610655,0.770789,0.253144,
     1               -0.389500,-0.703000,-0.968133/
c::::::::::::::
c mag/arc-sec =24
c::::::::::::::
        data sbx21/  3.464285,1.903381,1.033812,0.494252,
     1               -0.203342,-0.525901,-0.828800/
c::::::::::::::
c mag/arc-sec =25
c::::::::::::::
        data sbx22/  3.821092,2.256431,1.326535,0.760529,
     1               0.017200,-0.299200,-0.639400/
c::::::::::::::
c mag/arc-sec =26
c::::::::::::::
        data sbx23/  4.221000,2.632000,1.699000,1.132000,
     1               0.286000,-0.051000,-0.408000/
c::::::::::::::
c mag/arc-sec =27
c::::::::::::::
        data sbx24/  4.610000,3.066000,2.132000,1.585000,
     1               0.652000,0.241000,-0.121000/
c*debug*
c*debug write (6,1) sbx
c*debug1          format (24(7(f8.4, 1x),/))

c find bounding sb indices

        isb1 = int(sb) -3
        if (isb1 .lt. 1) isb1 = 1
        if (isb1 .gt. 23) isb1 = 23
        isb2 = isb1 +1
```

Table F.5 (cont.)

```
              i =7
              ang = alog10(angle)
              if (ang.le.angx(1)) go to 100
              if (ang.ge.angx(7)) go to 200
      c
      c       find angx values bounding ang
      c
              do 10 i = 1,6
                      if ((angx(i).lt.ang).and.(angx(i+1).ge.ang)) go to 15
      10      continue
      c
      c       interpolate along the angle direction.
      c
      c       x1 is the contrast interpolted in the angle direction
      c                       at the first bounding sb value
      c       x2 is the contrast interpolted in the angle direction
      c                       at the second bounding sb value
      c
      15      x1 = (ang-angx(i)) / (angx(i+1)-angx(i)) *
           1      (sbx(i+1,isb1)-sbx(i,isb1)) + sbx(i,isb1)

              x2 = (ang-angx(i)) / (angx(i+1)-angx(i)) *
           1      (sbx(i+1,isb2)-sbx(i,isb2)) + sbx(i,isb2)
              go to 500
      c
      c       ang less than beginning of table, so extraploate
      c
      100     x1 = sbx(1,isb)
              x2 = sbx(1,isb2)
              go to 500
      c
      c       ang greater than last in table, so extraploate
      c
      200     x1 = (ang-angx(6)) / (angx(7)-angx(6)) *
           1      (sbx(7,isb1)-sbx(6,isb1)) + sbx(6,isb1)
              x2 = (ang-angx(6)) / (angx(7)-angx(6)) *
           1      (sbx(7,isb2)-sbx(6,isb2)) + sbx(6,isb2)
      c
      c       Now interpolate (or extrapolate) along sb direction
      c       compute thrcon from x1, x2
      c
      500     if (sb .lt. 4.0) then
                      x = x1
              else if (sb .ge. 27.0) then
                      x = (sb-27.0)*(x2-x1) + x2
              else
                      x = (sb-real(int(sb))) * (x2-x1) + x1
              endif

              if (x.gt.37.0) x=37.0
              if (x.lt.-37.0) x= -37.0
```

APPENDIX F: OPTIMUM DETECTION MAGNIFICATIONS FOR DEEP-SKY OBJECTS

Table F.5 (*cont.*)

```
              thrcon = 10.0 ** x

c*debug:
c*debug  write (6,504) i, isbl
c*debug504         format (' i=',i6, '   isbl=',i6)
c*debug  iia = i+1
c*debug  if (iia .gt. 7) iia = 7
c*debug          write (6,505) i,ang, angx(i), angx(iia), sbx(i,isbl),
c*debug      1              sbx(iia,isbl), x
c*debug505         format (1x,
c*debug      1  '     ang           angx(i)   angx(i+1)  sbx(i,isbl)
    sbx(i+1,isbl)',
c*debug      2  '    x',/, 1x, i2, 6(f7.3, 1x), /, 1x, 72(1h-))

          return
          end

c----------------------------------------------------

          real*4 function opvang (sb)
          implicit integer*4 (i-n)
c
c       this function computes the optimum visual angle
c       of an object given the background surface brightness (sb)
c
c       sbx = surface brightness data for 18 values of optimum
c             visual angle.
c       opv = log base 10 optimum visual angles for sbx values
c
c       sbx and opv are used to interpolate to the opvang given
c                 by sb.
c
c
          real*4 sbx(18), opv(18)

          data sbx /4.0,9.0,11.0,13.0,14.0,15.0,16.0,17.0,18.0,
      1              19.0,20.0,21.0,22.0,23.0,24.0,25.0,26.0,27.0/

          data opv /0.97,1.01,1.04,1.07,1.11,1.14,1.20,1.26,1.36,
      1              1.49,1.62,1.71,1.79,1.83,1.86,1.90,1.96,2.07/

          if (sb.le.sbx(1)) go to 100
          if (sb.ge.sbx(18)) go to 200
c
c       find sbx values bounding sb
c
          do 10 i = 1,17
          if ((sbx(i).lt.sb).and.(sbx(i+1).ge.sb)) go to 15
10        continue
c
c       interpolate
c
```

Table F.5 (cont.)

```
15         x = (sb-sbx(i)) / (sbx(i+1)-sbx(i)) *
      1          (opv(i+1)-opv(i)) + opv(i)
           go to 500
c
c          sb less than beginning of table
c
100        x = opv(1)
           go to 500
c
c          sb greater than last in table
c
200        x = (sb-sbx(17)) / (sbx(18)-sbx(17)) *
      1          (opv(18)-opv(17)) + opv(17)
c
c          compute opvang from x
c
500        if (x.gt.37.0) x=37.0
           if (x.lt.-37.0) x= -37.0

c*debug          write (6,505) i,sb, sbx(i), sbx(i+1), opv(i), opv(i+1), x
c*debug505       format (1x,
c*debug     1  '   sb     sbx(i)  sbx(i+1)  opv(i)   opv(i+1)      x',
c*debug     2   /, 1x, i2, 6(f7.3, 1x), /, 1x, 72(1h-))

           opvang = 10.0 ** x
c
c limit opvang to 360.0 arc-minutes.
c
           if (opvang .gt. 360.0) opvang = 360.0

           return
           end
```

APPENDIX F: OPTIMUM DETECTION MAGNIFICATIONS FOR DEEP-SKY OBJECTS

TELEDOM examples

Each example shows the user input in bold.

Example 1: NGC 134:

***** Program Teleodm: Telescope Optimum Detection Magnification *****

Enter the telescope aperture in inches
8.
Enter the total magnitude of the object
11.4
Enter the minimum and maximum sizes of the object in arc-minutes
5.,1.

Telescope aperture: 8.0 inches
Minimum usable magnification: 27x
Faintest star: 15.2

Object size: 5.00 by   1.00 arc minutes
surface brightness (magnitudes / sq. arc-sec.):
   object with no telescope: 21.8
   reduction due to ODM: 4.6
   object in telescope at ODM: 26.2
   background in telescope at ODM: 28.8

ODM: 187x

Log object contrast: 1.0
Log threshold contrast: .6
The object is detectable

TELEDOM examples (cont.)

Example 2: Sculptor System:

***** Program Teleodm: Telescope Optimum Detection Magnification *****

Enter the telescope aperture in inches
8.
Enter the total magnitude of the object
8.8
Enter the minimum and maximum sizes of the object in arc-minutes
60,60

Telescope aperture: 8.0 inches
Minimum usable magnification: 27x
Faintest star: 15.2

Object size: 60.00 by   60.00 arc minutes
surface brightness (magnitudes / sq. arc-sec.):
    object with no telescope: 26.3
    reduction due to ODM: .4
    object in telescope at ODM: 24.5
    background in telescope at ODM: 24.6

ODM: 27x

Log object contrast: -.8
Log threshold contrast: -1.2
The object is detectable

***** Program Teleodm: Telescope Optimum Detection Magnification *****

Enter the telescope aperture in inches

Bibliography

Astronomical Almanac. US Government Printing Office, Washington, DC.

Bartley, S. H. (1951). The Psychophysiology of Vision. In *Handbook of Experimental Psychology* (S.S. Stevens, Ed.) Wiley, New York, 921-984.

Blackwell, H. R. (1946). Contrast Thresholds of the Human Eye *J. Opt Soc Amer.* **36** 624-643.

Bowen, K. P. (1984). Vision and the Amateur Astronomer. *Sky & Telscope* April, 321-324.

Crossier, W. J. and A. H. Holway (1939). Theory and Measurement of Visual Mechanisms I. A Visual Discriminometer. II. Threshold Stimulus Intensity and Retinal Position. *J. Gen. Physiol.*, **22**, 341-364.

Curtis, H. D. (1901). On the Limits of Averted Vision. *Lick Obs Bull* **2** 67-69.

Di Cicco, D. (1979). Filters to Pierce the Nightime Veil. *Sky & Telescope.* March, 231-236.

Ferris, T. (1982). *Galaxies.* Stewart, Tabori and Chang, New York.

Graham, C. H., N. R. Bartlett, J. L. Brown, Y. Hsia. C. G. Mueller and L. A. Riggs. (1965). *Vision and Visual Perception.* Wiley, New York.

De Groot, S. G., J. M. Dodge and J. A. Smith. (1952). Factors in Night Vision Sensitivity: the Effect of Brightness. MRL Rep. 194, **11**, 1-17.

Hecht, S., C. Haig and G. Wald (1935). The Dark Adaptation of Retinal Fields of Different Size and Location. *J. Gen. Physiol.*, **19**, 321-339.

Hirshfield, A. and R. W. Sinnott (1985). *Sky Catalog 2000.0, Volume 2, Double Stars, Variable Stars and Nonstellar Objects.* Sky Publishing Corp., Cambridge, Mass., and Cambridge University Press, Cambridge, England, 385pp.

Hoag, A. A., H. L. Johnson, B. Iriarte, R. J. Mitchell, K. L. Hallam and S. Sharpless (1961). Photometry of Stars in Galactic Cluster Fields, *Pub. US Naval Obs, Second Series,* **17** 349-542.

Kingslake, R. (1965). *Applied Optics and Optical Engineering, Volume 1. Light: Its Generation and Modification,* Academic Press, New York and London.

Mallas, J. H. and Kreimer, E. (1978) *The Messier Album,* Sky Publishing Corp., Cambridge, Mass., **216** pp.

Middleton, W. E. K. (1958). *Vision Through The Atmosphere.* Univ. of Toronto Press, Toronto.

O'Dell, C. R. (1965). Photoelectric Spectrophotometry of Gaseous Nebulae II. The Reflection Nebulae Around Merope. *Astrophy. J.* **142** 604-608.

Roach, F. E. and Jamnick, P. M. (1958). The Sky and the Eye. *Sky & Telescope.* February, 164-167.

Schweizer, F. (1976). Photometric Studies of Spiral Structure. I. The Disks and Arms of Six Sb and Sc Galaxies. *Astrophy. J. Suppl.* **31**, 313-332.

Sinnott, R. W. (1984). Taming Our Chaotic Calendar. *Sky & Telescope,* May, 454-455.

Stoltzmann, D. E. (1983). Resolution Criteria for Diffraction Limited Telescopes. *Sky & Telescope,* February, 176-181.

Index

aberration, **19**, 20, 22, 24, 26, 29, 48
absorption
   of starlight, **268**, 279
   in refractor 50
age
   effect on pupil size, 17
   effect on exit pupil choice, 30
air mass, 55, **268**, **271**, 278
airglow, 31
Airy disk, 31
altazimuth, 23
altitude, **271**, 278
apparent diameter, **28**, 279
apparent field, 26, **28**, 279
astigmatism, **20**, 22
astronomical east, 23, 41, **67**, 249
astronomical north, 23, 38, 42, 44, 45, **67**
astronomical south, 23, 38, 41, 42, 43, **67**
astronomical west, 23, 42, 43, 44, 45, **67**
atmospheric dispersion, 270
atmospheric extinction, 50, **268**
atmospheric refraction, 270
atmospheric seeing, **30**, 31
aurora, 31
averted vision, **12**, 17
azimuth, **271**, 278

Barlow lens, **26**, 27
barrel distortion, 20
blind spot, **4**, **5**, 6, 18

candela, **5**, 7
caring for optics, 46–7
Cassegrain (reflector or telescope), **22**, 24
catadioptric telescope, 23
celestial sphere, **38**, 39
chromatic aberration, **20**, 22
cleaning optics, 46–7
color, 15
color vision, 15
coma, **20**, 22
compound telescope, 23
cone cells (or cones), 4, 6, 8, 9
constellation, 38, 40, 41, 42, 245, 246, 283, 284
continuum radiation, 32

contrast, 7, 8, 11, **12**, **13**, 18, 22, 31, **59**, 60, 65, 279, 285, **321**
   increase with nebula filters, 36
contrast discrimination, 7
Coordinated Universal Time, *see* Universal Time
cornea, **4**, 5, 24
critical visual angle 8, 13, **14**

Dall–Kirkham (telescope), 23
dark adaptation, **6**, 10
dark country sky, 11, 49, 62, 67, 285
declination, **38**, 40
definitions, **67**, 279–81
dew, 47
diffraction disk (or pattern), **30**, 31, 49, 279
diffuse nebula, 32 (*for specific objects, see also individual objects in Chapter 7, Appendix E and Appendix F*)
distortion, 20
Dobsonian (mount), **24**, 25
Doppler shift, 188, *see also* red shift

east (astronomical), *see* astronomical east
emission nebula, 15, **32** (*for specific objects, see also individual objects in Chapter 7, Appendix E and Appendix F*)
equatorial, **23**, 24, 39, 41
Erfle (eyepiece), **26**, 27
excellent skies, 67
exit pupil, 25, 26, **29**, 30, 279
extinction, 50, **268**
eye lens, **4**, 5, 17, 24
eye limiting magnitude, 16, **49–53**, 280 (*see also* faintest star)
eye relief, **25**, 26
eyepiece, **19**, 24–9, 56, 279
eyepiece maximum usable focal length, **28**, 29, 279

faintest star (or limiting magnitude), 16, **49–53**, 55, 67, 249, 280, 318
field curvature, 20
field lens, **25**, 26
field of view, 25, **28**, 42, 44–6, 279, 280

352

INDEX

filter, 31–8
   ultraviolet-filtering glasses, 17
   interference, 33, 34
finder, 24, 40, 41, **44–7**
finder magnification ratio (FMR), 45, 46
flashlight, 55, 56
focal length, **19**, **24**, **28**, 279
focal ratio, *see* f/ratio
forbidden line, **32**, 33
fork (mount), **23**, 24
fovea, **4**, 5, 10, 12, 17, 18
f/ratio, **19**, 20, **22**, 24, 25, 28, 29, 37

ganglion cells, 4
German equatorial (mount), **23**, 24
good skies, 67
Gregorian calendar, 275
Gregorian (telescope), 23

high power, **30**, **31**, 50, **67**
HII association (or region), 77
hour angle, 270, 279
Huygens (eyepiece), **25**, 27

illuminance, 5
interference filter, 33, 34
iris, **4**, 5, 26, 30 (*see also* pupil)

Julian centuries, **268**, 279
Julian Day, **269**, 279

Kellner (eyepiece), **25**, 27

large amateur telescope, 67
large telescope, 67
light pollution, 31–8
light-pollution filter, 31–8
limiting magnitude, *see* faintest star
linear size of an image (in the focal plane), **28**, 280
Local group, **68**, 74, 77, 287
low power, 17, 28, 44, 46, 47, **67**
lowest usable power, *see* minimum usable magnification
Lumen, **5**, 7
luminance, 5

M1 (NGC 1952), **95–7**, 292, 293
M2 (NGC 7089), **232–3**, 315, 316
M3 (NGC 5272), 306, 307
M4 (NGC 6121), **184–5**, 308, 309
M5 (NGC 5904), 308, 309
M6 (NGC 6405), 310, 311
M7 (NGC 6475), 310, 311
M8 (NGC 6523), 44, **192–5**, 310, 311
M9 (NGC 6333), 310, 311
M10 (NGC 6254), 308, 309
M11 (NGC 6705), **205–7**, 312, 313
M12 (NGC 6218), 308, 309
M13 (NGC 6205), **186–7**, 308, 309
M14 (NGC 6402), 310, 311
M15 (NGC 7078), 45, **230–1**, 314, 315
M16 (NGC 6611), **197–9**, 310, 311
M17 (NGC 6618), **201–2**, 310, 311
M18 (NGC 6613), 310, 311
M19 (NGC 6273), 308, 309
M20 (NGC 6514), **188–91**, 310, 311
M21 (NGC 6531), 310, 311
M22 (NGC 6656), 312, 313
M23 (NGC 6494), **258–9**, 310, 311
M24, 310, 311
M25 (IC 4725), 312, 313
M26 (NGC 6694), 312, 313
M27 (NGC 6853), **210–11**, 312, 313
M28 (NGC 6626), 310, 311
M29 (NGC 6913), 314, 315
M30 (NGC 7099), 314, 315
M31 (NGC 224), **68–71**, 77, 286, 287
M32 (NGC 221), **68–71**, 286, 287
M33 (NGC 598), **77–9**, 286, 287
M34 (NGC 1039), 288, 289
M35 (NGC 2168), 292, 293
M36 (NGC 1960), 292, 293
M37 (NGC 2099), 292, 293
M38 (NGC 1912), 292, 293
M39 (NGC 7092), 314, 315
M40, 300, 301
M41 (NGC 2287), 41, 294, 295
M42 (NGC 1976), 15, **98–105**, 292, 293
M43 (NGC 1982), **98–105**, 292, 293
M44 (NGC 2632), 294, 295
M45, **90–4**, 290, 291
M46 (NGC 2437), **115–17**, 294, 295
M47 (NGC 2422), **256–7**, 294, 295
M48 (NGC 2548), 294, 295
M49 (NGC 4472), 302, 303
M50 (NGC 2323), 294, 295
M51 (NGC 5194), 43, **59–63**, **176–81**, 306, 307
M52 (NGC 7654), 316, 317
M53 (NGC 5024), 306, 307
M54 (NGC 6715), 312, 313
M55 (NGC 6809), 312, 313
M56 (NGC 6779), 312, 313
M57 (NGC 6720), **208–9**, 312, 313
M58 (NGC 4579), 302, 303
M59 (NGC 4621), 304, 305
M60 (NGC 4649), 304, 305
M61 (NGC 4303), 300, 301
M62 (NGC 6266), 308, 309
M63 (NGC 5055), **166–7**, 306, 307
M64 (NGC 4826), **164–6**, 304, 305
M65 (NGC 3623), **135–7**, 298, 299
M66 (NGC 3627), **135–7**, 298, 299
M67 (NGC 2682), **118–19**, 294, 295
M68 (NGC 4590), 302, 303
M69 (NGC 6637), 312, 313

INDEX

M70 (NGC 6681), 312, 313
M71 (NGC 6838), 312, 313
M72 (NGC 6981), 314, 315
M73 (NGC 6994), 314, 315
M74 (NGC 628), **80–1**, 288, 289
M75 (NGC 6864), 314, 315
M76 (NGC 650–1), **82**, 288, 289
M77 (NGC 1068), **86**, 288, 289
M78 (NGC 2068), 110–11, 292, 293
M79 (NGC 1904), 292, 293
M80 (NGC 6093), 308, 309
M81 (NGC 3031), **124–5**, 296, 297
M82 (NGC 3034), **126–7**, 296, 297
M83 (NGC 5236), **182–3**, 306, 307
M84 (NGC 4374), **146–9**, 300, 301
M85 (NGC 4382), 300, 301
M86 (NGC 4406), **146–9**, 300, 301
M87 (NGC 4486), **153–5**, 169, 296, 297
M88 (NGC 4501), 302, 303
M89 (NGC 4552), 302, 303
M90 (NGC 4569), **158–9**, 302, 303
M91 (NGC 4548), 302, 303
M92 (NGC 6341), 310, 311
M93 (NGC 2447), 294, 295
M94 (NGC 4736), **162–3**, 304, 305
M95 (NGC 3351), 296, 297
M96 (NGC 3368), **128–9**, 296, 297
M97 (NGC 3587), **137**, 298, 299
M98 (NGC 4192), 300, 301
M99 (NGC 4254), **142**, 300, 301
M100 (NGC 4321), **144–5**, 300, 301
M101 (NGC 5457), 306, 307
M102, doesn't exist
M103 (NGC 581), 286, 287
M104 (NGC 4594), **160–1**, 304, 305
M105 (NGC 3379), **130–1**, 296, 297
M106 (NGC 4258), **143**, 300, 301
M107 (NGC 6171), 308, 309
M108 (NGC 3556), **132**, 298, 299
M109 (NGC 3992), **138–9**, 298, 299
M110 (NGC 205), **68–71**, 286, 287
magnification, 2, **11**, **13**, 18, **19**, 20, 26, **28**, **29**, 30, 31, 44, 45, 46, 48, **49**, 50, 51, 54, 56, 59, 60, 63, 64, 66, 67, 249, 279, 280, **285**, **318**, 319–22, 349, 350
magnification limit, **30**, **31**, 242, 243
magnitude, **5**, **6**, 7–18, 34–7, 42, 43, 45, **49**, **53**, 55, 59, 60, 62, 64, 67, **249**, 268, 279, 280, 284, 285, 318, 321
maximum usable focal length, **28**, 279
medium amateur telescope, 67
medium power, 67
medium size telescope, 67
medium telescope, 67
meridian, **39**, 270, 279
Merope, **90**, 91, 93, 94, 197
Merope nebula, **90**, 91, 106, 197, 215, 291, 317
Milky Way, 44, 65, 68, 83, 215, 317

minimum optimum detection magnification (MDM), **318**, **285**
minimum usable magnification, **30**, 49, 63, 285
minimum useful power, *see* minimum usable magnification
moderate skies, 67
mount, *see* telescope mountings *or individual mounting names*

Nagler (eyepiece), **26**, 27
nebular filter, 31–8
neutron star, 97
Newtonian reflector (*or* Newtonian telescope), **22**, 23, 24, 25
north (astronomical), *see* astronomical north

optics, cleaning, 46–7
optimum detection magnification (ODM), 280, 285, **318**, 321, 323–39, 349, 350
optimum magnified visual angle (OMVA), 2, **11**, **13**, **15**, **60**, 64, 280, 319–21
orthoscopic (eyepiece), **25**, 26, 27
oxygen
  and the eye, 16
  in nebulae, 32, 98, 208

Petzval curvature, 20
pincushion distortion, 20
planetary nebula, 2, 15, 31, 32, *see also individual objects in Chapter 7, Appendix E and Appendix F*
plate scale, 50, **56**, 280
Plossl (eyepiece), 25, **26**, 27
point source, **8**, 11, 13, **14**, 30, 49
polar alignment, 41–3
poor skies, 67
precession, 40
pulsar, 95
pupil, **5**, 6, 17, **25**, 26, **29–30**, 279, *see also* iris
Purkinje effect, 15

quasar, 284

Ramsden (eyepiece), 25
Rayleigh
  resolution limit, 30
  scattering, **32**, 90
red shift, 142, 146, 160
reflection nebula, **32**, 90, 110
reflector, 22, *see also* telescope
refractor, 21–2, *see also* telescope
resolution
  of eye, **11**, 65, *see also* critical visual angle
  of telescope, 22, **30**, **31**
retina, 4, 6, 14, 64
rhodopsin, 6
right ascension, **38**, 40
Ritchey–Chretien (telescope), 23
rod cells (or rods), **4**, 6, 8, 9, 55

INDEX

Schmidt–Cassegrain (telescope), 21, **23**
seeing, **30**, **31**, 48, 50
setting circles, **41**, 283
Seyfert galaxy, 86, 87
shock excitation, 218
sidereal time, 268, 270, 280
sky surface brightness, **11**, 49, 50, 51, 67, **285**
small amateur telescope, 67
small telescope, 67
smoking, 16
south (astronomical), *see* astronomical south
spherical aberration, **20**, 23
star atlas, 38, 48, **246**, 247
star-hopping, **40**, 41–5
stellar magnitude, **5**, 7, 8, 268, 280
sunglasses, 17
Supercluster, 146
surface brightness, **5**, 6, 7, 10–16, 28, 31, 49, 50, 51, 53, 59, 62, 63, 66, 67, 279, 280, 285, 318, 319, 320, 322, 349, 350, *see also individual objects in Chapter 7, Appendix E and Appendix F*
surface brightness reduction, **49**, 59, 280, **318**, 319, 349, 350

telescope
    eyepieces, 24–7, 29
    finders, 44–6
    mountings, 23–5
    transmission factor, **49**, 280, 285
    types, 20–5, 29
    transmission factor, *see* telescope transmission factor
transmittance
    of eye, 17
    of filters, 34–7
transparency, 11, 17
true angular diameter, **28**, 279
true field (of view), 28, 29, 44–6, 56

ultraviolet radiation (or light), 17, 98, 208, 210, 227, 235
Universal Time (UT), 54, 66, 268–71, 280

very high power, 31, 60, 67
very low power, 60, 67
very poor skies, 67
viewing distance, 50, **56**, **66**, 249–67
Virgo Cluster (of Galaxies), 146, 153, 158, 160, 164
visual purple, 6

west (astronomical), *see* astronomical west
Wolf–Rayet star, 212
WWV, 280

zenith, 24, **39**, 55, 66, 268, 269, 270, 280, 281
zenith angle, 268–71, 280, 281
zonal aberration, 20